Universitext

Springer Science+Business Media, LLC

Universitext

Editors (North America): S. Axler, F.W. Gehring, and K.A. Ribet

Aksoy/Khamsi: Nonstandard Methods in Fixed Point Theory
Andersson: Topics in Complex Analysis
Aupetit: A Primer on Spectral Theory
Berberian: Fundamentals of Real Analysis
Booss/Bleecker: Topology and Analysis
Borkar: Probability Theory: An Advanced Course
Böttcher/Silbermann: Introduction to Large Truncated Toeplitz Matrices
Carleson/Gamelin: Complex Dynamics
Cecil: Lie Sphere Geometry: With Applications to Submanifolds
Chae: Lebesgue Integration (2nd ed.)
Charlap: Bieberbach Groups and Flat Manifolds
Chern: Complex Manifolds Without Potential Theory
Cohn: A Classical Invitation to Algebraic Numbers and Class Fields
Curtis: Abstract Linear Algebra
Curtis: Matrix Groups
DiBenedetto: Degenerate Parabolic Equations
Dimca: Singularities and Topology of Hypersurfaces
Edwards: A Formal Background to Mathematics I a/b
Edwards: A Formal Background to Mathematics II a/b
Foulds: Graph Theory Applications
Friedman: Algebraic Surfaces and Holomorphic Vector Bundles
Fuhrmann: A Polynomial Approach to Linear Algebra
Gardiner: A First Course in Group Theory
Gårding/Tambour: Algebra for Computer Science
Goldblatt: Orthogonality and Spacetime Geometry
Gustafson/Rao: Numerical Range: The Field of Values of Linear Operators and Matrices
Hahn: Quadratic Algebras, Clifford Algebras, and Arithmetic Witt Groups
Holmgren: A First Course in Discrete Dynamical Systems
Howe/Tan: Non-Abelian Harmonic Analysis: Applications of $SL(2, R)$
Howes: Modern Analysis and Topology
Humi/Miller: Second Course in Ordinary Differential Equations
Hurwitz/Kritikos: Lectures on Number Theory
Jennings: Modern Geometry with Applications
Jones/Morris/Pearson: Abstract Algebra and Famous Impossibilities
Kannan/Krueger: Advanced Analysis
Kelly/Matthews: The Non-Euclidean Hyperbolic Plane
Kostrikin: Introduction to Algebra
Luecking/Rubel: Complex Analysis: A Functional Analysis Approach
MacLane/Moerdijk: Sheaves in Geometry and Logic
Marcus: Number Fields
McCarthy: Introduction to Arithmetical Functions
Meyer: Essential Mathematics for Applied Fields
Mines/Richman/Ruitenburg: A Course in Constructive Algebra
Moise: Introductory Problems Course in Analysis and Topology
Morris: Introduction to Game Theory
Polster: A Geometrical Picture Book
Porter/Woods: Extensions and Absolutes of Hausdorff Spaces
Ramsay/Richtmyer: Introduction to Hyperbolic Geometry
Reisel: Elementary Theory of Metric Spaces
Rickart: Natural Function Algebras

(continued after index)

Albrecht Böttcher   Bernd Silbermann

Introduction to Large Truncated Toeplitz Matrices

With 62 Figures

Springer

Albrecht Böttcher
Fakultät für Mathematik
Technische Universität Chemnitz
Chemnitz, 09107
Germany

Bernd Silbermann
Fakultät für Mathematik
Technische Universität Chemnitz
Chemnitz, 09107
Germany

Mathematics Subject Classification (1991): 15-02, 47B35

Library of Congress Cataloging-in-Publication Data
Böttcher, Albrecht.
Introduction to large truncated Toeplitz matrices / Albrecht Böttcher, Bernd Silbermann.
p. cm. — (Universitext)
Includes bibliographical references and index.
ISBN 978-0-387-98570-1 ISBN 978-1-4612-1426-7 (eBook)
DOI 10.1007/978-1-4612-1426-7
1. Toeplitz matrices. I. Silbermann, Bernd, 1941– II. Title.
QA188.B67 1998
512.9'434—dc21 98-9923

Printed on acid-free paper.

Production managed by Allan Abrams; manufacturing supervised by Jeffrey Taub.
Photocomposed copy provided from the authors' LaTeX files.

9 8 7 6 5 4 3 2 1

ISBN 978-0-387-98570-1

Preface

Toeplitz matrices have been enjoying immense popularity for many decades. They are easily defined (as matrices constant along the parallels to the main diagonal), they emerge in a variety of problems of quite different natures, they cause interesting and difficult questions, and they usually lead to beautiful results. On the one hand, Toeplitz matrices are easy enough to serve as ideal illustrations of various abstract results and methods of linear algebra and functional analysis, and on the other hand, they are sufficiently nontrivial and therefore have the potential to create new concepts, techniques, insights, and, of course, to raise new questions.

Figuratively speaking, the theory of Toeplitz matrices has grown to a grandiose city. Specialists know the parts of this city well and are about to reconstruct and expand it, but beginners and amateurs often have problems with finding their way in the labyrinth. This book is addressed to the latter group of people. It is intended as a guide to three main roads of this city, whose names are:

pseudospectra,
singular values,
eigenvalues,

and the book also contains glimpses at several side-streets.

We consider large finite Toeplitz matrices as truncations of infinite Toeplitz matrices and hence, we study properties of an individual large finite Toeplitz matrix by embedding it into the sequence of the truncations (finite sections) of an infinite Toeplitz matrix.

The three roads mentioned start at a place that bears the name

stability.

Given an infinite Toeplitz matrix A, let $\{A_n\}_{n=1}^{\infty}$ stand for the sequence of its $n \times n$ truncations. The central problem of the entire theory is as follows: if A induces an invertible operator, are the finite sections A_n invertible for all sufficiently large n, say for $n \geq n_0$, and are the norms of the inverses, $\|A_n^{-1}\|$, bounded from above by a finite constant independent of $n \geq n_0$? If this is the case, the sequence $\{A_n\}$ is said to be stable. Properties of infinite Toeplitz matrices (including invertibility criteria) are studied in Chapter 1 and the stability of sequences of truncated Toeplitz matrices is the topic of Chapter 2.

Chapters 3,4, and 5 deal with pseudospectra, singular values, and eigenvalues of the truncated matrices A_n, respectively. The investigation of the pseudospectra of A_n is heavily based on the ability of computing the limit of $\|A_n^{-1}\|$ as n goes to infinity. The computation of this limit is in turn a nice application of the theory of C^*-algebras. The asymptotic distribution of the singular values of A_n is intimately tied in with asymptotic Moore-Penrose inversion of the matrices A_n. Finally, the main results on the asymptotic behavior of the eigenvalues of A_n are all derived from asymptotic formulas for the traces and determinants of Toeplitz-like matrices. Thus, we could also name our three main roads:

condition numbers,
Moore-Penrose inversion,
traces and determinants,

respectively.

The matrices one encounters nowadays in applications are often not Toeplitz matrices but block Toeplitz matrices. Many results on Toeplitz matrices can be extended to block Toeplitz matrices, although this is usually a hard job. In accordance with the purpose of this text, we focus our attention on scalar Toeplitz matrices. However, in Chapter 6 we describe some of the phenomena caused by block Toeplitz matrices and cite several results, referring for proofs to the literature.

In Chapter 7, we exhibit some results on Toeplitz operators on Banach spaces and acquaint the reader with certain techniques employed in this field. We remark that the Banach space theory of Toeplitz operators with piecewise continuous symbols is more beautiful than the corresponding Hilbert space theory! Moreover, we will exemplify in Chapter 7 that there are Hilbert space results which can be most easily understood by passing to Banach spaces.

We suppose that knowledge of the basic facts of functional analysis in conjunction with some patience and persistence should suffice to enable a reading of the bulk of the text. Of course, the sights along the three roads mentioned will represent our taste, and some important topics are not treated at all. We would be happy if we could nevertheless convey to the reader an idea of what is known and of what is going on in the extensive and beautiful field of large Toeplitz matrices.

Acknowledgments. We wish to express our especially sincere gratitude to Sylvia Böttcher for the production of the LaTeX masters of the book and to Harald Heidler for making the majority of the computer pictures. We are greatly indebted to Torsten Ehrhardt and Steffen Roch for proof-reading the entire manuscript and for improving it by many useful remarks.

Albrecht Böttcher
Chemnitz, December 1997 Bernd Silbermann

Contents

1

Infinite Matrices

1.1 Boundedness and Invertibility

The purpose of this section is to fix some standard notations and to recall some terminology.

Bounded linear operators. Given a Banach space X, we denote by $\mathcal{B}(X)$ the collection of all bounded linear operators on X. The *norm* of an operator $A \in \mathcal{B}(X)$ is defined in the usual way:

$$\|A\| := \sup_{x \neq 0} \|Ax\|/\|x\|. \tag{1.1}$$

We say that a sequence $\{A_n\}_{n=1}^{\infty}$ of operators $A_n \in \mathcal{B}(X)$ *converges uniformly* (or in the norm) to an operator $A \in \mathcal{B}(X)$ if $\|A_n - A\| \to 0$ as $n \to \infty$, and the sequence $\{A_n\}_{n=1}^{\infty}$ is said to *converge strongly* to $A \in \mathcal{B}(X)$ if $\|A_n x - Ax\| \to 0$ as $n \to \infty$ for every $x \in X$.

Infinite matrices. In what follows we are mainly concerned with the case where X is the Hilbert space $l^2(J)$ and J stands for the integers $\mathbf{Z}$ or the natural numbers $\mathbf{N} = \{1, 2, 3, \ldots\}$. An orthonormal basis in $l^2(J)$ is constituted by the elements $\{e_j\}_{j \in J}$ where e_j is the sequence (or the vector) whose jth entry is 1 and the remaining entries of which are zero. Thus, with every operator $A \in \mathcal{B}(l^2(J))$ we may associate the infinite matrix $(a_{jk})_{j,k \in J}$ given by

$$a_{jk} = (Ae_k, e_j). \tag{1.2}$$

Thinking of $l^2(J)$ as a space of infinite columns, we can describe the action of A on $l^2(J)$ as multiplication by the infinite matrix $(a_{jk})_{j,k\in J}$: we have

$$Ax = \left(\begin{array}{ccc|ccc} \cdots & \cdots & \cdots & \cdots & \cdots & \cdots \\ \cdots & a_{-2,-2} & a_{-2,-1} & a_{-2,0} & a_{-2,1} & \cdots \\ \cdots & a_{-1,-2} & a_{-1,-1} & a_{-1,0} & a_{-1,1} & \cdots \\ \hline \cdots & a_{0,-2} & a_{0,-1} & a_{00} & a_{01} & \cdots \\ \cdots & a_{1,-2} & a_{1,-1} & a_{10} & a_{11} & \cdots \\ \cdots & \cdots & \cdots & \cdots & \cdots & \cdots \end{array}\right) \left(\begin{array}{c} \vdots \\ x_{-2} \\ x_{-1} \\ \hline x_0 \\ x_1 \\ \vdots \end{array}\right) = \left(\begin{array}{c} \vdots \\ y_{-2} \\ y_{-1} \\ \hline y_0 \\ y_1 \\ \vdots \end{array}\right) = y$$

and

$$Ax = \begin{pmatrix} a_{11} & a_{12} & a_{13} & \cdots \\ a_{21} & a_{22} & a_{23} & \cdots \\ a_{31} & a_{32} & a_{33} & \cdots \\ \cdots & \cdots & \cdots & \end{pmatrix} \begin{pmatrix} x_1 \\ x_2 \\ x_3 \\ \vdots \end{pmatrix} = \begin{pmatrix} y_1 \\ y_2 \\ y_3 \\ \vdots \end{pmatrix} = y$$

with $y_j = \sum_{k\in J} a_{jk}x_k$ in the cases $J = \mathbf{Z}$ and $J = \mathbf{N}$, respectively.

Every operator $A \in \mathcal{B}(l^2(J))$ can be represented by an infinite matrix as above, but not every infinite matrix defines a bounded operator on $l^2(J)$. We say that an infinite matrix $(a_{jk})_{j,k\in J}$ generates a bounded operator on $l^2(J)$ (or simply that the infinite matrix *is* a bounded operator on $l^2(J)$) if there exists an $A \in \mathcal{B}(l^2(J))$ such that (1.2) is valid for all $j, k \in J$. Equivalently, the infinite matrix $(a_{jk})_{j,k\in J}$ generates a bounded operator on $l^2(J)$ if and only if there exists a constant $M < \infty$ such that for every $x \in l^2(J)$ the following hold:

(i) the series $y_j := \sum_{k\in J} a_{jk}x_k$ converge for all $j \in J$;

(ii) $y := \{y_j\}_{j\in J}$ belongs to $l^2(J)$;

(iii) $\|y\| \le M\|x\|$;

here, of course, $\|\cdot\|$ is nothing but the norm in $l^2(J)$. The smallest constant M for which this is true equals the norm $\|A\|$ of the operator $A \in \mathcal{B}(l^2(J))$ induced by the given infinite matrix.

Banach algebras. A (complex) Banach space $\mathcal{A}$ with an associative and distributive multiplication is called a *Banach algebra* if $\|ab\| \le \|a\|\,\|b\|$ for all $a, b \in \mathcal{A}$. If a Banach algebra has a unit element (which is also frequently called the identity), then this element is usually denoted by e, 1, or I. We always require that $\|e\| = \|1\| = \|I\| = 1$.

Invertibility and spectrum. Let $\mathcal{A}$ be a Banach algebra with the unit element e. An element $a \in \mathcal{A}$ is said to be *invertible* (in $\mathcal{A}$) if there is an element $b \in \mathcal{A}$ such that $ab = ba = e$. The element b is uniquely determined whenever it exists; in that case it is denoted by a^{-1} and is called the *inverse* of a. The *spectrum* of an element $a \in \mathcal{A}$ is defined as the set

$$\operatorname{sp} a := \operatorname{sp}_{\mathcal{A}} a := \{\lambda \in \mathbf{C} : a - \lambda e \ \text{ is not invertible in } \ \mathcal{A}\}.$$

It is well known that the spectrum of an element of a Banach algebra with identity is always a nonempty compact subset of the complex plane $\mathbf{C}$ which is contained in the disk $\{\lambda \in \mathbf{C} : |\lambda| \leq \|a\|\}$.

If X is a Banach space, then $\mathcal{B}(X)$ is a Banach algebra with obvious algebraic operations and the norm (1.1). The unit of $\mathcal{B}(X)$ is the identity operator I. An operator $A \in \mathcal{B}(X)$ is called *invertible* if it is invertible as an element of $\mathcal{B}(X)$, and the *spectrum* of A is simply the set $\mathrm{sp}_{\mathcal{B}(X)} A$.

1.2 Laurent Matrices

Suppose we are given a sequence $\{a_n\}_{n=-\infty}^{\infty}$ of complex numbers and A is the infinite matrix

$$A = \left(\begin{array}{ccc|cccc} \dots & \dots & \dots & \dots & \dots & \dots & \dots \\ \dots & a_0 & a_{-1} & a_{-2} & a_{-3} & a_{-4} & \dots \\ \dots & a_1 & a_0 & a_{-1} & a_{-2} & a_{-3} & \dots \\ \hline \dots & a_2 & a_1 & a_0 & a_{-1} & a_{-2} & \dots \\ \dots & a_3 & a_2 & a_1 & a_0 & a_{-1} & \dots \\ \dots & a_4 & a_3 & a_2 & a_1 & a_0 & \dots \\ \dots & \dots & \dots & \dots & \dots & \dots & \dots \end{array}\right). \tag{1.3}$$

Such matrices, that is, doubly-infinite matrices which are constant along the diagonals, are called *Laurent matrices.* When does A generate a bounded operator on $l^2(\mathbf{Z})$? The answer is given by the following well-known result.

Theorem 1.1. *The Laurent matrix* (1.3) *generates a bounded operator on* $l^2(\mathbf{Z})$ *if and only if there is a function* $a \in L^\infty(\mathbf{T})$ *such that* $\{a_n\}_{n=-\infty}^{\infty}$ *is the sequence of the Fourier coefficients of* a*:*

$$a_n = \frac{1}{2\pi} \int_0^{2\pi} a(e^{i\theta}) e^{-in\theta}\, d\theta \quad (n \in \mathbf{Z}).$$

Here and in what follows, $\mathbf{T}$ stands for the complex unit circle $\{z \in \mathbf{C} : |z| = 1\}$. The function $a \in L^\infty(\mathbf{T})$ (or better: the equivalence class of $L^\infty(\mathbf{T})$ containing a) whose existence is ensured by this theorem is determined uniquely. We therefore denote the matrix (1.3) as well as the bounded operator generated by this matrix on $l^2(\mathbf{Z})$ by $L(a)$. The function a is in this context usually referred to as the *symbol* of the matrix or the operator $L(a)$.

Multiplication operators. Let $L^p := L^p(\mathbf{T})$ $(1 \leq p \leq \infty)$ be the usual Lebesgue spaces on $\mathbf{T}$ and denote by $\|\cdot\|_p$ the norm in L^p. For $a \in L^\infty$, the multiplication operator

$$M(a) : L^2 \to L^2, \;\; f \mapsto af$$

is obviously bounded. Clearly, $\|M(a)\| \le \|a\|_\infty$. It is not difficult to show that actually $\|M(a)\| = \|a\|_\infty$ and that $M(a)$ is not bounded if a does not belong to L^∞. Denote by

$$\Phi : L^2 \to l^2(\mathbf{Z}), \quad f \mapsto \{f_n\}_{n\in\mathbf{Z}}$$

the operator which sends a function to the sequence of its Fourier coefficients. The operator Φ is bijective and, by Parseval's equality,

$$\sqrt{2\pi}\|\Phi f\|_2 = \|f\|_2 \quad \text{for all} \quad f \in L^2.$$

The basic properties of Laurent matrices follow from the fact that they are the matrix representations of multiplication operators on L^2 with respect to the orthonormal basis

$$\left\{\frac{1}{\sqrt{2\pi}}e^{in\theta}\right\}_{n\in\mathbf{Z}}.$$

In other words, we have

$$L(a) = \Phi M(a)\Phi^{-1}. \tag{1.4}$$

For example, this equality at once implies the "if" portion of Theorem 1.1; it is the key to the proof of the "only if" part as well. Also notice that (1.4) shows that

$$L(a)L(b) = L(ab) \quad \text{for all} \quad a, b \in L^\infty \tag{1.5}$$

and

$$\|L(a)\| = \|a\|_\infty \quad \text{for all} \quad a \in L^\infty. \tag{1.6}$$

Essential range. The space L^∞ is a Banach algebra under pointwise operations and the norm $\|\cdot\|_\infty$. For $a \in L^\infty$, we denote by $\mathcal{R}(a)$ the spectrum of a as an element of L^∞. Equivalently, we may define $\mathcal{R}(a)$ as the spectrum of the multiplication operator $M(a)$ on L^2. Finally, it is not hard to see that

$$\mathcal{R}(a) = \big\{\lambda \in \mathbf{C} : \big|\{t \in \mathbf{T} : |a(t) - \lambda| < \varepsilon\}\big| > 0 \ \forall\, \varepsilon > 0\big\},$$

where $|E|$ stands for the Lebesgue measure of E. The (nonempty and compact) set $\mathcal{R}(a)$ is called the *essential range* of a. Notice that alteration of a on a set of measure zero does not change $\mathcal{R}(a)$.

Theorem 1.2. *If $a \in L^\infty$, then $\operatorname{sp} L(a) = \mathcal{R}(a)$. If $0 \notin \mathcal{R}(a)$, then the inverse of $L(a)$ is $L(a^{-1})$.*

This follows easily from (1.4) and (1.5). Under some additional hypotheses, Theorem 1.2 was first proved by Otto Toeplitz [169].

We now discuss a few concrete symbol classes.

Example 1.3: band matrices. If $L(a)$ is a band matrix, i.e. if $a_n = 0$ for $|n| > N$, then the symbol a is a trigonometric polynomial:

$$a(t) = \sum_{n=-N}^{N} a_n t^n \quad (t = e^{i\theta} \in \mathbf{T}).$$

In particular, the symbol of the matrix

$$\left(\begin{array}{cccc|cccc} \dots & \dots & \dots & \dots & \dots & \dots & \dots & \dots \\ \dots & 1 & 2 & 1 & 0 & 0 & 0 & \dots \\ \dots & 0 & 1 & 2 & 1 & 0 & 0 & \dots \\ \hline \dots & 0 & 0 & 1 & 2 & 1 & 0 & \dots \\ \dots & 0 & 0 & 0 & 1 & 2 & 1 & \dots \\ \dots & 0 & 0 & 0 & 0 & 1 & 2 & \dots \\ \dots & \dots & \dots & \dots & \dots & \dots & \dots & \dots \end{array}\right) \tag{1.7}$$

is

$$a(e^{i\theta}) = 2 + e^{i\theta} + e^{-i\theta} = 2 + 2\cos\theta.$$

It follows from Theorem 1.2 that the spectrum of the operator given on $l^2(\mathbf{Z})$ by (1.7) is the line segment $[0, 4]$. Some more interesting symbols of Toeplitz band matrices are in Figure 1. ■

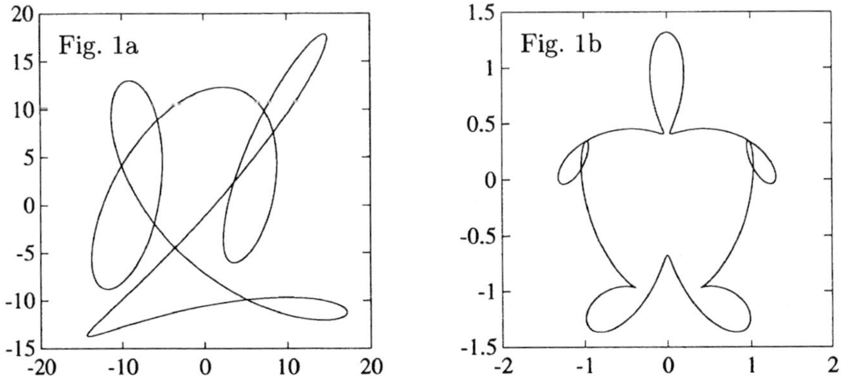

In Figures 1a and 1b we plotted $\mathcal{R}(a)$ for two trigonometric polynomials a.

Example 1.4: rational symbols. The restriction of a rational function a to $\mathbf{T}$ belongs to L^∞ if and only if the function a has no poles on $\mathbf{T}$. Such symbols define Laurent matrices whose entries decay as a geometric sequence. For example, the symbol of the matrix

$$\left(\begin{array}{cccc|cccc}
\cdots & \cdots & \cdots & \cdots & \cdots & \cdots & \cdots & \cdots \\
\cdots & \beta & 1 & \alpha & \alpha^2 & \alpha^3 & \alpha^4 & \cdots \\
\cdots & \beta^2 & \beta & 1 & \alpha & \alpha^2 & \alpha^3 & \cdots \\
\hline
\cdots & \beta^3 & \beta^2 & \beta & 1 & \alpha & \alpha^2 & \cdots \\
\cdots & \beta^4 & \beta^3 & \beta^2 & \beta & 1 & \alpha & \cdots \\
\cdots & \beta^5 & \beta^4 & \beta^3 & \beta^2 & \beta & 1 & \cdots \\
\cdots & \cdots & \cdots & \cdots & \cdots & \cdots & \cdots & \cdots
\end{array}\right) \qquad (|\alpha| < 1,\ |\beta| < 1)$$

is (see Figure 2)

$$a(t) = 1 + \sum_{n=1}^{\infty} \alpha^n t^{-n} + \sum_{n=1}^{\infty} \beta^n t^n = 1 + \frac{\alpha}{t-\alpha} + \frac{\beta}{1/t - \beta} \qquad (t \in \mathbf{T}). \blacksquare$$

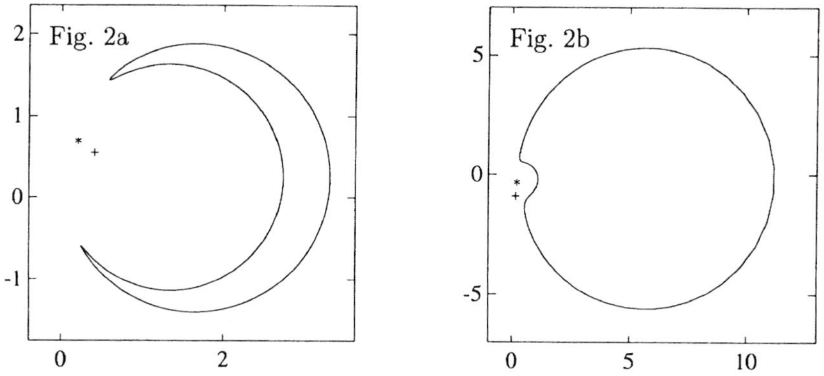

Figures 2a and 2b show $\mathcal{R}(a)$ for two symbols as in Example 1.4. The corresponding points α and β are marked by $*$ and $+$, respectively.

Example 1.5: symbols in the Wiener algebra. If $\sum_{n\in\mathbf{Z}} |a_n| < \infty$, then the matrix $L(a)$ has the symbol

$$a(t) = \sum_{n\in\mathbf{Z}} a_n t^n \qquad (t = e^{i\theta} \in \mathbf{T}).$$

The set of all such functions is denoted by $W := W(\mathbf{T})$ and is called the *Wiener algebra*. It is a Banach algebra with pointwise algebraic operations and the norm

$$\|a\|_W := \sum_{n\in\mathbf{Z}} |a_n|.$$

Clearly, W is contained in the Banach algebra $C := C(\mathbf{T})$ of all (complex-valued) continuous functions on $\mathbf{T}$ with the maximum norm. *Wiener's theorem* says that if $a \in W$ and a has no zeros on $\mathbf{T}$, then $a^{-1} = 1/a \in W$.

Thus, by Theorem 1.2, the inverse of an invertible Laurent operator with a symbol in the Wiener algebra is again a Laurent operator with a symbol in the Wiener algebra. ■

Example 1.6: continuous symbols. The essential range of a function $a \in C$ is its image $a(\mathbf{T})$ and hence, $\operatorname{sp} L(a) = a(\mathbf{T})$.

There are many functions in $C \setminus W$, but these are sometimes difficult to identify in terms of their Fourier coefficients. To get an idea of the problem, let us look at a special class of functions. Suppose $\{b_n\}_{n=2}^{\infty}$ is a sequence of positive numbers *converging monotoneously to zero* and consider the series

$$\sum_{n=2}^{\infty} b_n \sin n\theta = \sum_{n=2}^{\infty} \frac{b_n}{2i}(e^{in\theta} - e^{-in\theta}) \qquad (e^{i\theta} \in \mathbf{T}). \tag{1.8}$$

The following result is well known (see, e.g., [68, Section 7.2.2]):

(i) the series (1.8) is the Fourier series of a function in C if and only if

$$b_n = o(1/n) \quad \text{as} \quad n \to \infty;$$

(ii) the series (1.8) is the Fourier series of a function in L^∞ if and only if

$$b_n = O(1/n) \quad \text{as} \quad n \to \infty.$$

In particular, the symbol of the Laurent matrix induced by $\{a_n\}_{n\in\mathbf{Z}}$ with

$$a_{-1} = a_0 = a_1 = 0, \qquad a_n = \frac{1}{n \log|n|} \qquad (|n| \geq 2)$$

belongs to $C \setminus W$, while the Laurent matrix defined by $\{a_n\}_{n\in\mathbf{Z}}$ with

$$a_{-1} = a_0 = a_1 = 0, \qquad a_n = \frac{\log|n|}{n} \qquad (|n| \geq 2)$$

does not generate a bounded operator on $l^2(\mathbf{Z})$. ■

Example 1.7: piecewise continuous symbols. A function $a \in L^\infty$ is said to be *piecewise continuous* if for every $t = e^{i\theta} \in \mathbf{T}$ the one-sided limits

$$a(t+0) := \lim_{\varepsilon \to 0+0} a\big(e^{i(\theta+\varepsilon)}\big), \qquad a(t-0) := \lim_{\varepsilon \to 0+0} a\big(e^{i(\theta-\varepsilon)}\big)$$

exist. We always think of the unit circle $\mathbf{T}$ as being oriented counterclockwise, which also accounts for the notations $a(t-0)$ and $a(t+0)$. The set of all piecewise continuous functions is denoted by $PC := PC(\mathbf{T})$. It is well known that PC is a closed subalgebra of L^∞. Functions in PC have at most countably many jumps, i.e., if $a \in PC$, then the set

$$\Lambda_a := \big\{t \in \mathbf{T} : a(t-0) \neq a(t+0)\big\}$$

is at most countable. Moreover, for each $\delta > 0$ the set

$$\{t \in \mathbf{T} : |a(t+0) - a(t-0)| > \delta\}$$

is finite. Given $a \in PC$, we always assume that a is continuous on $\mathbf{T} \setminus \Lambda_a$. Thus,

$$\mathcal{R}(a) = \bigcup_{t \in \mathbf{T} \setminus \Lambda_a} \{a(t)\} \cup \bigcup_{t \in \Lambda_a} \{a(t-0),\, a(t+0)\}. \tag{1.9}$$

Theorem 1.2 tells us that (1.9) is the spectrum of $L(a)$.

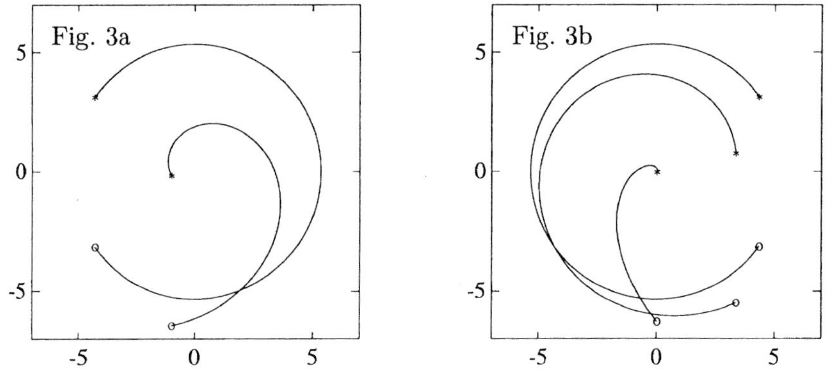

Figure 3a shows the set $\mathcal{R}(\psi_\gamma)$ for $\gamma = 0.8$ and $\gamma = 0.8 + 0.3i$, in Figure 3b we plotted $\mathcal{R}(\psi_\gamma)$ for $\gamma = -0.8$, $\gamma = -0.8 + 0.1i$, and $\gamma = -0.8 + i$. The points $\psi_\gamma(1+0)$ and $\psi_\gamma(1-0)$ are marked by $*$ and $\circ$, respectively.

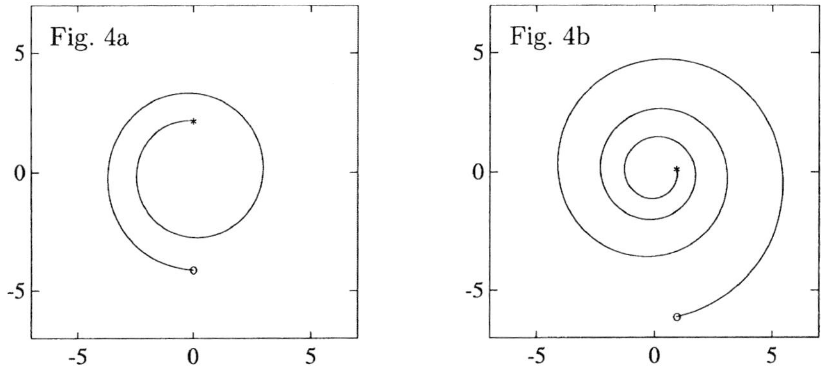

In Figure 4a we see the set $\mathcal{R}(\psi_\gamma)$ for $\gamma = -1.5 + 0.1i$ and in Figure 4b we have $\mathcal{R}(\psi_\gamma)$ for $\gamma = 3.25 + 0.3i$. The points $\psi_\gamma(1+0)$ and $\psi_\gamma(1-0)$ are again marked by $*$ and $\circ$, respectively. Figure 4b convincingly indicates that $\mathcal{R}(\psi_\gamma)$ is a piece of a logarithmic spiral. Thus, we encounter once more the phenomenon that logarithmic spirals are everywhere (see [42], [32]).

To have a concrete example, pick $\gamma \in \mathbf{C}\setminus\mathbf{Z}$ and put

$$\psi_\gamma(e^{i\theta}) = \frac{\pi}{\sin \pi\gamma} e^{i\pi\gamma} e^{-i\gamma\theta}, \qquad \theta \in [0, 2\pi). \tag{1.10}$$

A moment's thought reveals that this is a function in PC with a single jump at $e^{i\theta} = 1$:

$$\psi_\gamma(1+0) = \frac{\pi}{\sin \pi\gamma} e^{i\pi\gamma}, \qquad \psi_\gamma(1-0) = \frac{\pi}{\sin \pi\gamma} e^{-i\pi\gamma}. \tag{1.11}$$

A straightforward computation gives

$$(\psi_\gamma)_n = \frac{1}{n+\gamma} \qquad (n \in \mathbf{Z}),$$

i.e.,

$$L(\psi_\gamma) = \left(\frac{1}{j-k+\gamma}\right)_{j,k=-\infty}^{\infty}$$

is a so-called Cauchy-Laurent matrix. By Theorems 1.1 and 1.2, the Laurent operator $L(\psi_\gamma)$ is bounded and

$$\|L(\psi_\gamma)\| = \frac{\pi}{|\sin \pi\gamma|} e^{\pi|\operatorname{Im}\gamma|},$$

$$\operatorname{sp} L(\psi_\gamma) = \left\{\frac{\pi}{\sin \pi\gamma} e^{i\pi\gamma} e^{-i\gamma\theta} : \theta \in [0, 2\pi]\right\}.$$

Figures 3 and 4 show plots of the essential range of ψ_γ. ■

Example 1.8: a glance at the abyss of L^∞. Every compact set $M \subset \mathbf{C}$ is the spectrum of some Laurent operator. Indeed, let $\{z_j\}_{j\in\mathbf{N}}$ be a countable dense subset of M, let $\{E_j\}_{j\in\mathbf{N}}$ be a sequence of pairwise disjoint arcs $E_j \subset \mathbf{T}$ whose union is all of $\mathbf{T}$, and define $a \in L^\infty$ as

$$a = \sum_{j\in\mathbf{N}} z_j \chi_{E_j},$$

where χ_E is the characteristic function of E. Then $\operatorname{sp} L(a) = \mathcal{R}(a) = M$. ■

1.3 Toeplitz Matrices

The *Toeplitz matrix* defined by a sequence $\{a_n\}_{n=-\infty}^{\infty}$ of complex numbers is the infinite matrix

$$A = \begin{pmatrix} a_0 & a_{-1} & a_{-2} & \dots \\ a_1 & a_0 & a_{-1} & \dots \\ a_2 & a_1 & a_0 & \dots \\ \dots & \dots & \dots & \dots \end{pmatrix}. \tag{1.12}$$

We henceforth abbreviate $l^2(\mathbf{N})$ to l^2.

Theorem 1.9. *The Toeplitz matrix* (1.12) *generates a bounded operator on l^2 if and only if there is a function $a \in L^\infty$ whose sequence of Fourier coefficients is the sequence $\{a_n\}_{n\in\mathbf{Z}}$.*

This theorem was established by Toeplitz [169] in 1911. Notice that Toeplitz' paper actually deals with Laurent matrices and that the main result of his paper is Theorem 1.2. However, Theorem 1.9 is proved in a footnote of [169] and it is this theorem which led to naming the matrices (1.12) after Toeplitz. Independently, Theorem 1.9 was also found by Brown and Halmos [43].

Full proofs of Theorem 1.9 are in [39, Theorem 2.7], [43], [95, Problem 194], for example. In what follows we only need the "if portion" of Theorem 1.9, which can be easily proved. Indeed, identify l^2 as a subspace of $l^2(\mathbf{Z})$ in the natural manner and denote by P the orthogonal projection of $l^2(\mathbf{Z})$ onto l^2. Then the operator A given by the matrix (1.12) can be identified with $PL(a)P$. This and the sufficiency portion of Theorem 1.1 show that A generates a bounded operator on l^2 whenever $a \in L^\infty$. From (1.6) we also infer that

$$\|A\| = \|PL(a)P\| \le \|L(a)\| = \|a\|_\infty. \tag{1.13}$$

The function $a \in L^\infty$ given by Theorem 1.9 is called the *symbol* of the Toeplitz matrix (1.12) and of the operator induced by this matrix on l^2. Throughout the following we denote (1.12) by $T(a)$.

Norm of a Toeplitz operator. It is well known that for every $a \in L^\infty$ the equality

$$\|T(a)\| = \|a\|_\infty \tag{1.14}$$

holds. The estimate $\|T(a)\| \le \|a\|_\infty$ is contained in (1.13). To prove the reverse inequality, denote by S_n $(n = 1, 2, 3, \ldots)$ the projection on $l^2(\mathbf{Z})$ given by

$$(S_n x)_k = \begin{cases} 0 & \text{if } k < -n, \\ x_k & \text{if } k \ge -n. \end{cases}$$

Obviously, $S_n \to I$ strongly on $l^2(\mathbf{Z})$. It follows that $S_n L(a) S_n$ converges strongly to $L(a)$, and because evidently $\|S_n L(a) S_n\| = \|T(a)\|$, we deduce from the Banach-Steinhaus theorem (see Theorem 2.1 below) that

$$\|L(a)\| \le \liminf_{n\to\infty} \|S_n L(a) S_n\| = \|T(a)\|.$$

Now (1.6) implies that $\|T(a)\| \ge \|a\|_\infty$.

Spectrum of a Toeplitz operator. The spectra of Laurent operators are completely described by Theorem 1.2. The determination of the spectra of Toeplitz operators is a much more difficult problem. In the case where

$a \in C$, we will identify $\operatorname{sp} T(a)$ in Section 1.6. Some more results on the spectra of Toeplitz operators will be discussed in Section 1.8.

The rest of this section is devoted to Coburn's lemma, which divides the problem of deciding whether a Toeplitz operator $T(a)$ is invertible into finding out whether $T(a)$ is Fredholm and into computing the index of $T(a)$.

Fredholmness and index. Let X be a Banach space and $A \in \mathcal{B}(X)$. The *kernel* and the *image* (= *range*) of A are defined by

$$\operatorname{Ker} A := \{x \in X : Ax = 0\}, \qquad \operatorname{Im} A := \{Ax : x \in X\}.$$

The operator A is said to be *Fredholm* if $\operatorname{Im} A$ is a closed subspace of X and the two numbers

$$\alpha(A) := \dim \operatorname{Ker} A, \qquad \beta(A) := \dim(X/\operatorname{Im} A)$$

are finite. The space $X/\operatorname{Im} A$ is also frequently referred to as the cokernel of A and denoted by $\operatorname{Coker} A$. If A is Fredholm, the index of A is defined as the integer

$$\operatorname{Ind} A := \alpha(A) - \beta(A).$$

Calkin algebra and essential spectrum. Let $\mathcal{K}(X)$ denote the set of all compact operators on a Banach space X. It is well known that $\mathcal{K}(X)$ is a closed two-sided ideal of the Banach algebra $\mathcal{B}(X)$. One can show that an operator $A \in \mathcal{B}(X)$ is Fredholm if and only if the coset $A + \mathcal{K}(X)$ is invertible in the quotient algebra $\mathcal{B}(X)/\mathcal{K}(X)$ (see, e.g., [87, Chapter 4, Theorem 7.1] or [59, Theorem 5.17]). The algebra $\mathcal{B}(X)/\mathcal{K}(X)$ is also referred to as the *Calkin algebra* of X. The *essential spectrum* $\operatorname{sp}_{\text{ess}} A$ of $A \in \mathcal{B}(X)$ is the spectrum of $A + \mathcal{K}(X)$ in $\mathcal{B}(X)/\mathcal{K}(X)$, that is,

$$\operatorname{sp}_{\text{ess}} A := \{\lambda \in \mathbf{C} : A - \lambda I \text{ is not Fredholm on } X\}.$$

Clearly, $\operatorname{sp}_{\text{ess}} A \subset \operatorname{sp} A$.

Hardy spaces. In Section 1.2 we saw that the basic properties of Laurent matrices result from the fact that they are canonical matrix representations of multiplication operators on L^2. To proceed in an analogous way in the context of Toeplitz operators, we need the *Hardy spaces* $H^2 := H^2(\mathbf{T})$ and $H^2_- := H^2_-(\mathbf{T})$. By definition,

$$H^2 := \{f \in L^2 : f_n = 0 \text{ for } n < 0\}, \quad H^2_- := \{f \in L^2 : f_n = 0 \text{ for } n \geq 0\},$$

where $\{f_n\}_{n \in \mathbf{Z}}$ is the sequence of the Fourier coefficients of f. The spaces H^2 and H^2_- are closed subspaces of L^2, and it is clear that L^2 decomposes into the orthogonal sum

$$L^2 = H^2 \oplus H^2_-.$$

Let P stand for the orthogonal projection of L^2 onto H^2. The functions

$$\left\{\frac{1}{\sqrt{2\pi}}e^{in\theta}\right\}_{n=0}^{\infty}$$

form an orthonormal basis in H^2, and it can be readily verified that if $a \in L^\infty$, then the matrix representation of the operator

$$H^2 \to H^2, \quad f \mapsto P(af) \tag{1.15}$$

is just the Toeplitz matrix $T(a)$. Notice that the operator (1.15) is the compression $PM(a)P$ of the multiplication operator $M(a)$ to H^2.

A crucial property of functions in H^2 is revealed by the *F. and M. Riesz theorem*: a function in H^2 vanishes either almost everywhere or almost nowhere on $\mathbf{T}$ (see, e.g., [59, Theorem 6.13]).

Theorem 1.10 (Coburn's Lemma). *Let $a \in L^\infty$ and suppose a does not vanish identically. Then $T(a)$ has a trivial kernel on l^2 or its image is dense in l^2. In particular, $T(a)$ is invertible if and only if $T(a)$ is Fredholm of index zero:*

$$\operatorname{sp} T(a) = \operatorname{sp}_{\text{ess}} T(a) \cup \big\{\lambda \in \mathbf{C} \setminus \operatorname{sp}_{\text{ess}} T(a) : \operatorname{Ind}\big(T(a) - \lambda I\big) \neq 0\big\}. \tag{1.16}$$

Proof. The adjoint operator of $T(a)$ is $T(\overline{a})$ where $\overline{a}(t) := \overline{a(t)}$ for $t \in \mathbf{T}$ and the bar denotes complex conjugation. Assume that $T(a)$ has a nontrivial kernel and that the image of $T(a)$ is not dense. The latter assumption implies that $T(\overline{a})$ has a nontrivial kernel. Hence, there are nonzero functions $f_+ \in H^2$ and $g_+ \in H^2$ such that $af_+ =: f_- \in H^2_-$ and $\overline{a}g_+ =: g_- \in H^2_-$ (recall (1.15)). By the Riesz brothers' theorem, $f_+ \neq 0$ and $g_+ \neq 0$ almost everywhere (a.e.) on $\mathbf{T}$. We have

$$\overline{g}_- f_+ = a\overline{g}_+ f_+ = af_+\overline{g}_+ = f_-\overline{g}_+ =: \varphi.$$

Obviously, $\varphi \in L^1$. Moreover, $\varphi_n = (\overline{g}_- f_+)_n = 0$ for $n \leq 0$ and $\varphi_n = (f_-\overline{g}_+)_n = 0$ for $n \geq 0$. Consequently, $\varphi = 0$. Since $f_+ \neq 0$ a.e. on $\mathbf{T}$, we conclude that $g_- = 0$ a.e. on $\mathbf{T}$, and since $g_- = \overline{a}g_+$ and $g_+ \neq 0$ a.e. on $\mathbf{T}$, it follows that $a = 0$ a.e. on $\mathbf{T}$. This contradicts the hypothesis of the theorem and shows that $T(a)$ has a trivial kernel or a dense range.

Now suppose $T(a)$ is Fredholm of index zero. Then $\operatorname{Ker} T(a) = \{0\}$ or $\operatorname{Im} T(a) = l^2$ by what was already proved. If $T(a)$ were not invertible, we had $\alpha(T(a)) > 0$ and $\beta(T(a)) = 0$ or $\alpha(T(a)) = 0$ and $\beta(T(a)) > 0$. In either case it would follow that $\operatorname{Ind} T(a) \neq 0$. Thus, $T(a)$ must be invertible. Finally, (1.16) results from the equality $T(a) - \lambda I = T(a - \lambda)$. ■

Theorem 1.10 was established by Coburn [49] and independently (and in a more general setting) also by Simonenko [161]. For continuous symbols, the theorem is already in Gohberg's paper [78].

1.4 Hankel Matrices

With each sequence $\{a_n\}_{n\in\mathbf{Z}}$ of complex numbers we associate two *Hankel matrices*

$$A = \begin{pmatrix} a_1 & a_2 & a_3 & \dots \\ a_2 & a_3 & \dots & \dots \\ a_3 & \dots & \dots & \dots \\ \dots & \dots & \dots & \dots \end{pmatrix}, \quad \tilde{A} = \begin{pmatrix} a_{-1} & a_{-2} & a_{-3} & \dots \\ a_{-2} & a_{-3} & \dots & \dots \\ a_{-3} & \dots & \dots & \dots \\ \dots & \dots & \dots & \dots \end{pmatrix}. \tag{1.17}$$

The problem of describing the sequences $\{a_n\}_{n\in\mathbf{Z}}$ for which A or $\tilde{A}$ generate a bounded operator on l^2 is more delicate than in the Laurent and Toeplitz cases. Its solution is given by the following theorem.

Theorem 1.11 (Nehari). *The matrix A (resp., $\tilde{A}$) generates a bounded operator on l^2 if and only if there is a function $b \in L^\infty$ such that $b_n = a_n$ (resp., $b_{-n} = a_{-n}$) for all $n \ge 1$.*

A proof of this result is in almost every text on Hankel and Toeplitz operators. We here confine ourselves to a few remarks.

Let $a \in L^\infty$ and let $\{a_n\}_{n\in\mathbf{Z}}$ be the sequence of the Fourier coefficients of a. We then denote by $H(a)$ the matrix A of (1.17) and by $H(\tilde{a})$ the matrix $\tilde{A}$ of (1.17). Thus, for each $a \in L^\infty$ we define two Hankel matrices $H(a)$ and $H(\tilde{a})$. However, note that if we assign a new function $\tilde{a} \in L^\infty$ to a by the formula

$$\tilde{a}(t) := a(1/t) \quad (t \in \mathbf{T}), \tag{1.18}$$

then $H(\tilde{a})$ is nothing but $H(c)$ with $c = \tilde{a}$.

If $a \in L^\infty$, then the boundedness of $H(a)$ and $H(\tilde{a})$ is immediate from Theorem 1.1, because $H(a)$ is the left lower quarter and $H(\tilde{a})$ is the right upper quarter of $L(a)$. To be more precise, define P, Q, J on $l^2(\mathbf{Z})$ as follows:

$$(Px)_k := \begin{cases} x_k & \text{for} \quad k \ge 0, \\ 0 & \text{for} \quad k < 0, \end{cases} \qquad (Qx)_k := \begin{cases} x_k & \text{for} \quad k < 0, \\ 0 & \text{for} \quad k \ge 0, \end{cases}$$

$$(Jx)_k := x_{-k-1}.$$

An easy computation shows that

$$H(a) = PL(a)QJ|\operatorname{Im} P \quad \text{and} \quad H(\tilde{a}) = JQL(a)P|\operatorname{Im} P,$$

which proves that $\|H(a)\|$ and $\|H(\tilde{a})\|$ do not exceed $\|L(a)\| = \|a\|_\infty$.

Caution. Let A be the left matrix of (1.17) and suppose A generates a bounded operator on l^2. Then, by Theorem 1.11, there is a $b \in L^\infty$ such that $A = H(b)$. However, this does not imply that there exists a $c \in L^\infty$ such that $c_n = 0$ for $n \le 0$ and $c_n = a_n$ for $n \ge 1$, i.e., the function c given formally by

$$c(t) = \sum_{n=1}^{\infty} a_n t^n \quad (t \in \mathbf{T})$$

need not belong to L^∞. To see this, consider the function

$$b(e^{i\theta}) = -i\theta, \quad \theta \in [0, 2\pi). \tag{1.19}$$

Clearly, $b \in PC \subset L^\infty$ and

$$H(b) = \begin{pmatrix} \frac{1}{1} & \frac{1}{2} & \frac{1}{3} & \dots \\ \frac{1}{2} & \frac{1}{3} & \dots & \dots \\ \frac{1}{3} & \dots & \dots & \dots \\ \dots & \dots & \dots & \dots \end{pmatrix},$$

but the function

$$e^{i\theta} + \frac{1}{2}e^{2i\theta} + \frac{1}{3}e^{3i\theta} + \dots = -\log(1 - e^{i\theta})$$

is not bounded.

Norm of a Hankel operator. Given $a \in L^\infty$, there are infinitely many different functions $b \in L^\infty$ such that $H(a) = H(b)$. One can show that

$$\|H(a)\| = \inf\{\|b\|_\infty : H(b) = H(a)\}. \tag{1.20}$$

The role played by Hankel matrices in Toeplitz theory is uncovered by the following simple but important result.

Proposition 1.12. *If $a, b \in L^\infty$, then*

$$T(ab) = T(a)T(b) + H(a)H(\tilde{b}).$$

Proof. With P, Q, J as above, this is nothing but the obvious identity

$$PabP = PaPbP + PaQbP = PaPPbP + PaQJJQbP. \blacksquare$$

Thus, unlike (1.5), the product of two Toeplitz matrices is in general not a Toeplitz matrix.

Notes. Theorem 1.11 and the equality (1.20) are due to Nehari [126]. Full proofs are in [39, Theorem 2.11], [127, Lecture VIII], [129, Chapter 3], [135, Chapter 1], and [147, Chapter 9]. Results like Proposition 1.12 have been used for more than fifty years; in the form cited here, Proposition 1.12 appeared in Widom's paper [185] for the first time.

1.5 Wiener-Hopf Factorization

Triangular Toeplitz matrices. Let

$$H^\infty := \{a \in L^\infty : a_n = 0 \text{ for } n < 0\}, \quad \overline{H^\infty} := \{a \in L^\infty : a_n = 0 \text{ for } n > 0\}.$$

The sets H^∞ and $\overline{H^\infty}$ are obviously closed subalgebras of the Banach algebra L^∞. Clearly, $H^\infty \cap \overline{H^\infty} = \mathbf{C}$, where $\mathbf{C}$ here refers to the constant functions. While $H^2 + H^2_- = L^2$ (recall Section 1.3), the sum $H^\infty + \overline{H^\infty}$ does not coincide with L^∞; for instance, function (1.19) can be shown to be not in $H^\infty + \overline{H^\infty}$.

If $a \in H^\infty$, then $H(\tilde{a}) = 0$ and $T(a)$ is lower triangular. Analogously, if $a \in \overline{H^\infty}$, then $H(a) = 0$ and $T(a)$ is upper triangular. These observations and Proposition 1.12 imply the following fact.

Proposition 1.13. *If* $a \in \overline{H^\infty}$, $b \in L^\infty$, $c \in H^\infty$, *then*

$$T(abc) = T(a)T(b)T(c). \blacksquare$$

This proposition is the origin of the so-called *Wiener-Hopf factorization.* Given $a \in L^\infty$, one looks for functions $a_- \in \overline{H^\infty}$ and $a_+ \in H^\infty$ such that $a = a_- a_+$. Provided we have found a_- and a_+, we can factorize $T(a) = T(a_-)T(a_+)$ by virtue of Proposition 1.13. If, in addition, a_- is invertible in $\overline{H^\infty}$ and a_+ is invertible in H^∞, then, again due to Proposition 1.13,

$$T(a_-^{-1})T(a_-) = T(a_-^{-1}a_-) = I = T(a_- a_-^{-1}) = T(a_-)T(a_-^{-1}),$$
$$T(a_+^{-1})T(a_+) = T(a_+^{-1}a_+) = I = T(a_+ a_+^{-1}) = T(a_+)T(a_+^{-1}),$$

which shows that $T(a)$ is invertible and that

$$T^{-1}(a) = T(a_+^{-1})T(a_-^{-1}).$$

Note that if $a \in H^\infty$ is invertible in L^∞, then a need not necessarily be invertible in H^∞ (example: $a(e^{i\theta}) = e^{i\theta}$).

Our next purpose is to construct a Wiener-Hopf factorization for functions in the Wiener algebra W (recall Example 1.5).

The group of invertible elements. Let $\mathcal{A}$ be a Banach algebra with identity element e. We denote by $G\mathcal{A}$ the collection of all invertible elements of $\mathcal{A}$. The set $G\mathcal{A}$ is a multiplicative group and an open subset of $\mathcal{A}$. Let $G_0\mathcal{A}$ stand for the connected component of $G\mathcal{A}$ which contains the identity. If $\mathcal{A}$ is *commutative*, then

$$a \in G_0\mathcal{A} \iff a = \exp(b) \text{ with some } b \in \mathcal{A} \tag{1.21}$$

(see, e.g., [59, Corollary 2.15]).

Winding number. The group $GC := GC(\mathbf{T})$ consists precisely of the functions in C which have no zeros on $\mathbf{T}$. For $a \in GC$, we denote by $\text{wind}(a, 0)$ the *winding number* of a with respect to the origin: every function $a \in GC$ may be written in the form $a = |a|e^{ic}$ where $c : \mathbf{T} \setminus \{1\} \to \mathbf{R}$ is continuous, and $\text{wind}(a, 0)$ is defined as the integer

$$\frac{1}{2\pi}\big(c(1-0) - c(1+0)\big). \tag{1.22}$$

Notice that c is unique up to an additive constant of the form $2k\pi$ $(k \in \mathbf{Z})$ and that (1.22) does not depend on the particular choice of c.

If $a \in C(\mathbf{T})$ and $\lambda \in \mathbf{C} \setminus a(\mathbf{T})$, we define the winding number $\text{wind}(a, \lambda)$ of a with respect to λ by $\text{wind}(a, \lambda) = \text{wind}(a - \lambda, 0)$ (see Figure 5).

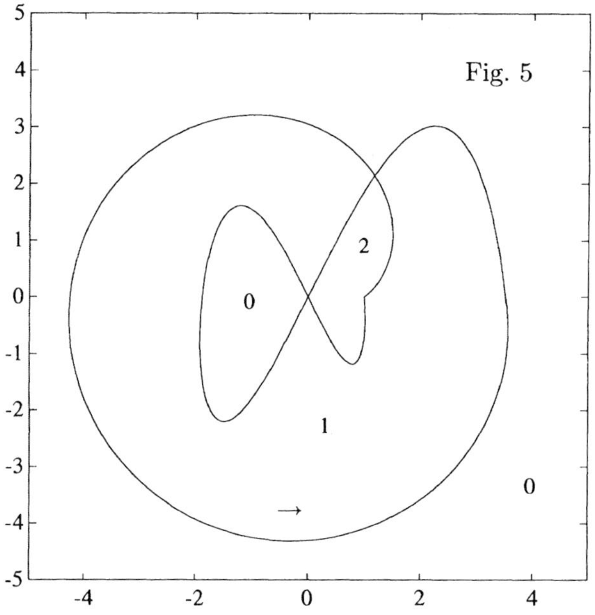

The "dolphin" curve in Figure 5 divides the plane into four regions. Figure 5 shows the winding number of the curve with respect to the points in these regions.

For $n \in \mathbf{Z}$, we define χ_n by $\chi_n(t) = t^n$ $(t \in \mathbf{T})$. Equivalently, $\chi_n(e^{i\theta}) = e^{in\theta}$. By the definition of the winding number, $\text{wind}(\chi_n, 0) = n$. It is easily seen that if $a, b \in GC$, then

$$\text{wind}(ab, 0) = \text{wind}(a, 0) + \text{wind}(b, 0).$$

Further, it is well known that

$$G_0C = \big\{a \in GC : \text{wind}(a, 0) = 0\big\}. \tag{1.23}$$

By Wiener's theorem, we have $GW = W \cap GC$. One can show that the analogue of (1.23) is also true in the Wiener algebra:

$$G_0 W = \{a \in GW : \text{wind}(a,0) = 0\}. \tag{1.24}$$

Theorem 1.14 (Wiener-Hopf factorization for Wiener functions). *Let* $a \in W \cap GC$ *and let* $\text{wind}(a,0) = \varkappa$. *Then*

$$a = a_- \chi_\varkappa a_+, \tag{1.25}$$

where $a_- \in W \cap G\overline{H^\infty}$ *and* $a_+ \in W \cap GH^\infty$.

The following proof shows how the factors $a_\pm$ can be found.

Proof. Since $\text{wind}(a\chi_{-\varkappa}, 0) = 0$, we deduce from (1.21) and (1.24) that $a\chi_{-\varkappa} = e^b$ with $b \in W$. Thus,

$$b(t) = \sum_{n\in\mathbf{Z}} b_n t^n \ (t \in \mathbf{T}) \quad \text{and} \quad \sum_{n\in\mathbf{Z}} |b_n| < \infty.$$

Define $b_+ \in W \cap H^\infty$ and $b_- \in W \cap \overline{H^\infty}$ by

$$b_+(t) = \sum_{n=0}^{\infty} b_n t^n, \quad b_-(t) = \sum_{n=-\infty}^{-1} b_n t^n \quad (t \in \mathbf{T}).$$

We have $b = b_- + b_+$ and hence (1.25) holds with $a_- := e^{b_-}$ and $a_+ := e^{b_+}$. Obviously, $a_+^{-1} = e^{-b_+} \in W \cap H^\infty$ and $a_-^{-1} = e^{-b_-} \in W \cap \overline{H^\infty}$. ■

Analytic Wiener functions. Put $W_+ := W \cap H^\infty$ and $W_- := W \cap \overline{H^\infty}$. The preceding proof worked for functions in W because $W = W_- + W_+$ (and it does not work for functions in L^∞ for several reasons, one being that $L^\infty \neq \overline{H^\infty} + H^\infty$).

Let $\mathbf{D} := \{z \in \mathbf{C} : |z| < 1\}$ be the open unit disk. If $a_+ \in W_+$, then a_+ can be extended to an analytic function in $\mathbf{D}$ by the formula

$$a_+(z) := \sum_{n=0}^{\infty} a_n z^n \quad (z \in \mathbf{D}),$$

where $\{a_n\}_{n=0}^{\infty}$ is the sequence of the Fourier coefficients of a. Analogously, a function $a_- \in W_-$ admits analytic continuation to $(\mathbf{C} \cup \{\infty\}) \setminus (\mathbf{D} \cup \mathbf{T})$ via

$$a_-(z) := \sum_{n=0}^{\infty} a_{-n} z^{-n} \quad (z \notin \mathbf{D} \cup \mathbf{T}).$$

One can show that if $a_\pm \in W_\pm$, then

$$a_+ \in W \cap GH^\infty \Leftrightarrow a_+ \in GW_+ \Leftrightarrow a_+(z) \neq 0 \ \forall z \in \mathbf{D} \cup \mathbf{T},$$
$$a_- \in W \cap G\overline{H^\infty} \Leftrightarrow a_- \in GW_- \Leftrightarrow a_-(z) \neq 0 \ \forall z \in \big(\mathbf{C} \cup \{\infty\}\big) \setminus \mathbf{D}.$$

Theorem 1.15 (M.G. Krein). *Suppose $a \in W$. The operator $T(a)$ is Fredholm on l^2 if and only if a has no zeros on $\mathbf{T}$. In that case*

$$\operatorname{Ind} T(a) = -\operatorname{wind}(a, 0).$$

In particular, $T(a)$ is invertible if and only if $a(t) \neq 0$ for all $t \in \mathbf{T}$ and $\operatorname{wind}(a,0) = 0$. *In the latter case, the inverse is*

$$T^{-1}(a) = T(a_+^{-1})T(a_-^{-1}),$$

where $a = a_- a_+$ is any factorization as in Theorem 1.14.

Proof. Assume a has no zeros on $\mathbf{T}$, put $\varkappa := \operatorname{wind}(a,0)$, and factorize a as in (1.25). Then $T(a) = T(a_-)T(\chi_\varkappa)T(a_+)$ due to Proposition 1.13, and the operators $T(a_\pm)$ are invertible by the remark after Proposition 1.13. Looking at the matrix of $T(\chi_\varkappa)$ it is readily seen that $T(\chi_\varkappa)$ is Fredholm and that

$$\alpha\big(T(\chi_\varkappa)\big) = \max\{-\varkappa, 0\}, \quad \beta\big(T(\chi_\varkappa)\big) = \max\{0, \varkappa\}.$$

Consequently, $\operatorname{Ind} T(\chi_\varkappa) = -\varkappa$. This implies that $T(a)$ is Fredholm and that

$$\operatorname{Ind} T(a) = \operatorname{Ind} T(a_-) + \operatorname{Ind} T(\chi_\varkappa) + \operatorname{Ind} T(a_+) = 0 + (-\varkappa) + 0 = -\varkappa.$$

Now assume the curve $a(\mathbf{T})$ passes through the origin but $T(a)$ is Fredholm. Let $\varkappa := \operatorname{Ind} T(a)$. On slightly perturbing a, we can produce functions $b, c \in W$ without zeros on $\mathbf{T}$ such that $\|a - b\|_\infty$ and $\|a - c\|_\infty$ are as small as desired and $|\operatorname{wind}(b,0) - \operatorname{wind}(c,0)| = 1$. As the property of being Fredholm and the index are stable under small perturbations, it follows that $T(b)$ and $T(c)$ are Fredholm and that $\operatorname{Ind} T(b) = \operatorname{Ind} T(c) = \varkappa$. However, from what was already proved we know that

$$|\operatorname{Ind} T(b) - \operatorname{Ind} T(c)| = |\operatorname{wind}(b,0) - \operatorname{wind}(c,0)| = 1.$$

This contradiction shows that $T(a)$ cannot be Fredholm.

The assertions concerning the invertibility and the inverse of $T(a)$ follow from Theorem 1.10 and the discussion after Proposition 1.13. ∎

Notes. The books [87] and [59] contain very readable expositions of the properties of $G\mathcal{A}$ and of Fredholm operators we used in this section.

What we call Wiener-Hopf factorization is a method which was developed by Gakhov [75], [76] (but see also Plemelj's paper [134]). Mark Krein [110] was the first to understand the Banach algebraic background of Wiener-Hopf factorization and to present the method in a crystal-clear manner. The results of this section are all due to him.

1.6 Continuous Symbols

Since Fredholmness is equivalent to invertibility modulo compact operators, Proposition 1.12 motivates the search for compactness criteria for Hankel operators. The following theorem provides such a criterion.

Theorem 1.16 (Hartman). *The matrix A of* (1.17) *generates a compact operator on l^2 if and only if there is a function $b \in C$ such that $b_n = a_n$ for all $n \geq 1$. Equivalently, if $a \in L^\infty$, then $H(a)$ is compact on l^2 if and only if*

$$a \in C + \overline{H^\infty} := \{c + g : c \in C,\ g \in \overline{H^\infty}\}.$$

This theorem is proved in many standard texts on Hankel and Toeplitz operators. In what follows, we merely need the sufficiency part of the theorem, which can be easily verified. Indeed, let $c \in C$ and $g \in \overline{H^\infty}$. Then $H(c+g) = H(c)$, and we are left with showing that $H(c)$ is compact. Let $\{\varphi_n\}_{n=1}^\infty$ be any sequence of trigonometric polynomials converging uniformly to c on $\mathbf{T}$; for instance, let φ_n be the nth Fejer-Cesaro mean of c. We then have

$$\|H(c) - H(\varphi_n)\| = \|H(c - \varphi_n)\| \leq \|c - \varphi_n\|_\infty = o(1),$$

and as the operators $H(\varphi_n)$ have finite rank, it follows that $H(c)$ is compact.

We are now in a position to describe the spectra of Toeplitz operators with continuous symbols. If $a \in C$, then $a(t)$ traces out a continuous closed oriented curve $a(\mathbf{T})$ as t traverses $\mathbf{T}$ in the counterclockwise direction. As in Section 1.5, given $\lambda \in \mathbf{C}\backslash a(\mathbf{T})$, we let $\operatorname{wind}(a,\lambda)$ stand for the winding number of the curve $a(\mathbf{T})$ about λ, that is, $\operatorname{wind}(a,\lambda) := \operatorname{wind}(a-\lambda, 0)$.

Theorem 1.17 (Gohberg). *Let $a \in C$. The operator $T(a)$ is Fredholm on the space l^2 if and only if a has no zeros on $\mathbf{T}$, in which case*

$$\operatorname{Ind} T(a) = -\operatorname{wind}(a, 0).$$

Equivalently,

$$\operatorname{sp}_{\text{ess}} T(a) = a(\mathbf{T}), \tag{1.26}$$

$$\operatorname{sp} T(a) = a(\mathbf{T}) \cup \{\lambda \in \mathbf{C}\backslash a(\mathbf{T}) : \operatorname{wind}(a,\lambda) \neq 0\}. \tag{1.27}$$

Proof. Suppose first that $a \in GC$. By Proposition 1.12,

$$T(a^{-1})T(a) = I - H(a^{-1})H(\tilde{a}),$$
$$T(a)T(a^{-1}) = I - H(a)H(\tilde{a}^{-1}),$$

and since all occurring Hankel operators are compact due to (the sufficiency portion of) Theorem 1.16, the operator $T(a^{-1})$ is an inverse of $T(a)$ modulo compact operators. This shows that $T(a)$ is Fredholm.

Let $\text{wind}(a, 0) = \varkappa$. Then a is homotopic to the function $\chi_\varkappa$ defined by $\chi_\varkappa(t) = t^\varkappa$ $(t \in \mathbf{T})$ within GC. Therefore

$$\text{Ind}\, T(a) = \text{Ind}\, T(\chi_\varkappa).$$

In the proof of Theorem 1.15 we observed that $\text{Ind}\, T(\chi_\varkappa) = -\varkappa$.

The index perturbation argument of the proof of Theorem 1.15 yields that necessarily $a \in GC$ if $T(a)$ is Fredholm.

Finally, since $T(a) - \lambda I = T(a - \lambda)$, we arrive at (1.26) and (1.27). ■

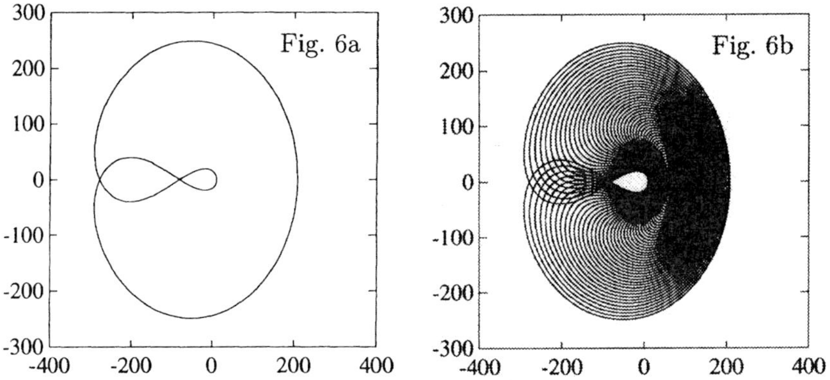

For a special symbol a, we see $\text{sp}_{\text{ess}}\, T(a)$ in Figure 6a, while $\text{sp}\, T(a)$ is indicated in Figure 6b.

Notes. Theorem 1.16 was established by Hartman [97]. Full proofs are in [39, Theorem 2.54], [127, Lecture VIII], [129, Chapter 3], [135, Chapter 1], or [147, Chapter 9].

Theorem 1.17 has a long history. That $T(a)$ is Fredholm and has the index $-\text{wind}(a, 0)$ whenever $a \in GC$ is more or less explicit in the works by F. Noether, S.G. Mikhlin, N.I. Muskhelishvili, F.D. Gakhov, V.V. Ivanov, M.G. Krein, A.P. Calderón, F. Spitzer, H. Widom, A. Devinatz, G. Fichera, and certainly others. In the form cited here, the theorem appeared in Gohberg's papers [77], [78]. Figure 6 illustrates Theorem 1.17.

1.7 Locally Sectorial Symbols

We now turn to Toeplitz operators with discontinuous symbols. The following result provides a useful upper estimate of the spectrum.

Theorem 1.18 (Brown-Halmos). *If* $a \in L^\infty$ *then*

$$\operatorname{sp} T(a) \subset \operatorname{conv} \mathcal{R}(a),$$

where $\operatorname{conv} \mathcal{R}(a)$ *stands for the convex hull of the essential range* $\mathcal{R}(a)$ *of the function* a.

Proof. Let $\lambda \in \mathbf{C} \setminus \operatorname{conv} \mathcal{R}(a)$ and put $b = a - \lambda$. Then $0 \notin \operatorname{conv} \mathcal{R}(b)$. Hence, there is a $\gamma \in \mathbf{T}$ such that $\gamma \operatorname{conv} \mathcal{R}(b) = \operatorname{conv} \mathcal{R}(\gamma b)$ is completely contained in the right open half-plane. This implies that we can find a $\delta > 0$ such that $\|1 - \delta\gamma b\|_\infty < 1$, whence

$$\|I - \delta\gamma T(b)\| = \|T(1 - \delta\gamma b)\| \le \|1 - \delta\gamma b\|_\infty < 1.$$

Consequently, $\delta\gamma T(b)$ and thus also $T(b) = T(a) - \lambda I$ is invertible. ■

Sectoriality. A function $a \in L^\infty$ is said to be *sectorial* if $0 \notin \operatorname{conv} \mathcal{R}(a)$. The previous theorem says that Toeplitz operators with sectorial symbols are invertible. It is easily seen (Figure 7) that $a \in L^\infty$ is sectorial if and only if

$$a \in GL^\infty \quad \text{and} \quad \operatorname{dist}(a/|a|, \mathbf{C}) < 1, \tag{1.28}$$

where $(a/|a|)(t) := a(t)/|a(t)|$, $\mathbf{C}$ stands for the constant functions, and the distance is measured in the L^∞ norm, i.e.,

$$\operatorname{dist}(f, \mathbf{C}) = \inf_{c \in \mathbf{C}} \|f - c\|_\infty.$$

Local sectoriality. Let $a \in L^\infty$ and $\tau \in \mathbf{T}$. Given a subarc $U \subset \mathbf{T}$, we denote by $L^\infty(U)$ the essentially bounded functions on U and we define $\mathcal{R}_U(a)$ as the essential range of the restriction $a|U$, that is, as the spectrum of $a|U$ in the Banach algebra $L^\infty(U)$. Finally, we let $\mathcal{U}_\tau$ denote the collection of all arcs $U \subset \mathbf{T}$ containing τ, and we set

$$\mathcal{R}_\tau(a) := \bigcap_{U \in \mathcal{U}_\tau} \mathcal{R}_U(a).$$

The set $\mathcal{R}_\tau(a)$ is called the *local essential range* of a at τ. The points in $\mathcal{R}_\tau(a)$ are also referred to as the *essential cluster points* of a at τ. If $a \in PC$, then obviously

$$\mathcal{R}_\tau(a) = \big\{a(\tau - 0),\ a(\tau + 0)\big\}. \tag{1.29}$$

The function a is called *locally sectorial at* $\tau \in \mathbf{T}$ if $0 \notin \operatorname{conv} \mathcal{R}_\tau(a)$ and it is said to be *locally sectorial on* $\mathbf{T}$ if $0 \notin \operatorname{conv} \mathcal{R}_\tau(a)$ for every $\tau \in \mathbf{T}$ (see Figure 8).

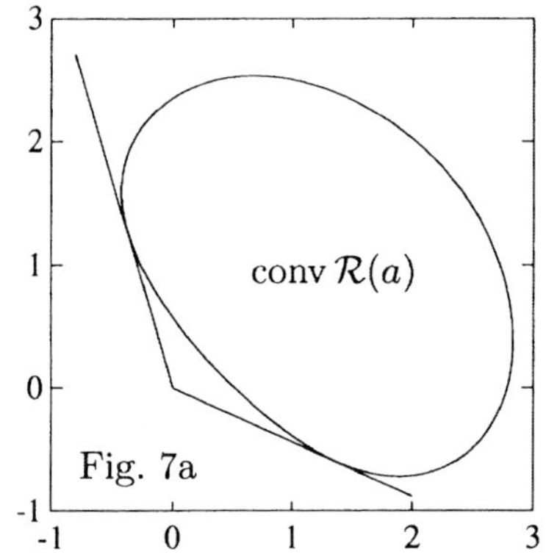

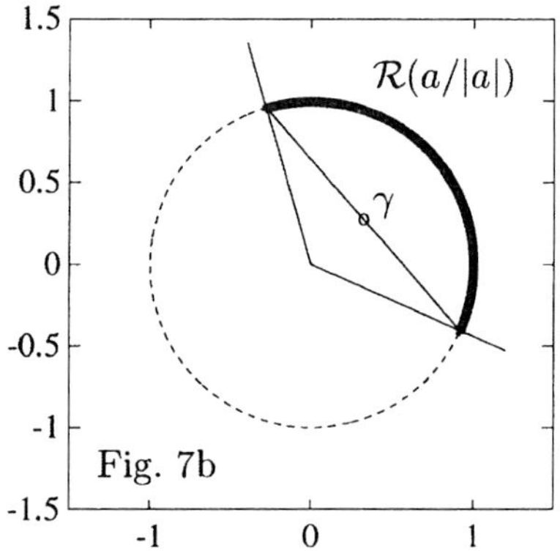

Figure 7a tells us why a function a is called sectorial if $0 \notin \operatorname{conv} \mathcal{R}(a)$: in that case $\mathcal{R}(a)$ is contained in some open sector with the vertex at the origin and with an opening less than π. If $\operatorname{conv} \mathcal{R}(a)$ is as in Figure 7a, then $\operatorname{dist}(a/|a|, \mathbf{C}) = \|a/|a| - \gamma\|_\infty$ with γ as in Figure 7b.

For $f \in L^\infty$ and $\tau \in \mathbf{T}$, put

$$\varrho_\tau(f) := \max_{\lambda \in \mathcal{R}_\tau(f)} |\lambda|, \qquad \operatorname{dist}_\tau(f, \mathbf{C}) := \inf_{c \in \mathbf{C}} \varrho_\tau(f - c).$$

Thus, $\varrho_\tau(f)$ is the "local L^∞ norm" of f at τ, while $\operatorname{dist}_\tau(f, \mathbf{C})$ is the "local distance" of f at τ to the constants. A little thought reveals that $a \in GL^\infty$ is locally sectorial at $\tau \in \mathbf{T}$ if and only if

$$\operatorname{dist}_\tau(a/|a|, \mathbf{C}) < 1. \tag{1.30}$$

Lemma 1.19. *If $f \in L^\infty$, then*

$$\operatorname{dist}(f, C) = \max_{\tau \in \mathbf{T}} \operatorname{dist}_\tau(f, \mathbf{C}), \tag{1.31}$$

where

$$\operatorname{dist}(f, C) := \inf\{\|f - c\|_\infty : c \in C\}.$$

The maximum in (1.31) *is attained at some $\tau_0 \in \mathbf{T}$.*

Proof. The estimate "$\geq$" in (1.31) is obvious. To prove the reverse estimate, fix $\varepsilon > 0$. For each $\tau \in \mathbf{T}$, there are a number $\gamma_\tau \in \mathbf{C}$ and an arc $U_\tau \in \mathcal{U}_\tau$ such that

$$\|f|U_\tau - \gamma_\tau\|_{L^\infty(U_\tau)} < \operatorname{dist}_\tau(f, \mathbf{C}) + \varepsilon.$$

Choose finitely many $U_j := U_{\tau_j}$ and $\gamma_j := \gamma_{\tau_j}$ such that $\mathbf{T}$ is covered by the union of the U_j's. There exist functions $\varphi_j \in C$ such that

$$0 \leq \varphi_j \leq 1, \quad \operatorname{supp} \varphi_j \subset U_j, \quad \sum_j \varphi_j = 1.$$

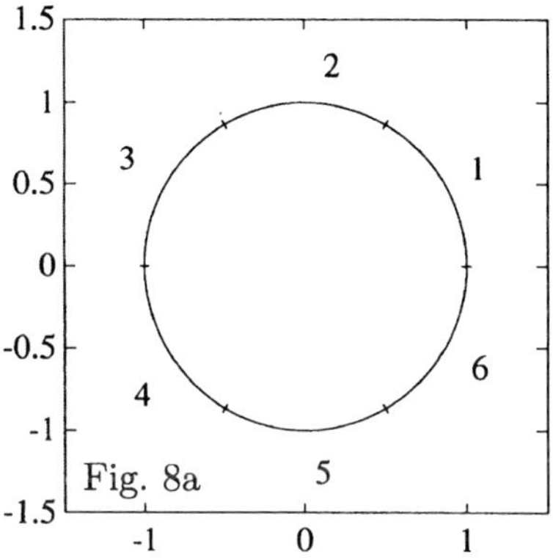

Fig. 8a

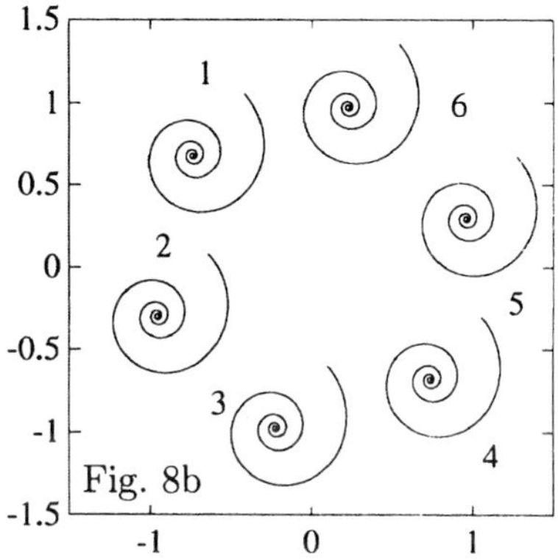

Fig. 8b

If $\mathbf{T}$ is divided into 6 arcs as in Figure 8a and if the images of these arcs are as in Figure 8b, then a is locally (but not globally) sectorial.

Since

$$\left\| f - \sum_j \gamma_j \varphi_j \right\|_\infty = \left\| f \sum_j \varphi_j - \sum_j \gamma_j \varphi_j \right\|_\infty$$

$$= \left\| \sum_j (f - \gamma_j) \varphi_j \right\|_\infty \le \max_j \| f | U_j - \gamma_j \|_{L^\infty(U_j)},$$

we get the estimate "$\le$" in (1.31).

The map $\tau \mapsto \varrho_\tau(f)$ is easily seen to be upper semicontinuous. This implies that the map $\tau \mapsto \operatorname{dist}_\tau(f, \mathbf{C})$ is also upper semicontinuous, by virtue of which the maximum in (1.31) is attained. ∎

Corollary 1.20. *If $a \in L^\infty$ is locally sectorial on $\mathbf{T}$, then there exist a function $c \in GC$ and a sectorial function $s \in GL^\infty$ such that $a = cs$.*

Proof. Put $u = a/|a|$. Combining (1.30) and Lemma 1.19, we obtain that $\operatorname{dist}(u, C) < 1$. Let $c \in C$ be any function such that $\|u - c\|_\infty < 1$. Because u is unimodular, we have $\|1 - u^{-1}c\|_\infty = \|u - c\|_\infty < 1$, which shows that $\sigma := u^{-1}c$ is sectorial. In particular, $c \in GC$. Furthermore, we can write $a = c|a|\sigma^{-1}$ and it is clear that $s := |a|\sigma^{-1}$ is sectorial together with σ. ∎

Theorem 1.21 (Simonenko). *If $a \in L^\infty$ is locally sectorial on $\mathbf{T}$, then $T(a)$ is Fredholm on l^2.*

Proof. Write $a = cs$ as in Corollary 1.20. From Proposition 1.12 we infer that $T(a) = T(c)T(s) + H(c)H(\tilde{s})$. The operator $H(c)$ is compact in view of the (sufficiency part of) Theorem 1.16, the operator $T(c)$ is Fredholm due to Theorem 1.17, and the operator $T(s)$ is invertible owing to Theorem 1.18. ∎

Winding number of a locally sectorial function. Let $a \in L^\infty$ be locally sectorial on $\mathbf{T}$ and let $a = cs$ with a function $c \in GC$ and a sectorial function s. From Theorem 1.17 and the proof of the preceding theorem we see that

$$\begin{aligned} -\text{wind}(c, 0) &= \operatorname{Ind} T(c) = \operatorname{Ind} T(c) + \operatorname{Ind} T(s) \\ &= \operatorname{Ind} T(c)T(s) = \operatorname{Ind} T(a). \end{aligned}$$

Thus, if $a = c_1 s_1 = c_2 s_2$ with $c_1, c_2 \in GC$ and with sectorial functions s_1, s_2, then $\text{wind}(c_1, 0) = -\operatorname{Ind} T(a) = \text{wind}(c_2, 0)$. In other words, the winding number $\text{wind}(c, 0)$ is independent of the special representation $a = cs$ as in Corollary 1.20 (also see Figure 9). We denote this number by $\text{wind}(a, 0)$ and so have the following result.

Theorem 1.22 (Simonenko). *If $a \in L^\infty$ is locally sectorial on $\mathbf{T}$, then*

$$\operatorname{Ind} T(a) = -\text{wind}(a, 0). \blacksquare$$

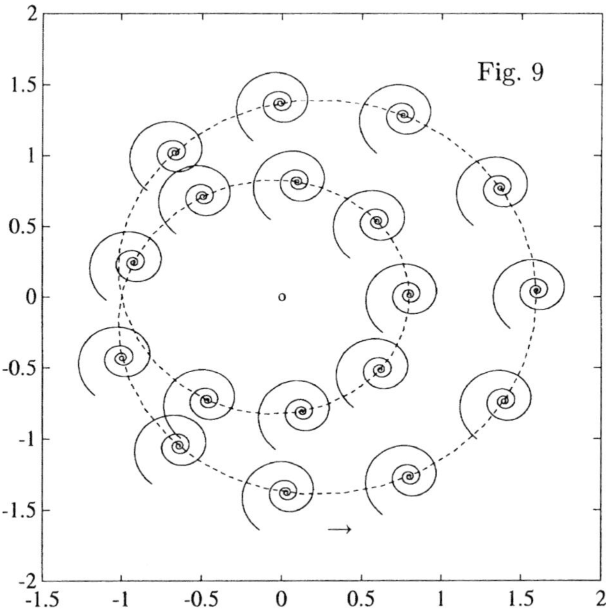

The winding number with respect to the origin of the locally sectorial symbol whose essential range is indicated in Figure 9 equals 2.

Notes. Theorem 1.18 appeared explicitly in the Brown-Halmos paper [43] for the first time, but it is implicit also in Simonenko's papers [158], [161]. Theorems 1.21 and 1.22 are Simonenko's [158], [161]. They were also obtained independently and by different methods by Devinatz [53], Douglas and Widom [63], and Douglas and Sarason [62]. One can show that $T(a)$ is Fredholm if a is locally sectorial in a much weaker sense than above, e.g., if a is locally sectorial over the fibers of QC (R. Douglas) or the maximal antisymmetric sets of $C+H^\infty$ (S. Axler); for more about this topic see [39, Sections 2.75–2.91].

1.8 Discontinuous Symbols

In this section we collect some results on the spectra of Toeplitz operators with discontinuous (but not necessarily locally sectorial) symbols. A detailed discussion of this topic is in [39].

Piecewise continuous symbols. Let first $a \in PC$ be a piecewise continuous function (recall Example 1.7). We denote by $a^{\#}(\mathbf{T})$ the closed continuous and naturally oriented curve which results from the essential range of a by filling in the line segment $[a(t-0),\, a(t+0)]$ between the endpoints $a(t-0)$ and $a(t+0)$ of each jump. For $\lambda \in \mathbf{C} \setminus a^{\#}(\mathbf{T})$, we let $\operatorname{wind}(a^{\#}, \lambda)$ stand for the winding number of $a^{\#}(\mathbf{T})$ with respect to λ (see Figure 10).

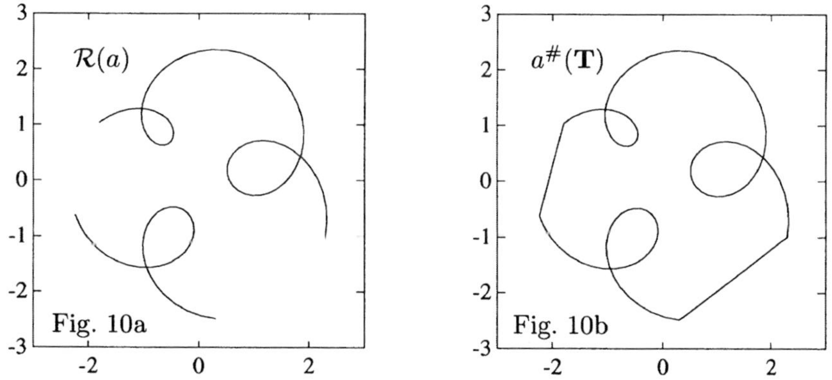

The essential range $\mathcal{R}(a)$ of a piecewise continuous function (Figure 10a) and the corresponding curve $a^{\#}(\mathbf{T})$ (Figure 10b).

The following beautiful result was discovered by many people, including Calderón, Spitzer, Widom, Devinatz, Gohberg, Krupnik, and Simonenko.

Theorem 1.23. *Let $a \in PC$. The operator $T(a)$ is Fredholm on l^2 if and only if $0 \notin a^{\#}(\mathbf{T})$. In that case*

$$\operatorname{Ind} T(a) = -\operatorname{wind}(a^{\#}, 0).$$

Thus,

$$\begin{aligned} \mathrm{sp}_{\mathrm{ess}}\, T(a) &= a^{\#}(\mathbf{T}), \\ \mathrm{sp}\, T(a) &= a^{\#}(\mathbf{T}) \cup \{\lambda \in \mathbf{C}\backslash a^{\#}(\mathbf{T}) : \mathrm{wind}(a^{\#}, \lambda) \neq 0\}. \end{aligned}$$

Proof. If $0 \notin a^{\#}(\mathbf{T})$ then, by (1.29), a is locally sectorial on $\mathbf{T}$. It is not difficult to verify that $\mathrm{wind}(a,0) = \mathrm{wind}(a^{\#},0)$, where $\mathrm{wind}(a,0)$ is understood as in Theorem 1.22. Therefore Theorems 1.21 and 1.22 imply the Fredholmness of $T(a)$ and the index formula. That $T(a)$ cannot be Fredholm if $0 \in a^{\#}(\mathbf{T})$ can be shown by the index perturbation argument of the proof of Theorem 1.15. ■

Example 1.24: Cauchy-Toeplitz matrices. For $\gamma \in \mathbf{C}\backslash\mathbf{Z}$, define $\psi_\gamma \in PC$ by (1.10) as in Section 1.2:

$$\psi_\gamma(e^{i\theta}) = (\pi/\sin\pi\gamma)e^{i\pi\gamma}e^{-i\gamma\theta}, \qquad \theta \in [0, 2\pi).$$

From (1.11) we infer that

$$0 \in \psi_\gamma^{\#}(\mathbf{T}) \iff e^{2\pi i\gamma} \in (-\infty, 0) \iff \mathrm{Re}\,\gamma - \frac{1}{2} \in \mathbf{Z}.$$

As θ moves from 0 to 2π, the argument of $e^{-i\gamma\theta}$ changes from 0 to $-2\pi\,\mathrm{Re}\,\gamma$. This shows that for $k \in \mathbf{Z}$,

$$k - \frac{1}{2} < \mathrm{Re}\,\gamma < k + \frac{1}{2} \implies \mathrm{wind}(\psi_\gamma^{\#}, 0) = -k$$

(see also Figure 11). Hence, Theorem 1.23 gives that

$$T(\psi_\gamma) \text{ is Fredholm of index } k \iff k - \frac{1}{2} < \mathrm{Re}\,\gamma < k + \frac{1}{2}.$$

In particular, $T(\psi_\gamma)$ is invertible if and only if $|\mathrm{Re}\,\gamma| < 1/2$. ■

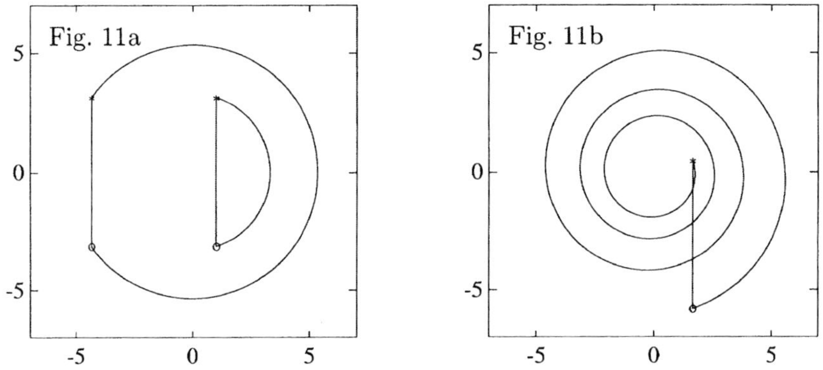

In Figure 11 we see $\mathrm{sp}_{\mathrm{ess}}\, T(\psi_\gamma)$ for $\gamma = 0.8$ and $\gamma = 0.4$ (Figure 11a) and for $\gamma = 3.25 + 0.2i$ (Figure 11b).

Two general results. Things are more complicated for symbols beyond PC, i.e., for symbols in $L^\infty \setminus PC$. We have the following two very useful theorems.

Theorem 1.25 (Hartman-Wintner). *If $a \in L^\infty$, then*

$$\mathcal{R}(a) \subset \operatorname{sp_{ess}} T(a).$$

From Theorems 1.18 and 1.25 we see that both $\operatorname{sp_{ess}} T(a)$ and $\operatorname{sp} T(a)$ are always included between $\mathcal{R}(a)$ and $\operatorname{conv} \mathcal{R}(a)$.

Theorem 1.26 (Douglas-Widom). *If a is in L^∞, then the two spectra $\operatorname{sp_{ess}} T(a)$ and $\operatorname{sp} T(a)$ are connected sets.*

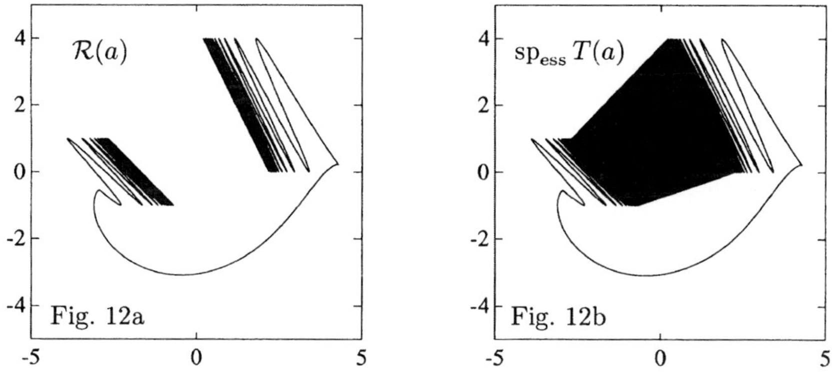

Figure 12a shows the essential range $\mathcal{R}(a)$ of a function a in PQC which has an oscillating discontinuity. The essential spectrum $\operatorname{sp_{ess}} T(a)$ is plotted in Figure 12b. It is clearly seen that $\operatorname{sp_{ess}} T(a)$ is a connected set. For PQC, see [39], for example.

Selfadjoint Toeplitz operators. The Toeplitz operator $T(a)$ $(a \in L^\infty)$ is selfadjoint if and only if $\overline{a}_n = a_{-n}$ for all $n \in \mathbf{Z}$, which is the case if and only if a is real-valued. Combining Theorems 1.18, 1.25, 1.26 we arrive at the following result.

Theorem 1.27 (Hartman-Wintner). *If $a \in L^\infty$ is real-valued, then*

$$\operatorname{sp_{ess}} T(a) = \operatorname{sp} T(a) = \operatorname{conv} \mathcal{R}(a).$$

There is a simple *direct proof* of this theorem. It suffices to consider the case where a is not constant. If $\lambda \notin \operatorname{conv} \mathcal{R}(a)$, then $a - \lambda$ is sectorial and hence $T(a) - \lambda I = T(a - \lambda)$ is invertible. So suppose $\lambda \in \operatorname{conv} \mathcal{R}(a)$, put $b = a - \lambda$, and assume $T(b)$ is Fredholm. Let $\operatorname{Ind} T(b) = \varkappa$. Since b is real-valued, we have

$$\operatorname{Ind} T(b) = \operatorname{Ind} T(\bar{b}) = \operatorname{Ind} T^*(b) = -\varkappa,$$

which implies that $\varkappa = 0$. Theorem 1.10 therefore shows that $T(b)$ is invertible. Let $x \in l^2$ be the solution of the equation $T(b)x = e_0$ where $(e_0)_1 = 1$ and $(e_0)_n = 0$ for $n \geq 2$. Denoting by $f \in H^2$ the function

$$f(t) = x_1 + x_2 t + x_3 t^2 + \dots \quad (t \in \mathbf{T}),$$

we see that $bf = 1 + g$ with $g \in H^2_-$ (recall (1.15)). Thus, if $n \geq 1$, then

$$\int_0^{2\pi} b(e^{i\theta})|f(e^{i\theta})|^2 e^{-in\theta}\, d\theta = \int_0^{2\pi} b(e^{i\theta}) f(e^{i\theta}) \overline{f(e^{i\theta})} e^{-in\theta}\, d\theta$$

$$= \int_0^{2\pi} \big(1 + g(e^{i\theta})\big) \overline{f(e^{i\theta})} e^{-in\theta}\, d\theta = 0.$$

Since $b|f|^2$ is real-valued, it follows that all Fourier coefficients with nonzero index of $b|f|^2$ must be equal to zero. Consequently, $b|f|^2 = (a - \lambda)|f|^2$ is some constant $c \in \mathbf{R}$. If $c = 0$, then $a = \lambda$ a.e. because $|f|^2 \neq 0$ a.e. by the F. and M. Riesz theorem. This case was excluded. Hence $c \neq 0$. If λ is an inner point of the segment $\operatorname{conv} \mathcal{R}(a)$, then $a - \lambda$ changes its sign and therefore $(a - \lambda)|f|^2$ cannot be a nonzero constant. Thus, every inner point of $\operatorname{conv} \mathcal{R}(a)$ belongs to $\operatorname{sp}_{\text{ess}} T(a)$. As $\operatorname{sp}_{\text{ess}} T(a)$ is a closed set, we conclude that $\operatorname{sp}_{\text{ess}} T(a) = \operatorname{conv} \mathcal{R}(a)$. ■

Triangular Toeplitz matrices. For such matrices we have the following result (also see Figure 13).

Theorem 1.28 (Wintner-Douglas). *Let $a \in H^\infty$. The operator $T(a)$ is invertible on l^2 if and only if $a^{-1} \in H^\infty$, and the operator $T(a)$ is Fredholm on l^2 if and only if $a^{-1} \in C + H^\infty := \{c + h : c \in C,\ h \in H^\infty\}$.*

General symbols. Recall that GH^∞ stands for the set of all functions $a \in H^\infty$ which are invertible in H^∞. Every function $a \in H^\infty$ can be analytically extended into the complex unit disk $\mathbf{D} := \{z \in \mathbf{C} : |z| < 1\}$. The analytic extension $\hat{a}$ of a is given by $\hat{a}(z) = \sum_{n=0}^\infty a_n z^n$ where $\{a_n\}_{n=0}^\infty$ is the sequence of the Fourier coefficients of a. One can show that

$$GH^\infty = \Big\{ a \in H^\infty : \inf_{z \in \mathbf{D}} |\hat{a}(z)| > 0 \Big\}.$$

The following theorem provides us with invertibility criteria for Toeplitz operators with general symbols in L^∞. Since always $\mathcal{R}(a) \subset \operatorname{sp} T(a)$, we may without loss of generality assume that a is invertible in L^∞.

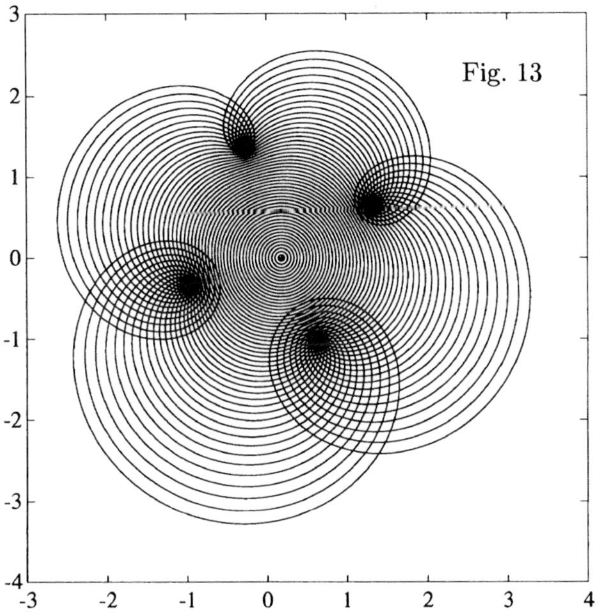

In Figure 13 we picked an analytic polynomial $a \in H^\infty$ and plotted $\hat{a}(r\mathbf{T})$ for $r = k/50$ $(k = 0, 1, \ldots, 50)$. Thus, Figure 13 indicates $\hat{a}(\mathbf{D}) = \operatorname{sp} T(a)$. Although the symbol at hand is continuous, Figure 13 nicely illustrates the invertibility part of Theorem 1.28. With respect to what follows in the forthcoming chapters, we also remark that the spectrum of every principal $n \times n$ section $T_n(a)$ of $T(a)$ is the singleton marked by $*$ in the center of Figure 13. Clearly, $\operatorname{sp} T_n(a)$ does not at all mimic $\operatorname{sp} T(a)$ as $n \to \infty$.

Theorem 1.29 (Widom-Devinatz). *Suppose $a \in GL^\infty$. Then the following are equivalent:*

(i) *$T(a)$ is invertible on l^2;*

(ii) *$T(a/|a|)$ is invertible on l^2;*

(iii) *$\operatorname{dist}_{L^\infty}(a/|a|, GH^\infty) < 1$, i.e., there exists an $h \in GH^\infty$ such that $\|a/|a| - h\|_\infty < 1$;*

(iv) *$a/|a| = hs$ where $h \in GH^\infty$ and $s \in L^\infty$ is sectorial.*

Note that the equivalence (i) $\Leftrightarrow$ (ii) tells us that invertibility of Toeplitz operators is solely determined by the "argument" of the symbol. Of course, (iii) and (iv) are in general difficult to check. In a sense, we can say that results like Theorems 1.17 or 1.23 represent verifiable necessary and sufficient conditions for (iii) and (iv) to be valid.

Notes. As for Theorem 1.25, we remark that Hartman and Wintner [98] showed that a is invertible in L^∞ if $T(a)$ is invertible, while Simonenko [161] observed that a is invertible in L^∞ if $T(a)$ is merely known to have closed range. The connectedness of $\operatorname{sp} T(a)$ was first proved by Widom [183], the connectedness of $\operatorname{sp}_{\mathrm{ess}} T(a)$ is due to Douglas [59, Theorem 7.45]. Theorem 1.27 was established in [98]. The H^∞ part of Theorem 1.28 is already in [192], the $C + H^\infty$ version of Theorem 1.28 was obtained by Douglas [58], [59, Corollary 7.34]. Theorem 1.29 goes back to Widom [181] and Devinatz [53]. Full proofs of the results of this section can also be found in [39]

2

Finite Section Method and Stability

2.1 Approximation Methods

Finite section method. Let $A = (a_{jk})_{j,k=1}^{\infty}$ be an infinite matrix and suppose A generates a bounded operator on l^2. In order to solve the equation $Ax = y$, i.e., the infinite linear system

$$\begin{pmatrix} a_{11} & a_{12} & a_{13} & \dots \\ a_{21} & a_{22} & a_{23} & \dots \\ a_{31} & a_{32} & a_{33} & \dots \\ \dots & \dots & \dots & \dots \end{pmatrix} \begin{pmatrix} x_1 \\ x_2 \\ x_3 \\ \vdots \end{pmatrix} = \begin{pmatrix} y_1 \\ y_2 \\ y_3 \\ \vdots \end{pmatrix} \tag{2.1}$$

we consider the truncated systems

$$\begin{pmatrix} a_{11} & \dots & a_{1n} \\ \vdots & & \vdots \\ a_{n1} & \dots & a_{nn} \end{pmatrix} \begin{pmatrix} x_1^{(n)} \\ \vdots \\ x_n^{(n)} \end{pmatrix} = \begin{pmatrix} y_1 \\ \vdots \\ y_n \end{pmatrix}. \tag{2.2}$$

To abbreviate notation, let P_n be the projection on l^2 acting by the rule

$$P_n : \{x_1, x_2, x_3, \dots\} \mapsto \{x_1, x_2, \dots, x_n, 0, 0, \dots\}. \tag{2.3}$$

The system (2.2) then takes the form

$$P_n A x^{(n)} = P_n y \quad (x^{(n)} \in \text{Im } P_n); \tag{2.4}$$

here and in what follows, we freely identify Im P_n, the image of P_n, with $\mathbf{C}^n$. Since $P_n x^{(n)} = x^{(n)}$ for $x^{(n)} \in \text{Im } P_n$, we might also write

$$P_n A P_n x^{(n)} = P_n y \quad (x^{(n)} \in \text{Im } P_n)$$

for (2.4). The matrix

$$A_n = \begin{pmatrix} a_{11} & \dots & a_{1n} \\ \vdots & & \vdots \\ a_{n1} & \dots & a_{nn} \end{pmatrix}$$

may be identified with the restriction of $P_n A P_n$ to Im P_n:

$$A_n = P_n A P_n | \text{Im } P_n.$$

For obvious reasons, the replacement of (2.1) by (2.2) is called the *finite section method.*

If $A = T(a)$ $(a \in L^\infty)$ is a Toeplitz matrix, the finite section method leads to solving the systems $T_n(a)x^{(n)} = P_n y$ where

$$T_n(a) := \begin{pmatrix} a_0 & a_{-1} & \dots & a_{-(n-1)} \\ a_1 & a_0 & \dots & a_{-(n-2)} \\ \vdots & \vdots & \ddots & \vdots \\ a_{n-1} & a_{n-2} & \dots & a_0 \end{pmatrix}. \tag{2.5}$$

The finite section method is a special case of more general approximation methods. For example, if $A = T(a)T(b)$ is the product of two Toeplitz matrices, two possible replacements of the equation $Ax = y$ are

$$P_n T(a) T(b) P_n x^{(n)} = P_n y \quad \text{and} \quad T_n(a) T_n(b) x^{(n)} = P_n y.$$

In the first case we are considering the finite section method, but in the second case we have something different.

General approximation sequences. Suppose we are given any sequence $\{A_n\}_{n=1}^\infty$ of $n \times n$ matrices A_n. On identifying $\mathbf{C}^n$ and Im P_n and on regarding A_n as $A_n P_n$, we can think of A_n as given on all of l^2. We call $\{A_n\}$ an *approximating sequence* for some operator $A \in \mathcal{B}(l^2)$ if $A_n P_n$ converges strongly to A on l^2, i.e., if

$$\|A_n x - Ax\| \quad (:= \|A_n P_n x - Ax\|) \quad \to 0 \quad \text{as} \quad n \to \infty$$

for every $x \in l^2$. Clearly, $\{P_n A P_n | \text{Im } P_n\}$ is always an approximating sequence for A.

We write $A \in \Pi\{A_n\}$ and say that the *approximation method* $\{A_n\}$ (and the finite section method in case $A_n = P_n A P_n | \text{Im } P_n$) *is applicable to* A if

(i) the matrices A_n are invertible for all sufficiently large n, say for $n \geq n_0$;

(ii) for every $y \in l^2$ the (unique) solutions $x^{(n)} \in \operatorname{Im} P_n$ of $A_n x^{(n)} = P_n y$ $(n \geq n_0)$ converge in l^2 to a solution $x \in l^2$ of the equation $Ax = y$.

The following theorem is also known as the uniform boundedness principle. It plays a fundamental role in numerical analysis.

Theorem 2.1 (Banach-Steinhaus). *If $\{A_n\}_{n=1}^{\infty}$ is any sequence of operators $A_n \in \mathcal{B}(l^2)$ such that $\{A_n x\}_{n=1}^{\infty}$ is a convergent sequence in l^2 for every $x \in l^2$, then $\sup_{n \geq 0} \|A_n\| < \infty$, the operator A defined by $Ax := \lim_{n\to\infty} A_n x$ is bounded on l^2, and*

$$\|A\| \leq \liminf_{n\to\infty} \|A_n\|.$$

A *proof* of this theorem is in every text on functional analysis. Here is a first application of the Banach-Steinhaus theorem.

Proposition 2.2. *Let $A \in \mathcal{B}(l^2)$ and let $\{A_n\}$ be an approximating sequence for A. Then $A \in \Pi\{A_n\}$ if and only if A is invertible, the matrices A_n are invertible for all sufficiently large n, and A_n^{-1} $(:= A_n^{-1} P_n)$ converges strongly to A^{-1}.*

Proof. The "if" portion is trivial. Now suppose $A \in \Pi\{A_n\}$. The "only if" part will follow as soon as we have shown that A is invertible. By the definition of $\Pi\{A_n\}$, the operator A is surjective. Since, also by the definition of $\Pi\{A_n\}$, the sequence $\{A_n^{-1} P_n y\}_{n \geq n_0}$ is convergent for every $y \in l^2$, Theorem 2.1 implies that

$$M := \sup_{n \geq n_0} \|A_n^{-1} P_n\| < \infty.$$

Thus, if $x \in l^2$ and $n \geq n_0$,

$$\begin{aligned} \|P_n x\| &= \|A_n^{-1} A_n P_n x\| = \|A_n^{-1} P_n A_n P_n x\| \\ &\leq \|A_n^{-1} P_n\| \, \|A_n P_n x\| \leq M \|A_n P_n x\|. \end{aligned} \tag{2.6}$$

As $A_n P_n x \to Ax$, it follows that $\|x\| \leq M\|Ax\|$ for all $x \in l^2$, which shows that A is injective. ■

Note that, in particular, no method $\{A_n\}$ is applicable (in the above sense) to A in case A is not invertible.

Stable approximation sequences. From the basic course in numerical analysis we know the principle

$$\text{convergence} = \text{approximation} + \text{stability}. \tag{2.7}$$

In the following we need not be worried about approximation, because we will always *require* that $\{A_n\}$ be an approximating sequence for A. What about stability? One says that a sequence $\{A_n\}_{n=1}^{\infty}$ of $n \times n$ matrices A_n is *stable* (or *uniformly invertible*) if A_n is invertible for all n large enough, for $n \ge n_0$ say, and

$$\sup_{n \ge n_0} \|A_n^{-1}\| < \infty. \tag{2.8}$$

Recall that $\|A_n^{-1}\| := \|A_n^{-1} P_n\|$. In order to avoid the inconvenient n_0, we put

$$\|A_n^{-1}\| = \infty \text{ in the case where } A_n \text{ is not invertible.}$$

With this convention, we may write (2.8) in the form

$$\limsup_{n \to \infty} \|A_n^{-1}\| < \infty. \tag{2.9}$$

If $\{A_n\}$ is a stable sequence, then the sequence $\{A_n^*\}$ of the adjoint matrices (operators) is also stable. However, if $\{A_n\}$ is an approximating sequence for A, then $\{A_n^*\}$ need not be an approximating sequence for the adjoint operator A^*. For example, if

$$A_n = \begin{pmatrix} 0 & \dots & 0 & 1 \\ 1 & \dots & 0 & 0 \\ \vdots & & \vdots & \vdots \\ 0 & \dots & 1 & 0 \end{pmatrix}, \quad A_n^* = \begin{pmatrix} 0 & 1 & \dots & 0 \\ \vdots & \vdots & & \vdots \\ 0 & 0 & \dots & 1 \\ 1 & 0 & \dots & 0 \end{pmatrix}, \tag{2.10}$$

then A_n converges strongly to the shift operator

$$T(\chi_1) : \{x_1, x_2, \dots\} \mapsto \{0, x_1, x_2, \dots\}$$

as $n \to \infty$, while A_n^* does not at all converge strongly.

Proposition 2.3. *Let $A \in \mathcal{B}(l^2)$ and let $\{A_n\}$ be an approximating sequence for A.*

(a) *If $\{A_n\}$ is stable, then the operator A is injective and* $\operatorname{Im} A$ *is a closed subspace of l^2.*

(b) *If $\{A_n\}$ is stable and, in addition, $\{A_n^*\}$ is an approximating sequence for A^*, then A is invertible.*

Proof. (a) From (2.8) and (2.6) we get $\|x\| \le M\|Ax\|$ for all $x \in l^2$. This implies that A is injective. Moreover, if $Ax_n \to y$, then

$$\|x_n - x_m\| \le \|A(x_n - x_m)\| = \|Ax_n - Ax_m\|,$$

hence $x_n \to x$ and thus $y = Ax$, which shows that $\operatorname{Im} A$ is closed.

(b) This follows from part (a) and the fact that our additional hypothesis yields the estimate $\|x\| \le M\|A^*x\|$ for all $x \in l^2$. ■

Note that if A_n is as in (2.10), then $\{A_n\}$ is stable but the strong limit $A = T(\chi_1)$ of A_n is not invertible. This reveals that in general the conclusion of Proposition 2.2(a) cannot be sharpened.

Here is what (2.7) states in our context.

Proposition 2.4. *Let $A \in \mathcal{B}(l^2)$ and let $\{A_n\}$ be an approximating sequence for A. Then $A \in \Pi\{A_n\}$ if and only if A is invertible and $\{A_n\}$ is a stable sequence.*

Proof. If $A \in \Pi\{A_n\}$, we deduce the invertibility of A and the stability of $\{A_n\}$ from Proposition 2.2 and Theorem 2.1.

Conversely, suppose A is invertible and $\{A_n\}$ is stable. Then for each $y \in l^2$,

$$\|A_n^{-1}P_n y - A^{-1}y\| \le \|A_n^{-1}P_n y - P_n A^{-1}y\| + \|P_n A^{-1}y - A^{-1}y\|,$$

the second term on the right goes to zero because $P_n \to I$ strongly, and the first term on the right is

$$\|A_n^{-1}(P_n y - A_n P_n A^{-1}y)\| \le M\|P_n y - A_n P_n A^{-1}y\| = o(1)$$

since $A_n P_n A^{-1} \to AA^{-1} = I$ strongly. ■

Propositions 2.2 and 2.4 do not tell us anything that might solve the question whether the finite section method (for example) is applicable to a given concrete operator. However, they reveal the heart of the problem: we have to study the stability of the sequence $\{P_n A P_n|\text{Im}\, P_n\}$.

Example 2.5. Let A be an infinite matrix which contains exactly one unit in every row and every column. Then A is a unitary operator on l^2 and hence invertible. If A is given by

$$A = \begin{pmatrix} 0 & 1 & 0 & 0 & \dots \\ 1 & 0 & 0 & 0 & \dots \\ 0 & 0 & 0 & 1 & \dots \\ 0 & 0 & 1 & 0 & \dots \\ \dots & \dots & \dots & \dots & \dots \end{pmatrix},$$

then A is such a matrix (note that A is a block diagonal Toeplitz matrix). The last row of $P_{2k+1}AP_{2k+1}|\text{Im}\, P_{2k+1}$ $(k = 0, 1, 2, \dots)$ consists only of zeros. Thus, although A is invertible, the matrices $P_n A P_n|\text{Im}\, P_n$ are not invertible whenever n is odd. By Proposition 2.2, the finite section method is not applicable to A. ■

Example 2.6. We now slightly modify the matrix of the previous example: let $\{\varepsilon_n\}_{n=0}^{\infty}$ be a sequence of numbers $\varepsilon_n \in (0, 1/2]$ converging to zero and put

$$A = \begin{pmatrix} \varepsilon_0 & 1 & 0 & 0 & \dots \\ 1 & \varepsilon_0 & 0 & 0 & \dots \\ 0 & 0 & \varepsilon_1 & 1 & \dots \\ 0 & 0 & 1 & \varepsilon_1 & \dots \\ \dots & \dots & \dots & \dots & \dots \end{pmatrix}$$

(this is a compactly perturbed block diagonal Toeplitz matrix). It is readily seen that

$$\left\| \begin{pmatrix} \varepsilon & 1 \\ 1 & \varepsilon \end{pmatrix}^{-1} \right\| = \frac{1}{1-\varepsilon^2} \left\| \begin{pmatrix} \varepsilon & -1 \\ -1 & \varepsilon \end{pmatrix} \right\| = \frac{1+\varepsilon}{1-\varepsilon^2} = \frac{1}{1-\varepsilon} \le 2$$

for $\varepsilon \in (0, 1/2]$. Thus, A is invertible and $\|A^{-1}\| \le 2$. Clearly, $A_n := P_n A P_n | \operatorname{Im} P_n$ is invertible for every $n \ge 1$. If n is even, then $\|A_n^{-1}\| \le 2$, while if $n = 2k+1$ ($k \ge 0$) is odd, then $\|A_n^{-1}\| \ge 1/\varepsilon_k$. Consequently, although A and all the truncations A_n are invertible, the norms $\|A_n^{-1}\|$ are not uniformly bounded. From Proposition 2.4 we infer that the finite section method is not applicable to A. ■

Treil's theorem. Let $a \in L^\infty$ and consider the infinite Toeplitz matrix $T(a)$ and its finite sections $T_n(a)$ ($n \ge 1$) given by (2.5). Since $\{T_n(a)\}$ and $\{T_n^*(a)\} = \{T_n(\overline{a})\}$ are approximating sequences for $T(a)$ and $T^*(a) = T(\overline{a})$, respectively, we obtain from Proposition 2.3 the implication

$$\{T_n(a)\} \text{ is stable } \Longrightarrow T(a) \text{ is invertible.}$$

Is the reverse implication also true? Here is the answer.

Theorem 2.7 (Treil). *There exist $a \in L^\infty$ such that $T(a)$ is invertible but $\{T_n(a)\}$ is not stable.*

Although the finite section method for Toeplitz operators has extensively been studied since the 1960s, this result was established by Treil [172] only in 1987. This fact uncovers the point of the matter: the construction of a symbol a as in Theorem 2.7 is rather difficult, because *for large classes of symbols a the invertibility of $T(a)$ indeed implies the stability of $\{T_n(a)\}$.*

We will not give a proof of Theorem 2.7 here; full proofs are in [172], [39, Theorem 7.92], and [28]. The latter paper contains explicitly given symbols a for which $T(a)$ is invertible but $\{T_n(a)\}$ is not stable: this is, for example, the case if a is the almost periodic function

$$a(e^{i\theta}) = \exp\left(i f\left(-\cot\frac{\theta}{2} \right) \right)$$

where f is 2π-periodic on $\mathbf{R}$ and $f(x) = 2|x|$ for $|x| \le \pi$. Note that

$$f(x) = \pi - \frac{8}{\pi}\left(\cos x + \frac{1}{3^2}\cos 3x + \frac{1}{5^2}\cos 5x + \dots \right), \qquad (2.11)$$

which shows that f, the "stretched argument" of a, has an absolutely convergent Fourier series (see also Figure 14).

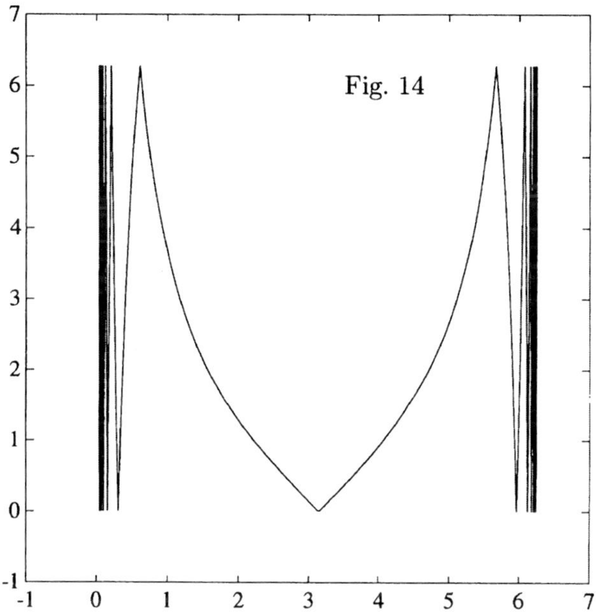

Figure 14 shows the function $(0, 2\pi) \to \mathbf{R}$, $\theta \mapsto f(-\cot(\theta/2))$ with f as in (2.11) and thus the argument of a discontinuous symbol a generating an invertible Toeplitz operator $T(a)$ for which $\{T_n(a)\}$ is not stable.

2.2 Continuous Symbols

As intimated in the end of the previous section, the sequence $\{T_n(a)\}$ is stable for large classes of invertible Toeplitz operators $T(a)$. In this section we prove that this is true if $a \in C$.

In what follows we will frequently make use of the following simple, well known, but basic lemma.

Lemma 2.8. *If K is a compact operator and $B_n \to B$ strongly, then $B_nK \to BK$ uniformly, i.e., $\|B_nK - BK\| \to 0$.*

Proof. Let $\varepsilon > 0$. Since K maps the closed unit ball $\{x : \|x\| \leq 1\}$ to a set whose closure is compact, there are $x_1, \ldots, x_N$ in the closed unit ball such that for every x on the unit sphere we can find an x_j satisfying $\|Kx - Kx_j\| \leq \varepsilon$. Clearly, the norm $\|B_nKx - BKx\|$ can be estimated from above by

$$\begin{aligned}\|B_n\|\,\|Kx - Kx_j\| + \|B_nKx_j - BKx_j\| + \|B\|\,\|Kx_j - Kx\| \\ \le \|B_n\|\varepsilon + \|B_nKx_j - BKx_j\| + \|B\|\varepsilon.\end{aligned}$$

Theorem 2.1 implies that $\|B_n\| \le M < \infty$ for all n, and because B_n converges strongly to B, the norms $\|B_nKx_j - BKx_j\|$ are less than ε for all j and all sufficiently large n. Thus, if n is large enough, then

$$\|B_nKx - BKx\| \le \big(M + 1 + \|B\|\big)\varepsilon = \big(M + 1 + \|B\|\big)\varepsilon\|x\|$$

whenever $\|x\| = 1$. ■

The following well-known fact has been shown by various authors to be extremely useful in several contexts. We learned it from A.V. Kozak in the late 1970s (private communication).

Lemma 2.9. *Suppose X is a linear space, P and Q are complementary projections on X (i.e., $P^2 = P$, $Q^2 = Q$, $P+Q = I$), and A is an invertible operator on X. Then the compression $PAP|\operatorname{Im} P$ is invertible on $\operatorname{Im} P$ if and only if the compression $QA^{-1}Q|\operatorname{Im} Q$ is invertible on $\operatorname{Im} Q$. In that case*

$$(PAP)^{-1}P = PA^{-1}P - PA^{-1}Q(QA^{-1}Q)^{-1}QA^{-1}P.$$

Proof. We have

$$\begin{aligned}&PAP\big(PA^{-1}P - PA^{-1}Q(QA^{-1}Q)^{-1}QA^{-1}P\big)\\ &\quad= PA(I - Q)A^{-1}P - PA(I - Q)A^{-1}Q(QA^{-1}Q)^{-1}QA^{-1}P\\ &\quad= P - PAQA^{-1}P - O + PAQA^{-1}P = P,\end{aligned}$$

and similarly we get $(PA^{-1}P - PA^{-1}Q(QA^{-1}Q)^{-1}QA^{-1}P)PAP = P$. ■

As a first consequence of Lemma 2.9 we record the following observation, which shows that the truncations of the inverse of Toeplitz matrices are always stable. Recall that P_n is defined by (2.3).

Theorem 2.10. *Let $a \in L^\infty$ and suppose the operator $T(a)$ is invertible. Then $\{P_nT^{-1}(a)P_n|\operatorname{Im} P_n\}$ is stable.*

Proof. Let $Q_n := I - P_n$. Since $Q_nT(a)Q_n|\operatorname{Im} Q_n$ has the same matrix as $T(a)$, we see that $Q_nT(a)Q_n|\operatorname{Im} Q_n$ is invertible for all $n \ge 1$ and that

$$\big\|\big(Q_nT(a)Q_n\big)^{-1}Q_n\big\| = \big\|T^{-1}(a)\big\|.$$

Lemma 2.9 therefore implies that $P_nT^{-1}(a)P_n|\operatorname{Im} P_n$ is invertible for all $n \ge 1$ and that the norm of the inverse is at most

$$\begin{aligned}\big\|P_nT(a)P_n - P_nT(a)Q_n\big(Q_nT(a)Q_n\big)^{-1}Q_nT(a)P_n\big\| \\ \le \|T(a)\| + \|T(a)\|\,\|T^{-1}(a)\|\,\|T(a)\|. \ ■\end{aligned}$$

The following theorem is a classical result of Gohberg and Feldman [80]. The proof given below is from [35, Section 3.10].

Theorem 2.11 (Gohberg-Feldman). *If $a \in C$ and $T(a)$ is invertible, then $\{T_n(a)\}$ is stable.*

Proof. By Proposition 1.12, $T(a)T(a^{-1}) = I - H(a)H(\tilde{a}^{-1})$ and thus,

$$T^{-1}(a) = T(a^{-1}) + T^{-1}(a)H(a)H(\tilde{a}^{-1}) =: T(a^{-1}) + K.$$

The operator K is compact due to (the sufficiency part of) Theorem 1.16. Since $Q_n \to 0$ strongly, we deduce from Lemma 2.8 that

$$\|Q_n K Q_n|\mathrm{Im}\, Q_n\| = \|Q_n K Q_n\| \le \|Q_n K\| = o(1),$$

whence

$$Q_n T^{-1}(a) Q_n|\mathrm{Im}\, Q_n = Q_n T(a^{-1}) Q_n|\mathrm{Im}\, Q_n + K_n \quad \text{with} \quad \|K_n\| = o(1).$$

The operator $Q_n T(a^{-1}) Q_n|\mathrm{Im}\, Q_n$ has the same matrix as $T(a^{-1})$, and as $T(a^{-1})$ is invertible together with $T(a)$ (Theorem 1.17), it follows that $Q_n T^{-1}(a) Q_n|\mathrm{Im}\, Q_n$ is invertible for all sufficiently large n and that for every $\varepsilon > 0$ there is an $n_0(\varepsilon)$ such that

$$\left\| \left(Q_n T^{-1}(a) Q_n\right)^{-1} Q_n \right\| < (1+\varepsilon)\|T^{-1}(a^{-1})\|$$

for all $n \ge n_0(\varepsilon)$. For these n we obtain from Lemma 2.9 that $T_n(a) = P_n T(a) P_n|\mathrm{Im}\, P_n$ is invertible and that $(P_n T(a) P_n)^{-1} P_n$ equals

$$P_n T^{-1}(a) P_n - P_n T^{-1}(a) Q_n \left(Q_n T^{-1}(a) Q_n\right)^{-1} Q_n T^{-1}(a) P_n.$$

Since the norm of this operator does not exceed

$$\|T^{-1}(a)\| + (1+\varepsilon)\|T^{-1}(a)\|\, \|T^{-1}(a^{-1})\|\, \|T^{-1}(a)\|,$$

we see that $\{T_n(a)\}$ is stable. ∎

2.3 Asymptotic Inverses

In addition to the projections P_n and $Q_n = I - P_n$, we need the operators W_n $(n \ge 1)$ which are defined on l^2 by

$$W_n : \{x_1, x_2, x_3, \ldots\} \mapsto \{x_n, x_{n-1}, \ldots, x_1, 0, 0, \ldots\}.$$

Obviously,

$$W_n^2 = P_n, \quad W_n P_n = P_n W_n = W_n, \quad W_n T_n(a) W_n = T_n(\tilde{a}).$$

Recall that $\tilde{a}(t) := a(1/t)$ $(t \in \mathbf{T})$. It can also be readily verified that W_n *converges weakly* to zero, i.e., $(W_n x, y) \to 0$ as $n \to \infty$ for every $x, y \in l^2$.

The following proposition is the finite section analogue of Proposition 1.12.

Proposition 2.12 (Widom). *If $a, b \in L^\infty$, then*

$$P_n T(a) Q_n T(b) P_n = W_n H(\tilde{a}) H(b) W_n, \tag{2.12}$$
$$T_n(ab) = T_n(a) T_n(b) + P_n H(a) H(\tilde{b}) P_n + W_n H(\tilde{a}) H(b) W_n. \tag{2.13}$$

Proof. Let $\chi_k(t) := t^k$ $(t \in \mathbf{T})$. Then for $n \geq 1$,

$$T(\chi_{-n}) : \{x_1, x_2, \ldots\} \mapsto \{x_{n+1}, x_{n+2}, \ldots\},$$
$$T(\chi_n) : \{x_1, x_2, \ldots\} \mapsto \{0, \ldots, 0, x_1, x_2, \ldots\},$$

the latter sequence containing n zeros. An easy computation shows that

$$W_n T(a) Q_n = P_n H(\tilde{a}) T(\chi_{-n}), \qquad Q_n T(b) W_n = T(\chi_n) H(b) P_n.$$

Since $T(\chi_{-n}) T(\chi_n) = I$, we therefore get

$$\begin{aligned} P_n T(a) Q_n T(b) P_n &= W_n W_n T(a) Q_n Q_n T(b) W_n W_n \\ &= W_n P_n H(\tilde{a}) T(\chi_{-n}) T(\chi_n) H(b) P_n W_n = W_n H(\tilde{a}) H(b) W_n, \end{aligned}$$

which is (2.12). By Proposition 1.12,

$$T_n(ab) = P_n T(ab) P_n = P_n T(a) T(b) P_n + P_n H(a) H(\tilde{b}) P_n,$$

and using (2.12) we obtain

$$\begin{aligned} P_n T(a) T(b) P_n &= P_n T(a) P_n T(b) P_n + P_n T(b) Q_n T(b) Q_n \\ &= T_n(a) T_n(b) + W_n H(\tilde{a}) H(b) W_n. \ \blacksquare \end{aligned}$$

Let $a \in L^\infty$ and suppose $T(a)$ is invertible. Then $a \in GL^\infty$ due to Theorem 1.25. Since $T(\tilde{a})$ is the transposed operator of $T(a)$, the operator $T(\tilde{a})$ is also invertible. Thus, the two operators

$$K(a) := T^{-1}(a) - T(a^{-1}) \quad \text{and} \quad K(\tilde{a}) := T^{-1}(\tilde{a}) - T(\tilde{a}^{-1}) \tag{2.14}$$

are well-defined bounded operators.

Lemma 2.13. *If $a \in L^\infty$ and $T(a)$ is invertible, then*

$$\begin{aligned} &\big(P_n T^{-1}(a) P_n + W_n K(\tilde{a}) W_n\big) T_n(a) \\ &\quad = P_n - P_n K(a) Q_n T(a) P_n - W_n K(\tilde{a}) Q_n T(\tilde{a}) W_n, \end{aligned} \tag{2.15}$$

$$\begin{aligned} &T_n(a) \big(P_n T^{-1}(a) P_n + W_n K(\tilde{a}) W_n\big) \\ &\quad = P_n - P_n T(a) Q_n K(a) P_n - W_n T(\tilde{a}) Q_n K(\tilde{a}) W_n. \end{aligned} \tag{2.16}$$

Proof. By (2.12) and Proposition 1.12,

$$\begin{aligned} P_n T(a^{-1}) Q_n T(a) P_n &= W_n H(\tilde{a}^{-1}) H(a) W_n \\ &= W_n \big(I - T(\tilde{a}^{-1}) T(\tilde{a})\big) W_n = W_n K(\tilde{a}) T(\tilde{a}) W_n \\ &= W_n K(\tilde{a}) W_n T(a) P_n + W_n K(\tilde{a}) Q_n T(\tilde{a}) W_n. \end{aligned}$$

Thus,

$$W_n K(\tilde{a}) W_n T_n(a) = P_n T(a^{-1}) Q_n T(a) P_n - W_n K(\tilde{a}) Q_n T(\tilde{a}) W_n. \tag{2.17}$$

On the other hand,

$$\begin{aligned} P_n T^{-1}(a) P_n T(a) P_n &= P_n - P_n T^{-1}(a) Q_n T(a) P_n \\ &= P_n - P_n K(a) Q_n T(a) P_n - P_n T(a^{-1}) Q_n T(a) P_n. \end{aligned} \tag{2.18}$$

Adding (2.17) and (2.18) we arrive at (2.15). The identity (2.16) results from (2.15) after passage to transposed matrices. ∎

Asymptotic inverses. Formulas (2.15) and (2.16) are useful if their right-hand sides can be shown to be of the form $P_n + D_n$ with $\|D_n\| \to 0$ as $n \to \infty$. In that case they provide an *asymptotic inverse* of the sequence $\{T_n(a)\}_{n=1}^{\infty}$, i.e., a sequence $\{B_n\}_{n=1}^{\infty}$ of $n \times n$ matrices B_n such that

$$\sup_{n \geq 1} \|B_n\| < \infty \tag{2.19}$$

and

$$\|B_n T_n(a) - P_n\| \to 0, \quad \|T_n(a) B_n - P_n\| \to 0 \quad \text{as } n \to \infty. \tag{2.20}$$

Note that if (2.19) and (2.20) are valid, then $\{T_n(a)\}$ is necessarily stable and

$$T_n^{-1}(a) = B_n + C_n \quad \text{with} \quad \|C_n\| \to 0 \text{ as } n \to \infty. \tag{2.21}$$

Indeed, if $B_n T_n(a) = P_n + D_n$ where $\|D_n\| \to 0$, then there is an n_0 such that $\|D_n\| < 1/2$ for all $n \geq n_0$. It follows that $P_n + D_n$ is invertible for $n \geq n_0$ and that $T_n^{-1}(a) = (P_n + D_n)^{-1} B_n$. Since $\|(P_n + D_n)^{-1}\| < 2$, we infer from (2.19) that

$$\|T_n^{-1}(a)\| < 2 \sup_{n \geq 1} \|B_n\|,$$

which proves the stability of $\{T_n(a)\}$ and also gives the representation (2.21) with $C_n = -D_n T_n^{-1}(a)$.

Clearly, (2.21) always implies (2.20), independently of whether (2.19) holds or not.

Here is an asymptotic inverse of $\{T_n(a)\}$ in case the symbol a is continuous.

Theorem 2.14 (Widom). *Let $a \in C$ and suppose $T(a)$ is invertible. Then for all sufficiently large n,*

$$\begin{aligned} T_n^{-1}(a) &= P_n T^{-1}(a) P_n + W_n K(\tilde{a}) W_n + C_n && (2.22)\\ &= T_n(a^{-1}) + P_n K(a) P_n + W_n K(\tilde{a}) W_n + C_n, && (2.23)\end{aligned}$$

where $K(a) = T^{-1}(a) - T(a^{-1})$ and $K(\tilde{a}) = T^{-1}(\tilde{a}) - T(\tilde{a}^{-1})$ are compact and $\|C_n\| \to 0$ as $n \to \infty$.

In particular, under the hypothesis of the theorem, (2.19), (2.20), and (2.21) are valid with

$$\begin{aligned} B_n &= P_n T^{-1}(a) P_n + W_n K(\tilde{a}) W_n\\ &= T_n(a^{-1}) + P_n K(a) P_n + W_n K(\tilde{a}) W_n. && (2.24)\end{aligned}$$

Proof. Consider the identities (2.15) and (2.16). By Proposition 1.12,

$$K(a) = H(a^{-1}) H(\tilde{a}) T^{-1}(a) \quad \text{and} \quad K(\tilde{a}) = H(\tilde{a}^{-1}) H(a) T^{-1}(\tilde{a}). \quad (2.25)$$

The continuity of a gives the compactness of all occurring Hankel operators (Theorem 1.16). Hence, $K(a)$ and $K(\tilde{a})$ are also compact. Since $Q_n \to 0$ strongly, Lemma 2.8 shows that $\|Q_n K(a)\| \to 0$ and $\|Q_n K(\tilde{a})\| \to 0$. Passing to adjoints we see that $\|K(a) Q_n\| \to 0$ and $\|K(\tilde{a}) Q_n\| \to 0$, too. Consequently, on defining B_n by (2.24) we deduce from Lemma 2.13 that (2.19) and (2.20) and thus also (2.21) hold. ∎

Second proof of Theorem 2.11. From (2.22) we see that if $a \in C$ and $T(a)$ is invertible, then $\{T_n(a)\}$ is stable. Moreover, it follows that for every $\varepsilon > 0$ there is an $n_0(\varepsilon)$ such that

$$\|T_n^{-1}(a)\| \le \|T^{-1}(a)\| + \|K(\tilde{a})\| + \varepsilon$$

for all $n \ge n_0(\varepsilon)$. ∎

Entries of the inverses. Of course, (2.22) and (2.23) provide information about the jk entry of $T_n^{-1}(a)$. Given a matrix A, we denote its jk by A_{jk} or $[A]_{jk}$. Clearly, $A_{jk} = (Ae_k, e_j)$ where e_n is the (finite or infinite) sequence having a unit at the nth position and zeros at all other positions.

If $a \in L^\infty$, $T(a)$ is invertible, and $\{T_n(a)\}$ is stable, then Propositions 2.2 and 2.4 imply that

$$\|T_n^{-1}(a) e_j - T^{-1}(a) e_j\| \to 0 \quad \text{as} \quad n \to \infty$$

for each j. Thus, the jth column of $T_n^{-1}(a)$ (extended by zeros to an element of l^2) converges in l^2 to the jth column of $T^{-1}(a)$. In particular,

$$\left[T_n^{-1}(a)\right]_{jk} = \left[T^{-1}(a)\right]_{jk} + o(1) \quad \text{as} \quad n \to \infty \qquad (2.26)$$

for each pair (j,k). Define $K(a)$ and $K(\tilde{a})$ by (2.14) and write

$$K(a) =: (g_{jk})_{j,k=1}^{\infty}, \qquad K(\tilde{a}) =: (h_{jk})_{j,k=1}^{\infty}.$$

Clearly,

$$\left[T^{-1}(a)\right]_{jk} = (a^{-1})_{j-k} + g_{jk}.$$

If a is smooth, then (2.26) can be improved. Here is an example of such a more precise estimate.

Theorem 2.15. *Let $a \in C$ and let $T(a)$ be invertible. In addition, suppose*

$$\sum_{n\in\mathbf{Z}} |n|^{\alpha}|a_n| < \infty \tag{2.27}$$

for some $\alpha > 0$. Then

$$\left[T_n^{-1}(a)\right]_{jk} = \left[T^{-1}(a)\right]_{jk} + h_{n+1-j,n+1-k} + o(1/n^{\alpha}) \quad \text{as} \quad n \to \infty \tag{2.28}$$

uniformly with respect to $(j,k) \in \mathbf{N}^2$. Further, for each bounded subset Ω of $\mathbf{N}^2$ there exists a constant $C_\Omega = C_\Omega(a)$ such that

$$|h_{n+1-j,n+1-k}| \le C_\Omega/n^{\alpha} \tag{2.29}$$

for all $(j,k) \in \Omega$ and all $n \ge 1$.

Proof. Let W^{α} denote the set of all functions $a \in C$ satisfying (2.27). It is well known that W^{α} is a Banach algebra with pointwise algebraic operations and the norm $\|a\| := \sum_{n\in\mathbf{Z}}(1+|n|)^{\alpha}|a_n|$ and that if $a \in W^{\alpha}$ and a has no zeros on $\mathbf{T}$, then $a^{-1} \in W^{\alpha}$.

Given a function $f \in L^{\infty}(\mathbf{T})$, we denote by $s_n f$ the $(n-1)$st partial sum of its Fourier series. If $f \in W^{\alpha}$, then

$$\|f - s_n f\|_{\infty} \le \sum_{|l|\ge n} |f_l| \le \frac{1}{n^{\alpha}} \sum_{|l|\ge n} |l|^{\alpha}|f_l| = o(1/n^{\alpha}).$$

Since $Q_n H(f) = Q_n H(f - s_n f)$, we get

$$\|Q_n H(a^{-1})\| \le \|a^{-1} - s_n a^{-1}\|_{\infty} = o(1/n^{\alpha}),$$
$$\|Q_n H(\tilde{a}^{-1})\| \le \|\tilde{a}^{-1} - s_n \tilde{a}^{-1}\|_{\infty} = o(1/n^{\alpha}).$$

Thus, by (2.25),

$$\|Q_n K(a)\| = o(1/n^{\alpha}), \qquad \|Q_n K(\tilde{a})\| = o(1/n^{\alpha}).$$

With B_n given by (2.24) we therefore deduce from (2.16) that

$$\|T_n(a)B_n - P_n\| = o(1/n^{\alpha}),$$

and since $\{T_n(a)\}$ is stable, it follows that $T_n^{-1}(a) = B_n + C_n$ with

$$\|C_n\| = \big\| - T_n^{-1}(a)\big(T_n(a)B_n - P_n\big)\big\| = o(1/n^\alpha).$$

As $[B_n]_{jk} = [T^{-1}(a)]_{jk} + h_{n+1-j,n+1-k}$, we see that (2.28) holds uniformly with respect to $(j,k) \in \mathbf{N}^2$.

By Proposition 1.12, $K(\tilde{a}) = T^{-1}(\tilde{a})H(\tilde{a})H(a^{-1})$. Hence

$$\begin{aligned}&\big|(W_n K(\tilde{a}) W_n e_k, e_j)\big| \\ &\quad = \big|(T^{-1}(\tilde{a})H(\tilde{a})H(a^{-1})e_{n+1-k}, e_{n+1-j})\big| \\ &\quad \le \|T^{-1}(\tilde{a})H(\tilde{a})\|\, \|H(a^{-1})e_{n+1-k}\|.\end{aligned}$$

Let $b := a^{-1}$. Then $b \in W^\alpha$ and thus,

$$M^2 := \sum_{n\in\mathbf{Z}} |n|^{2\alpha}|b_n|^2 < \infty.$$

Consequently, for $n \ge k$,

$$\begin{aligned}\|H(a^{-1})e_{n+1-k}\|^2 &= \sum_{l\ge 1} |b_{n-k+l}|^2 \\ &\le \frac{1}{(n+1-k)^{2\alpha}} \sum_{l\ge 1}(n-k+l)^{2\alpha}|b_{n-k+l}|^2 \le \frac{M^2}{(n+1-k)^{2\alpha}},\end{aligned}$$

whence

$$|h_{n+1-j,n+1-k}| \le M/(n+1-k)^\alpha. \tag{2.30}$$

For $n \ge j$, passage to transposed operators gives

$$|h_{n+1-j,n+1-k}| \le M/(n+1-j)^\alpha. \tag{2.31}$$

Clearly, each of the estimates (2.30) and (2.31) implies the last assertion of the theorem. ■

Notes. Results similar to Theorem 2.15 can be found in [136, pp. 106–107]. All other results of this section are from Widom's paper [185].

2.4 The Gohberg-Feldman Approach

The proofs of Theorems 2.11 and 2.14 given above break hopelessly down in case a is a piecewise continuous function: one can show that functions in $\overline{H^\infty}$ cannot have jumps, so that $H(a)$ is never compact if $a \in PC\backslash C$ (recall Theorem 1.16). In this section we present the approach of Gohberg and Feldman [80] to the stability of the finite sections of Toeplitz operators with piecewise continuous symbols.

The following theorem shows that the applicability of the finite section method is stable under compact perturbations.

Theorem 2.16. *Let $A \in \mathcal{B}(l^2)$ and suppose $\{P_n A P_n\}$ is stable. If $K \in \mathcal{K}(l^2)$ and $A + K$ is invertible, then $\{P_n(A+K)P_n\}$ is stable.*

Proof. Put $A_n = P_n A P_n | \operatorname{Im} P_n$ and notice first that

$$P_n(A+K)P_n = P_n A P_n (I + A_n^{-1} P_n K P_n) P_n$$

for all sufficiently large n. Since $A_n^{-1} P_n \to A^{-1}$ strongly due to Propositions 2.2 and 2.4, we infer from Lemma 2.8 that $A_n^{-1} P_n K$ converges uniformly to $A^{-1}K$. The operator $I + A^{-1}K = A^{-1}(A+K)$ is invertible; let us put $\|(I + A^{-1}K)^{-1}\| =: 1/\varepsilon$. We then have for every $x \in l^2$ the estimate

$$\begin{aligned} \varepsilon \|P_n x\| &\le \|(I + A^{-1}K)P_n x\| \\ &\le \|(I + A_n^{-1} P_n K) P_n x\| + \|A_n^{-1} P_n K - A^{-1} K\| \, \|P_n x\|, \end{aligned}$$

and the second term on the right is less than $(\varepsilon/2)\|P_n x\|$ if only n is large enough. For these n,

$$\begin{aligned} (\varepsilon/2)\|P_n x\| &\le \|(I + A_n^{-1} P_n K) P_n x\| \\ &\le \|A_n^{-1}\| \, \|(A_n + P_n K P_n) P_n x\| = \|A_n^{-1}\| \, \|P_n (A+K) P_n x\|, \end{aligned}$$

and as $\|A_n^{-1}\| \le M$ with some $M < \infty$, we arrive at the inequality

$$(\varepsilon/2M)\|P_n x\| \le \|P_n(A+K)P_n x\|. \tag{2.32}$$

It follows that $P_n(A+K)P_n|\operatorname{Im} P_n$ is injective and thus invertible. Moreover, (2.32) also yields the estimate $\|(P_n(A+K)P_n)^{-1}\| \le 2M/\varepsilon$, which proves that $\{P_n(A+K)P_n\}$ is stable. ∎

Perturbing Toeplitz operators. The idea of the Gohberg-Feldman approach is best understood for symbols in the Wiener algebra W. So assume $a \in W$ and $T(a)$ is invertible. Let

$$a = a_- a_+, \qquad a_- \in GW_-, \qquad a_+ \in GW_+,$$

be a Wiener-Hopf factorization of a (recall Theorems 1.14 and 1.15). We then have $T(a) = T(a_-)T(a_+)$, but the point is to consider the (invertible) operator

$$A := T(a_+)T(a_-).$$

We have

$$P_n A P_n = P_n T(a_+) P_n T(a_-) P_n + P_n T(a_+) Q_n T(a_-) P_n,$$

and as $P_n T(a_+) Q_n = Q_n T(a_-) P_n = 0$, it follows that

$$P_n A P_n = T_n(a_+) T_n(a_-).$$

Consequently, P_nAP_n is invertible for all $n \geq 1$ and

$$(P_nAP_n)^{-1} = T_n(a_-^{-1})T_n(a_+^{-1})$$

(note that $T_n(b)T_n(c) = T_n(bc)$ if $T_n(b)$ and $T_n(c)$ are both upper-triangular or both lower-triangular). Because

$$\|T_n(a_-^{-1})T_n(a_+^{-1})\| \leq \|T(a_-^{-1})\| \, \|T(a_+^{-1})\|,$$

we arrive at the conclusion that $\{P_nAP_n\}$ is stable.

We finally pass from A to $T(a)$. By Proposition 1.12,

$$T(a) = T(a_+)T(a_-) + H(a_+)H(\tilde{a}_-) = A + K,$$

where $K := H(a_+)H(\tilde{a}_-)$ is compact. Theorem 2.16 tells us that $\{T_n(a)\}$ is stable.

In order to extend the reasoning outlined above, we need the following simple analogue of the Brown-Halmos theorem (Theorem 1.18).

Proposition 2.17. *If $a \in L^\infty$, then*

$$\operatorname{sp} T_n(a) \subset \operatorname{conv} \mathcal{R}(a) \quad \text{for all} \quad n \geq 1. \tag{2.33}$$

If $a \in L^\infty$ is sectorial, then $d := \operatorname{dist}\big(0, \operatorname{conv} \mathcal{R}(a)\big) > 0$ and

$$\|T_n^{-1}(a)\| \leq \frac{1}{d}\left(1 + \sqrt{1 - \frac{d^2}{\|a\|_\infty^2}}\right) < \frac{2}{d} \quad \text{for all} \quad n \geq 1. \tag{2.34}$$

Proof. Fix $\lambda \in \mathbf{C} \setminus \operatorname{conv} \mathcal{R}(a)$ and put $b := a - \lambda$. There is a $\gamma \in \mathbf{T}$ such that the set $\gamma \operatorname{conv} \mathcal{R}(b)$ is contained in the set $\{z \in \mathbf{C} : \operatorname{Re} z \geq d, |z| \leq \|b\|_\infty\}$. Multiplying the latter set by $\delta := d/\|b\|_\infty^2$ we obtain a subset of the disk

$$\{z \in \mathbf{C} : |z - 1| < r\} \quad \text{where} \quad r := \sqrt{1 - d^2/\|b\|_\infty^2}.$$

Hence,

$$\|\delta\gamma T_n(b) - I\| = \|T_n(\delta\gamma b - 1)\| \leq \|\delta\gamma b - 1\|_\infty < r < 1. \tag{2.35}$$

This implies the invertibility of $T_n(b) = T_n(a) - \lambda I$ for all $n \geq 1$ and thus gives (2.33). Moreover, from (2.35) we get

$$\|T_n^{-1}(b)\| \leq \frac{\delta|\gamma|}{1 - r} = \frac{\delta(1 + r)}{1 - r^2} = \frac{1}{d}\left(1 + \sqrt{1 - \frac{d^2}{\|b\|_\infty^2}}\right),$$

and letting $\lambda = 0$ (i.e., $b = a$), we arrive at (2.34). ∎

Theorem 2.18 (Gohberg-Feldman). *Let $a \in L^\infty$ be locally sectorial on $\mathbf{T}$ and suppose $T(a)$ is invertible. Then $\{T_n(a)\}$ is stable.*

Proof. By Corollary 1.20, we have $a = cs$ with $c \in GC$ and a sectorial function $s \in GL^\infty$. We can approximate c by functions in the Wiener algebra (e.g., by its Fejer-Cesaro means) as closely as desired. Thus, given any $\varepsilon > 0$ there are $d \in W$ and $\varphi \in C$ such that $c = d(1+\varphi)$ and $\|\varphi\|_\infty < \varepsilon$. If ε is small enough, then $r := (1+\varphi)s$ is sectorial together with s. Hence, $a = dr$ with $d \in GW$ and a sectorial function r.

By Proposition 1.12, $T(a) = T(d)T(r) + H(d)H(\tilde{r})$. The operator $H(d)$ is compact (Theorem 1.16) and the operator $T(r)$ is invertible (Theorem 1.18). Consequently, $T(d)$ must be Fredholm of index zero. From Theorems 1.14 and 1.15 we therefore deduce that d admits a Wiener-Hopf factorization

$$d = d_- d_+, \qquad d_- \in GW_-, \qquad d_+ \in GW_+.$$

In summary, $a = d_- d_+ r = d_+ r d_-$.

Now put $A := T(d_+)T(r)T(d_-)$. Since $P_n T(d_+) Q_n = 0$ and $Q_n T(d_-) P_n = 0$, we obtain

$$P_n A P_n = P_n T(d_+) P_n T(r) P_n T(d_-) P_n = T_n(d_+) T_n(r) T_n(d_-).$$

The operators $T_n(d_+)$ and $T_n(d_-)$ have the uniformly bounded inverses $T_n(d_+^{-1})$ and $T_n(d_-^{-1})$, respectively, while $\{T_n(r)\}$ is stable due to Proposition 2.17. This implies that $\{P_n A P_n\}$ is stable.

Again by Proposition 1.12,

$$\begin{aligned} T(a) &= T(d_+)T(rd_-) + H(d_+)H(\tilde{r}\tilde{d}_-) \\ &= T(d_+)T(r)T(d_-) + T(d_+)H(r)H(\tilde{d}_-) + H(d_+)H(\tilde{r}\tilde{d}_-), \end{aligned}$$

and since $H(\tilde{d}_-)$ and $H(d_+)$ are compact, Theorem 2.14 shows that $\{T_n(a)\}$ is stable. ■

Corollary 2.19 (Gohberg-Feldman). *If $a \in PC$ and $T(a)$ is invertible, then $\{T_n(a)\}$ is stable.*

Proof. Theorem 1.23 in conjunction with (1.29) says that if $a \in PC$ and $T(a)$ is invertible (or even only Fredholm), then a is locally sectorial on $\mathbf{T}$. The assertion is therefore immediate from Theorem 2.16. ■

2.5 Algebraization of Stability

We now develop another approach to the stability problem for the finite sections of Toeplitz matrices. This approach allows us to give alternative proofs of Theorems 2.11 and 2.18. At first glance, the machinery constructed in the following seems to be unduly heavy, but this machinery will prove to be of deciding importance in the forthcoming chapters.

The big algebra. The idea of the approach is to build a Banach algebra $\mathcal{G}$ such that

$$\{A_n\} \text{ is stable } \iff \text{ something is invertible in } \mathcal{G}.$$

To begin with, let $\mathcal{F}$ be the set of all sequences $\{A_n\} = \{A_n\}_{n=1}^{\infty}$ of operators (matrices) $A_n \in \mathcal{B}(\operatorname{Im} P_n)$ ($\cong \mathbf{C}^{n\times n}$) for which

$$\|\{A_n\}\| := \sup_{n\geq 1} \|A_n\| < \infty. \tag{2.36}$$

The set $\mathcal{F}$ with the algebraic operations

$$\{A_n\} + \{B_n\} := \{A_n + B_n\}, \ \ \lambda\{A_n\} := \{\lambda A_n\}, \ \ \{A_n\}\{B_n\} := \{A_n B_n\}$$

and the norm (2.36) is easily seen to be a Banach algebra. An element $\{A_n\} \in \mathcal{F}$ is invertible in $\mathcal{F}$ if and only if A_n is invertible for all $n \geq 1$ and $\sup_{n\geq 1} \|A_n^{-1}\| < \infty$. Clearly, invertibility in $\mathcal{F}$ is not equivalent to stability, but it has undoubtedly something to do with it.

Now denote by $\mathcal{N}$ the subset of $\mathcal{F}$ consisting of all $\{C_n\} \in \mathcal{F}$ such that $\|C_n\| \to 0$ as $n \to \infty$. It is not difficult to show that $\mathcal{N}$ is a closed two-sided ideal of $\mathcal{F}$ and hence we may consider the quotient algebra $\mathcal{F}/\mathcal{N}$.

Proposition 2.20. *A sequence $\{A_n\} \in \mathcal{F}$ is stable if and only if the coset $\{A_n\} + \mathcal{N}$ is invertible in $\mathcal{F}/\mathcal{N}$.*

Proof. If $\{A_n\} + \mathcal{N}$ is invertible in $\mathcal{F}/\mathcal{N}$, then there is a sequence $\{B_n\} \in \mathcal{F}$ such that $B_n A_n = P_n + C_n$ with $\|C_n\| \to 0$. If $\|C_n\| < 1/2$, then $P_n + C_n$ is invertible (note that P_n is the identity operator on $\operatorname{Im} P_n$) and $(P_n + C_n)^{-1} B_n$ is the inverse of A_n. Since

$$\|(P_n + C_n)^{-1} B_n\| \leq (1 - \|C_n\|)^{-1} \|B_n\| < 2\|B_n\|,$$

it follows that $\{A_n\}$ is stable.

Conversely, let $\{A_n\}$ be stable. Suppose A_n is invertible for $n \geq n_0$. Define $B_n = A_n^{-1}$ for $n \geq n_0$ and $B_n = 0$ for $n < n_0$. Then $\{B_n\} \in \mathcal{F}$ and $\{B_n\}\{A_n\} - \{P_n\}$ as well as $\{A_n\}\{B_n\} - \{P_n\}$ belong to $\mathcal{N}$. ■

Note that if $\{B_n\} + \mathcal{N} \in \mathcal{F}/\mathcal{N}$ is the inverse of $\{A_n\} + \mathcal{N}$ if and only if $\{B_n\}$ is an asymptotic inverse of $\{A_n\}$ in the sense of Section 2.3:

$$\sup_{n\geq 1} \|B_n\| < \infty, \qquad \|A_n B_n - P_n\| \to 0, \qquad \|B_n A_n - P_n\| \to 0.$$

The modified algebra. Now suppose $a \in C$ and $T(a)$ is invertible. Proposition 2.12 then gives the formula

$$T_n(a^{-1}) T_n(a) = P_n - P_n H(a^{-1}) H(\tilde{a}) P_n - W_n H(\tilde{a}^{-1}) H(a) W_n \tag{2.37}$$

and a similar formula for $T_n(a)T_n(a^{-1})$. Note that all occurring Hankel operators are compact. If sequences of the form $\{P_nKP_n + W_nLW_n\}$ with compact operators K and L would belong to $\mathcal{N}$, then Proposition 2.20 would imply that $\{T_n(a)\}$ is stable. However, these sequences clearly do not belong to $\mathcal{N}$. So let us replace $\mathcal{N}$ by the set

$$\mathcal{J} := \big\{\{P_nKP_n + W_nLW_n + C_n\}_{n=1}^{\infty} : K, L \in \mathcal{K}(l^2),\ \|C_n\| \to 0\big\}.$$

Then $\{T_n(a^{-1})\}$ is an inverse of $\{T_n(a)\}$ modulo $\mathcal{J}$ and it would be nice to have a Banach algebra $\mathcal{S} \subset \mathcal{F}$ with the following properties:

- $\mathcal{J}$ is a closed two-sided ideal of $\mathcal{S}$;
- $\{T_n(a)\} \in \mathcal{S}$ for all $a \in L^{\infty}$;
- invertibility in the quotient algebra $\mathcal{S}/\mathcal{J}$ has something to do with invertibility in $\mathcal{F}/\mathcal{N}$, i.e., with stability.

Let $\mathcal{S}$ be the collection of all $\{A_n\} \in \mathcal{F}$ for which there are two operators A and $\tilde{A}$ in $\mathcal{B}(l^2)$ such that

$$A_n \to A, \qquad A_n^* \to A^*, \qquad W_nA_nW_n \to \tilde{A}, \qquad W_nA_n^*W_n \to \tilde{A}^*,$$

where the asterisk refers to the adjoint operator and $\to$ denotes strong convergence.

We now show that $\mathcal{S}$ enjoys the properties required above. First of all, it is easily seen that $\mathcal{S}$ is a closed subalgebra of $\mathcal{F}$.

Lemma 2.21. *$\mathcal{J}$ is a closed two-sided ideal of $\mathcal{S}$.*

Proof. Obviously, $\mathcal{J}$ is a selfadjoint linear subspace of $\mathcal{S}$. We now prove that $\mathcal{J}$ is closed. Let

$$\{A_n\} = \{P_nKP_n + W_nLW_n + C_n\} \in \mathcal{J}. \tag{2.38}$$

Since W_n converges weakly to zero and L is compact, it follows that LW_n and thus also W_nLW_n converges strongly to zero. Consequently, $A_n \to K$ and $W_nA_nW_n \to L$ strongly, whence

$$\|K\| \le \liminf_{n\to\infty} \|A_n\|, \qquad \|L\| \le \liminf_{n\to\infty} \|A_n\|,$$

by Theorem 2.1. Thus, if $\{A_n^{(j)}\} \subset \mathcal{J}$ is a Cauchy sequence, then so are $\{K^{(j)}\} \subset \mathcal{K}(l^2)$ and $\{L^{(j)}\} \subset \mathcal{K}(l^2)$. We conclude that there are $K, L \in \mathcal{K}(l^2)$ such that $\|K^{(j)} - K\| \to 0$ and $\|L^{(j)} - L\| \to 0$ as $j \to \infty$, which implies almost at once that there is a sequence $\{A_n\} \in \mathcal{J}$ such that $\|\{A_n^{(j)}\} - \{A_n\}\| \to 0$ as $j \to \infty$. This proves that $\mathcal{J}$ is closed.

If $\{A_n\} \in \mathcal{J}$ is given by (2.38) and $\{B_n\}$ is any sequence in $\mathcal{S}$, then

$$B_nA_n = P_n(B_nP_nK)P_n + W_n(W_nB_nW_nL)W_n + B_nC_n,$$

and since $B_n P_n K \to BK$ and $W_n B_n W_n L \to \tilde{B}L$ uniformly (Lemma 2.8), it results that

$$B_n A_n = P_n BKP_n + W_n \tilde{B} L W_n + C_n'$$

with $\{C_n'\} \in \mathcal{N}$. Hence $\{B_n\}\{A_n\} \in \mathcal{J}$. Passing to adjoints we see that $\{A_n\}\{B_n\}$ also belongs to $\mathcal{J}$. ■

Lemma 2.22. *If $a \in L^\infty$, then $\{T_n(a)\} \in \mathcal{S}$, the strong limits of $T_n(a)$ and $W_n T_n(a) W_n$ being $T(a)$ and $T(\tilde{a})$, respectively.*

Proof. This is immediate from the equalities

$$T_n^*(a) = T_n(\overline{a}), \quad T^*(a) = T(\overline{a}), \quad W_n T_n(a) W_n = T_n(\tilde{a}). \ ■$$

Theorem 2.23. *Let $\{A_n\} \in \mathcal{S}$ and denote the strong limits of A_n and $W_n A_n W_n$ by A and $\tilde{A}$, respectively. Then the following are equivalent:*

(i) *$\{A_n\}$ is stable;*

(ii) *A and $\tilde{A}$ are invertible operators and the coset $\{A_n\} + \mathcal{J}$ is invertible in the algebra $\mathcal{S}/\mathcal{J}$.*

Proof. (i) ⇒ (ii). Suppose $\{A_n\}$ is stable. Since $A_n^* \to A^*$ strongly by the definition of $\mathcal{S}$, we deduce from Proposition 2.3 that A is invertible. As

$$\|(W_n A_n W_n)^{-1}\| = \|W_n A_n^{-1} W_n\| = \|A_n^{-1}\|,$$

the sequence $\{W_n A_n W_n\}$ is also stable. Again by the definition of the algebra $\mathcal{S}$, $W_n A_n W_n \to \tilde{A}$ and $(W_n A_n W_n)^* \to \tilde{A}^*$ strongly, which, once more by Proposition 2.3, implies that $\tilde{A}$ is invertible.

From Proposition 2.4 we now deduce that

$$A \in \Pi\{A_n\}, \quad \tilde{A} \in \Pi\{W_n A_n W_n\}, \quad A^* \in \Pi\{A_n^*\}, \quad \tilde{A}^* \in \Pi\{W_n A_n^* W_n\}.$$

Suppose A_n is invertible for $n \ge n_0$. Put $B_n = A_n^{-1}$ for $n \ge n_0$ and $B_n = 0$ for $n < n_0$. Using Proposition 2.2 we obtain

$$B_n \to A^{-1}, \quad W_n B_n W_n \to \tilde{A}^{-1}, \quad B_n^* \to (A^*)^{-1}, \quad W_n B_n^* W_n \to (\tilde{A}^*)^{-1}.$$

This shows that $\{B_n\} \in \mathcal{S}$. Since

$$\{B_n\}\{A_n\} - \{P_n\} \in \mathcal{N} \subset \mathcal{J}, \quad \{A_n\}\{B_n\} - \{P_n\} \in \mathcal{N} \subset \mathcal{J},$$

the element $\{B_n\} + \mathcal{J}$ is the inverse of $\{A_n\} + \mathcal{J}$.

(ii) ⇒ (i). Let $\{B_n\} + \mathcal{J}$ be the inverse of $\{A_n\} + \mathcal{J}$ and denote by B and $\tilde{B}$ the strong limits of B_n and $W_n B_n W_n$, respectively. We have

$$A_n B_n = P_n + P_n K P_n + W_n L W_n + C_n$$

with $K, L \in \mathcal{K}(l^2)$ and $\|C_n\| \to 0$. Taking into account that $W_n L W_n$ and $W_n K W_n$ converge strongly to zero, we get $AB = I + K$ and $\tilde{A}\tilde{B} = I + L$, which shows that

$$S := A^{-1} - B = -A^{-1}K \quad \text{and} \quad T := \tilde{A}^{-1} - \tilde{B} = -\tilde{A}^{-1}L$$

are compact. Put

$$R_n := B_n + P_n S P_n + W_n T W_n.$$

Then $\{R_n\} \in \mathcal{S} \subset \mathcal{F}$ and

$$A_n R_n = P_n + P_n(K + A_n P_n S)P_n + W_n(L + W_n A_n W_n T)W_n + C_n.$$

Lemma 2.8 shows that this is

$$P_n + P_n(K + AS)P_n + W_n(L + \tilde{A}T)W_n + C_n',$$

where $\|C_n'\| \to 0$. As

$$K + AS = L + \tilde{A}T = 0,$$

we see that $A_n R_n = P_n + C_n'$. Hence, $\{A_n\} + \mathcal{N}$ is invertible from the right. Analogously one can prove that $\{A_n\} + \mathcal{N}$ is invertible from the left. Proposition 2.20 therefore yields the stability of $\{A_n\}$. ∎

In the course of the previous proof we showed that if $\{B_n\} + \mathcal{J}$ is the inverse of $\{A_n\} + \mathcal{J}$ in $\mathcal{S}/\mathcal{J}$, then

$$R_n = B_n + P_n(A^{-1} - B)P_n + W_n(\tilde{A}^{-1} - \tilde{B})W_n \tag{2.39}$$

is an asymptotic inverse of A_n, that is,

$$\sup_{n\geq 1} \|R_n\| < \infty, \qquad \|R_n A_n - P_n\| \to 0, \qquad \|A_n R_n - P_n\| \to 0.$$

Corollary 2.24. *Let $a \in L^\infty$ and suppose $T(a)$ is invertible. Then for $\{T_n(a)\}$ to be stable it is necessary and sufficient that $\{T_n(a)\} + \mathcal{J}$ be invertible in $\mathcal{S}/\mathcal{J}$.*

Proof. This follows from Lemma 2.22, Theorem 2.23, and the fact that $T(a)$ is invertible if and only if so is the (transposed) operator $T(\tilde{a})$. ∎

Third proof of Theorem 2.11. If $a \in GC$, then (2.37) and its analogue for $T_n(a)T_n(a^{-1})$ shows that $\{T_n(a^{-1})\} + \mathcal{J}$ is the inverse of $\{T_n(a)\} + \mathcal{J}$. Theorem 2.11 is therefore an immediate consequence of Corollary 2.24. ∎

Moreover, Theorem 2.14 is now seen to result straightforwardly from (2.39).

The real strength of Theorem 2.23 will be revealed in the forthcoming sections. This theorem reduces the stability problem for sequences in $\mathcal{S}$ to invertibility in the algebra $\mathcal{S}/\mathcal{J}$, and unlike the algebra $\mathcal{F}/\mathcal{N}$, the algebra $\mathcal{S}/\mathcal{J}$ is so nice that it can be studied with the help of so-called local principles.

Notes. Proposition 2.20 is the starting point of the local approach to projection methods developed by Kozak [109]. The approach exhibited here (including the algebra $\mathcal{S}$, the ideal $\mathcal{J}$, and Theorem 2.23) is from [154].

2.6 Local Principles

Homomorphisms and isomorphisms. Given two Banach algebras $\mathcal{A}$ and $\mathcal{B}$, a map $\varphi : \mathcal{A} \to \mathcal{B}$ is called a *Banach algebra homomorphism* if φ is a bounded linear operator and

$$\varphi(ab) = \varphi(a)\varphi(b) \quad \text{for all} \quad a, b \in \mathcal{A}.$$

Bijective Banach algebra homomorphisms are referred to as *Banach algebra isomorphisms.*

Commutative Banach algebras. Let $\mathcal{A}$ be a commutative Banach algebra with identity element e. The Banach algebra homomorphisms of $\mathcal{A}$ into $\mathbf{C}$ which send e to 1 are called the *multiplicative linear functionals* of $\mathcal{A}$. Let $\mathcal{M}$ denote the set of all maximal ideals of $\mathcal{A}$ and let M stand for the set of all multiplicative linear functionals of $\mathcal{A}$. One can show that the map $M \to \mathcal{M}$, $\varphi \mapsto \operatorname{Ker} \varphi$ is bijective. Therefore no distinction is usually made between multiplicative linear functionals and maximal ideals.

The formula $\hat{a}(m) = m(a)$ $(m \in M)$ assigns a function $\hat{a} : M \to \mathbf{C}$ to each $a \in \mathcal{A}$. This function is referred to as the *Gelfand transform* of a. Let $\hat{A}$ be the set $\{\hat{a} : a \in \mathcal{A}\}$. The *Gelfand topology* on M is the coarsest (weakest) topology on M which makes all functions $\hat{a} \in \hat{A}$ continuous. The set M equipped with the Gelfand topology is called the *maximal ideal space* of $\mathcal{A}$. One can show that M is a compact Hausdorff space. The map

$$\Gamma : \mathcal{A} \to C(M), \ a \mapsto \hat{a}$$

is called the *Gelfand map* of A.

Theorem 2.25 (Gelfand). *Let $\mathcal{A}$ be a commutative Banach algebra with identity element and let M be the maximal ideal space of $\mathcal{A}$. An element $a \in \mathcal{A}$ is invertible if and only if $\hat{a}(m) \neq 0$ for all $m \in M$.*

In words: the Gelfand map is a Banach algebra homomorphism of $\mathcal{A}$ into $C(M)$ which preserves spectra. It can be shown that Γ actually has the norm 1, i.e., $\|\hat{a}\|_\infty \le \|a\|$ for every $a \in \mathcal{A}$. A proof of Theorem 2.25 is in every textbook on Banach algebras. ■

Example 2.26: singly generated algebras. A Banach algebra $\mathcal{A}$ with identity element e is said to be singly generated by an element $c \in \mathcal{A}$ if the smallest closed subalgebra of $\mathcal{A}$ containing e and c coincides with all of $\mathcal{A}$. One can show that in this case the maximal ideal space of $\mathcal{A}$ is homeomorphic to $\operatorname{sp} c$ (with the topology induced from the embedding $\operatorname{sp} c \subset \mathbf{C}$) and that the Gelfand map can be given by

$$\Gamma : \mathcal{A} \to C(\operatorname{sp} c), \ \big(\Gamma f(c)\big)(\lambda) = f(\lambda)$$

for every polynomial f. ■

Example 2.27: Wiener algebras. It turns out that the multiplicative linear functionals of the Wiener algebra W are the maps

$$\varphi_\tau : W \to \mathbf{C},\ a \mapsto a(\tau) \quad (\tau \in \mathbf{T}).$$

Thus, the maximal ideal space $M(W)$ of W can be identified with $\mathbf{T}$ and the Gelfand map is then nothing but the embedding $\Gamma : W \to C(\mathbf{T})$. Hence, Theorem 2.25 immediately yields Wiener's theorem: $a \in W$ is invertible in W if and only if $a(\tau) \neq 0$ for all $\tau \in \mathbf{T}$.

Analogously, the Gelfand maps of W_+ and W_- are given by

$$\Gamma : W_+ \to C(\overline{\mathbf{D}}), \quad \left(\Gamma \sum_{n=0}^{\infty} a_n \chi_n\right)(z) = \sum_{n=0}^{\infty} a_n z^n,$$

$$\Gamma : W_- \to C(\overline{\mathbf{D}}), \quad \left(\Gamma \sum_{n=0}^{\infty} a_n \chi_{-n}\right)(z) = \sum_{n=0}^{\infty} a_n z^n,$$

where

$$\overline{\mathbf{D}} := \{z \in \mathbf{C} : |z| \leq 1\}.$$

Theorem 2.25 therefore implies the invertibility criteria for $W_\pm$ cited in Section 1.5. ■

Example 2.28: the simplest Toeplitz algebra. Let $\mathbf{A}(C)$ be the smallest closed subalgebra of $\mathcal{B}(l^2)$ containing the set $\{T(a) : a \in C\}$, i.e., all Toeplitz operators with continuous symbols. It is easily seen that every finite-rank operator and thus every compact operator belongs to $\mathbf{A}(C)$ (see, e.g., [39, p. 155]). Consequently,

$$\mathcal{D} := \{T(c) + K : c \in C, K \in \mathcal{K}(l^2)\} \subset \mathbf{A}(C).$$

Let $c \in C$ and $K \in \mathcal{K}(l^2)$. By Theorem 1.17, the spectral radius of $T(c) + \mathcal{K}(l^2)$ in the Calkin algebra is $\|c\|_\infty$. Therefore

$$\|c\|_\infty \leq \operatorname{dist}\big(T(c), \mathcal{K}(l^2)\big) \leq \|T(c) + K\|, \tag{2.40}$$

which shows that $\mathcal{D}$ is a closed subset of $\mathcal{B}(l^2)$. This observation together with Proposition 1.12 and the fact that Hankel operators with continuous symbols are compact shows that in fact $\mathcal{D} = \mathbf{A}(C)$, i.e.,

$$\mathbf{A}(C) = \{T(c) + K : c \in C,\ K \in \mathcal{K}(l^2)\}. \tag{2.41}$$

Abbreviate the coset $T(c) + \mathcal{K}(l^2)$ to $T^\pi(c)$. From (2.41) we infer that

$$\mathbf{A}(C)/\mathcal{K}(l^2) = \{T^\pi(c) : c \in C\}. \tag{2.42}$$

Denote the Banach algebra (2.42) by $\mathbf{A}^\pi(C)$. The algebra $\mathbf{A}^\pi(C)$ is commutative (again by Proposition 1.12 and the compactness of Hankel operators with continuous symbols). Using Theorem 1.17 one can easily verify that the multiplicative linear functionals of $\mathbf{A}^\pi(C)$ are the maps $\varphi_\tau : T^\pi(c) \mapsto c(\tau)$ $(\tau \in \mathbf{T})$. Hence, we can identify the maximal ideal space of $\mathbf{A}^\pi(C)$ with $\mathbf{T}$ and can write the Gelfand map as

$$\Gamma : \mathbf{A}^\pi(C) \to C(\mathbf{T}), \quad T^\pi(c) \mapsto c.$$

This map Γ is readily seen to be even an isometric Banach algebra isomorphism (recall (2.40)). Therefore, we henceforth tacitly identify $\mathbf{A}^\pi(C)$ and $C := C(\mathbf{T})$. ■

Local principles. Theorem 2.25 associates with every element a of a unital commutative Banach algebra a collection of numbers, $\{\hat{a}(m)\}_{m\in M}$, in terms of which we can decide whether the given element is invertible or not. The idea behind so-called local principles is to associate with an element of a non-commutative unital Banach algebra a set of simpler objects which can answer for invertibility of the given element. One concrete realization of this strategy is the following theorem.

The *center* of a Banach algebra $\mathcal{A}$ is the set of all $z \in \mathcal{A}$ such that $za = az$ for all $a \in \mathcal{A}$. Note that the center and every subalgebra of the center are automatically commutative.

Theorem 2.29 (Allan-Douglas). *Let $\mathcal{A}$ be a Banach algebra with identity element e and let Z be a closed subalgebra of the center of $\mathcal{A}$ which contains e. Denote the maximal ideal space of Z by Ω, and for each maximal ideal $\omega \in \Omega$, let J_ω be the smallest closed two-sided ideal of $\mathcal{A}$ which contains the set ω. Then an element $a \in \mathcal{A}$ is invertible in $\mathcal{A}$ if and only if the coset $a + J_\omega$ is invertible in $\mathcal{A}/J_\omega$ for every $\omega \in \Omega$.*

A proof of this theorem is in [39, Theorem 1.34], for example. ■

We remark that if $J_\omega = \mathcal{A}$, then we consider $a + J_\omega$ as invertible in $\mathcal{A}/J_\omega$ by definition. The algebra $\mathcal{A}/J_\omega$ is referred to as the *local algebra* of $\mathcal{A}$ at $\omega \in \Omega$, the spectrum of $a + J_\omega$ in $\mathcal{A}/J_\omega$ is called the *local spectrum* of a at $\omega \in \Omega$, and every element $a_\omega \in \mathcal{A}$ for which $a_\omega + J_\omega = a + J_\omega$ is said to be a *local representative* of a at ω.

If $\mathcal{A}$ itself is commutative, we can take $Z = \mathcal{A}$, and since $\mathcal{A}/J_\omega = \mathcal{A}/\omega$ is isomorphic to $\mathbf{C}$ (Gelfand-Mazur theorem), Theorem 2.29 goes over into Theorem 2.25. Clearly, the larger the center of an algebra $\mathcal{A}$ is the finer we can localize in $\mathcal{A}$ using Theorem 2.29. In case the center is trivial, i.e., equal to $\{\lambda e : \lambda \in \mathbf{C}\}$, Theorem 2.29 merely says that a is invertible if and only if a is invertible.

Example 2.30: the local essential range. Let $\mathcal{A}$ be the (commutative) Banach algebra L^∞ and put $Z = C$. The maximal ideal space of Z is $\mathbf{T}$: the maximal ideal associated with $\tau \in \mathbf{T}$ is the set $\{c \in C : c(\tau) = 0\}$. The corresponding ideal $J_\tau \subset L^\infty$ is the closure of the set of all finite sums of the form

$$\sum_j c_j f_j \quad \text{with} \quad c_j \in C, \; c_j(\tau) = 0, \; f_j \in L^\infty.$$

One can show that actually

$$J_\tau = \{cf : c \in C, \; c(\tau) = 0, \; f \in L^\infty\} \tag{2.43}$$

(see, e.g., [31, Proposition 8.6]).

If $a \in PC$, then obviously $a + J_\tau = a_\tau + J_\tau$ where $a_\tau \in PC$ is any function such that $a_\tau(\tau \pm 0) = a(\tau \pm 0)$. This easily implies that the spectrum of $a + J_\tau$ is the set $\{a(\tau - 0), a(\tau + 0)\}$. In the general case, $a \in L^\infty$, it is not difficult to see that the local spectrum of a at τ, i.e., the spectrum of $a + J_\tau$, is just the set $\mathcal{R}_\tau(a)$ introduced in Section 1.7. Moreover, the norm $\|a + J_\tau\|$ is nothing but the number $\varrho_\tau(a)$ we encountered in Section 1.7. In the case at hand, Theorem 2.29 simply says that a function $a \in L^\infty$ is invertible in L^∞ if and only if $0 \notin \mathcal{R}_\tau(a)$ for every $\tau \in \mathbf{T}$. ■

Example 2.31: operators of local type. Consider the Calkin algebra $\mathcal{B}^\pi := \mathcal{B}(l^2)/\mathcal{K}(l^2)$ and write $A^\pi := A + \mathcal{K}(l^2)$ for $A \in \mathcal{B}(l^2)$. The operators in

$$\Lambda := \{A \in \mathcal{B}(l^2) : AT(c) - T(c)A \in \mathcal{K}(l^2) \;\text{ for all }\; c \in C\}$$

are called *operators of local type*. Clearly, Λ contains $\mathcal{K}(l^2)$, and by virtue of Proposition 1.12 and the compactness of Hankel operators with continuous symbols, every Toeplitz operator $T(a)$ with $a \in L^\infty$ belongs to Λ. Put $\Lambda^\pi := \Lambda/\mathcal{K}(l^2)$. By the definition of Λ, the algebra $C = \mathbf{A}^\pi(C)$ of Example 2.28 is contained in the center of Λ^π.

The algebra Λ^π is obviously inverse closed in $\mathcal{B}^\pi$: if $A \in \Lambda$ and A^π is invertible in $\mathcal{B}^\pi$, then the inverse of A^π belongs to Λ^π. Hence, an operator $A \in \Lambda$ is Fredholm if and only if A^π is invertible in Λ^π. We can therefore employ Theorem 2.29 with $\mathcal{A} = \Lambda^\pi$ and $\mathcal{Z} = C = \mathbf{A}^\pi(C)$ to study Fredholmness of operators of local type.

The ideal $J_\tau \subset \Lambda^\pi$ corresponding to $\tau \in \mathbf{T} = M(C)$ is the closure of the set of all finite sums

$$\sum_j T^\pi(c_j) B_j^\pi \quad \text{with} \quad c_j \in C, \; c_j(\tau) = 0, \; B_j \in \Lambda.$$

Again it can be shown that in fact

$$J_\tau = \{T^\pi(c)B^\pi : c \in C, \; c(\tau) = 0, \; B \in \Lambda\}. \tag{2.44}$$

Put $A_\tau^\pi := A^\pi + J_\tau$. We so infer from Theorem 2.29 that if $A \in \Lambda$, then A is Fredholm if and only if A_τ^π is invertible in Λ^π/J_τ for every $\tau \in \mathbf{T}$. ■

Theorem 2.32. *Let $a \in L^\infty$. For $\tau \in \mathbf{T}$, put $T_\tau^\pi(a) := T^\pi(a) + J_\tau$ where J_τ is given by* (2.44). *Then*

$$\mathcal{R}_\tau(a) \subset \operatorname{sp} T_\tau^\pi(a) \subset \operatorname{conv} \mathcal{R}_\tau(a).$$

The right inclusion of this theorem easily follows from the definition of the local essential range $\mathcal{R}_\tau(a)$: if $0 \notin \operatorname{conv} \mathcal{R}_\tau(a)$, then there is a neighborhood $U \in \mathcal{U}_\tau$ such that $0 \notin \operatorname{conv} \mathcal{R}_U(a)$ (recall Section 1.7), hence the function given by

$$a_U(t) := \begin{cases} a(t) & \text{for } t \in U, \\ \lambda_0 \in \operatorname{conv} \mathcal{R}_U(a) & \text{for } t \in \mathbf{T} \setminus U \end{cases}$$

induces an invertible Toeplitz operator (Theorem 1.18), and since $T_\tau^\pi(a) = T_\tau^\pi(a_U)$, it results that $T_\tau^\pi(a)$ is invertible. The left inclusion of Theorem 2.32 is less trivial; a proof is in [39, Corollary 3.64]. ■

Second proof of Theorem 1.21. If $a \in L^\infty$ is locally sectorial on $\mathbf{T}$, then $0 \notin \operatorname{conv} \mathcal{R}_\tau(a)$ for every $\tau \in \mathbf{T}$. Hence, by Theorem 2.32, $T_\tau^\pi(a)$ is invertible for every $\tau \in \mathbf{T}$. From Theorem 2.29 and Example 2.31 we therefore deduce that $T(a)$ is Fredholm. ■

Notes. The reasoning of Example 2.27 goes back to I.M. Gelfand and was one of the first triumphs of the theory of Banach algebras. The results of Example 2.28 are due to Gohberg [79] and Coburn [50]. Theorem 2.29 was established by Allan [1]. In the case of C^*-algebras, Theorem 2.29 was independently discovered by Douglas [59, Theorem 7.47], who was also the first to realize the relevancy of this theorem in operator theory. Operators of local type were introduced by Simonenko [160]. He also developed a local principle for their investigation. This local principle was subsequently essentially simplified and generalized by Gohberg and Krupnik [87] and Kozak [109]. These local principles are all more or less equivalent to Theorem 2.29. Local Toeplitz operators, i.e., the cosets $T_\tau^\pi(a)$, are an invention of Douglas [60]. Douglas also asked whether $\operatorname{sp} T_\tau^\pi(a)$ is a connected set for every $a \in L^\infty$. This problem is still open. That the ideal J_τ is of the form (2.43), (2.44) was probably first observed by Semenyuta and Khevelev [150].

2.7 Localization of Stability

We now apply Theorem 2.29 to the algebra $\mathcal{S}/\mathcal{J}$. Our first objective is to obtain a finite section analogue of Example 2.28. Let $\mathbf{S}(C)$ denote the smallest closed subalgebra of $\mathcal{S}$ (or $\mathcal{F}$) containing the set $\{\{T_n(c)\} : c \in C\}$.

Proposition 2.33. *The algebra* $\mathbf{S}(C)$ *coincides with the set*

$$\big\{\{T_n(c) + P_n K P_n + W_n L W_n + C_n\} : c \in C;\ K, L \in \mathcal{K}(l^2);\ \|C_n\| \to 0\big\}.$$

Proof. The set under consideration is a closed subalgebra of $\mathcal{S}$ (recall the proofs of Lemma 2.21 and Theorem 2.23 and take into account (2.40)).

One can also show that this set is contained in $\mathbf{S}(C)$ (see [39, Proposition 7.27]). This implies the assertion. ∎

From Example 2.28 and Proposition 2.33 we see that $\mathcal{J}$ is a closed two-sided ideal of $\mathbf{S}(C)$. Put

$$\{A_n\}^\pi := \{A_n\} + \mathcal{J} \ \left(\{A_n\} \in \mathcal{S}\right), \qquad \mathbf{S}^\pi(C) = \mathbf{S}(C)/\mathcal{J}.$$

By Proposition 2.33, $\mathbf{S}^\pi(C) = \{\{T_n(c)\}^\pi : c \in C\}$. Proposition 2.12 and the fact that Hankel operators with continuous symbols are compact show that $\mathbf{S}^\pi(C)$ is commutative.

Proposition 2.34. *The maximal ideal space of* $\mathbf{S}^\pi(C)$ *can be identified with* $\mathbf{T}$ *and the Gelfand map is given by*

$$\Gamma : \mathbf{S}^\pi(C) \to C(\mathbf{T}), \ \left\{T_n(c)\right\}^\pi \mapsto c.$$

Proof. The map $\Phi : \mathbf{S}(C) \to \mathbf{A}^\pi(C)$, $\{A_n\} \mapsto (s\text{–}\lim_{n\to\infty} A_n)^\pi$ is a surjective Banach algebra homomorphism. From Proposition 2.33 we infer that $\operatorname{Ker}\Phi = \mathcal{J}$. This shows that $\mathbf{S}^\pi(C)$ is isomorphic to $\mathbf{A}^\pi(C)$. Example 2.28 completes the proof. ∎

We henceforth freely identify $\mathbf{S}^\pi(C)$ and C.

In analogy to operators of local type, we define Σ as the set of all $\{A_n\} \in \mathcal{S}$ satisfying $\{A_n T_n(c) - T_n(c)A_n\} \in \mathcal{J}$ for all $c \in C$. By Proposition 2.12 and (the sufficiency part of) Theorem 1.16, Σ contains $\{T_n(a)\}$ for every $a \in L^\infty$. Let $\Sigma^\pi := \Sigma/\mathcal{J}$. Then $C = \mathcal{S}^\pi(C)$ is contained in the center of Σ^π. Obviously, if $\{A_n\} \in \Sigma$ and $\{A_n\}^\pi$ is invertible in $\mathcal{S}/\mathcal{J}$, then $\{A_n\}^\pi$ is also invertible in $\Sigma^\pi = \Sigma/\mathcal{J}$.

For $\tau \in \mathbf{T}$, let J_τ be the smallest closed two-sided ideal of Σ^π containing $\{\{T_n(c)\}^\pi : c \in C, \ c(\tau) = 0\}$. One can show that

$$J_\tau = \left\{\left\{T_n(c)B_n\right\}^\pi : c \in C, \ c(\tau) = 0, \ \{B_n\} \in \Sigma\right\}.$$

Denote by $\{A_n\}_\tau^\pi$ the coset $\{A_n\}^\pi + J_\tau$ in the local algebra Σ^π/J_τ.

Theorem 2.35. *Let* $\{A_n\} \in \Sigma$. *Then* $\{A_n\}^\pi$ *is invertible in* $\mathcal{S}/\mathcal{J}$ *if and only if* $\{A_n\}_\tau^\pi$ *is invertible in* Σ^π/J_τ *for every* $\tau \in \mathbf{T}$.

Proof. Immediate from Theorem 2.29 and the above discussion. ∎

Theorem 2.36. *If* $a \in L^\infty$, *then for every* $\tau \in \mathbf{T}$,

$$\mathcal{R}_\tau(a) \subset \operatorname{sp}\left\{T_n(a)\right\}_\tau^\pi \subset \operatorname{conv} \mathcal{R}_\tau(a).$$

The right inclusion can be verified as the corresponding inclusion of Theorem 2.32 (also recall Proposition 2.17), the left inclusion follows from [39, Corollary 3.64]. ∎

Second proof of Theorem 2.18. If $a \in L^\infty$ is locally sectorial on $\mathbf{T}$, then $\{T_n(a)\}_\tau^\pi$ is invertible for every $\tau \in \mathbf{T}$ by virtue of the right inclusion of Theorem 2.36, and hence $\{T_n(a)\}^\pi$ is invertible in $\mathcal{S}/\mathcal{J}$ due to Theorem 2.35. If, in addition, $T(a)$ is invertible, it results from Corollary 2.24 that $\{T_n(a)\}$ is stable. ■

Note. In this section we followed [154] and [36].

3

Norms of Inverses and Pseudospectra

3.1 C^*-Algebras

C^*-algebras are especially nice Banach algebras. A map $a \mapsto a^*$ of a Banach algebra $\mathcal{A}$ onto itself is called an *involution* if

$$a^{**} = a, \qquad (a+b)^* = a^* + b^*, \qquad (ab)^* = b^* a^*, \quad (\lambda a)^* = \overline{\lambda} a^*$$

for all $a, b \in \mathcal{A}$ and all $\lambda \in \mathbf{C}$. A *C^*-algebra* is a Banach algebra with an involution such that

$$\|aa^*\| = \|a\|^2 \quad \text{for all } a \in \mathcal{A}. \tag{3.1}$$

Examples. If X is a Hausdorff space, then $C(X)$ is a C^*-algebra. The algebra L^∞ as well as their subalgebras C and PC are C^*-algebras. In these cases the involution is given by complex conjugation, $a \mapsto \overline{a}$.

Although complex conjugation is an involution on the Wiener algebra W, it does not make W into a C^*-algebra: if $a(t) = -t^{-1} + 1 + t$ ($t \in \mathbf{T}$), then $(a\overline{a})(t) = -t^{-2} + 3 - t^2$, whence $\|a\overline{a}\|_W = 5$ and $\|a\|_W^2 = 9$.

If H is a Hilbert space, then the algebras $\mathcal{B}(H)$ and $\mathcal{K}(H)$ of all bounded and compact linear operators on H are C^*-algebras with passage to the adjoint operator as the involution. The involution $\{A_n\}^* := \{A_n^*\}$ makes the algebras $\mathcal{F}$ and $\mathcal{S}$ introduced in Section 2.5 into C^*-algebras.

A subset of a C^*-algebra is said to be selfadjoint if it is invariant under the involution. Clearly, every closed and selfadjoint subalgebra of a C^*-algebra is itself a C^*-algebra. The algebra $\mathcal{F}_c$ of all sequences $\{A_n\} \in \mathcal{F}$ which have a strong limit is not a selfadjoint subalgebra of $\mathcal{F}$ (recall the example before

Proposition 2.3: there $A_n \to 0$ while A_n^* does not at all converge strongly). However, the algebra $\mathcal{F}_{cc}$ of all sequences $\{A_n\} \in \mathcal{F}$ for which there exists an operator $A \in \mathcal{B}(l^2)$ such that $A_n \to A$ and $A_n^* \to A$ (strongly) is a C^*-subalgebra of the C^*-algebra $\mathcal{F}$.

If $\mathcal{A}$ is a C^*-algebra and J is a closed two-sided ideal of $\mathcal{A}$, then J is automatically selfadjoint and $\mathcal{A}/J$ provided with the involution $(a+J)^* := a^* + J$ and the usual quotient norm is a C^*-algebra. In particular, $\mathcal{F}/\mathcal{N}$ and $\mathcal{S}/\mathcal{J}$ are C^*-algebras.

The algebras $\mathbf{A}(C)$, $\mathbf{A}^\pi(C)$ (Section 2.6) and $\mathbf{S}(C)$, $\mathbf{S}^\pi(C)$ (Section 2.7) are also C^*-algebras with natural involutions.

Inverse closedness. If $\mathcal{A}$ is a Banach algebra with identity and $\mathcal{B}$ is a closed subalgebra of $\mathcal{A}$ which contains the identity, then for an element $b \in \mathcal{B}$ the spectrum in $\mathcal{B}$ may be larger than its spectrum in $\mathcal{A}$ (example: $\mathrm{sp}_{L^\infty} \chi_1 = \mathbf{T}$, $\mathrm{sp}_{H^\infty} \chi_1 = \overline{\mathbf{D}}$). As the following well-known result shows, this cannot happen for C^*-algebras.

Proposition 3.1. *If $\mathcal{A}$ is a C^*-algebra with identity and $\mathcal{B}$ is a C^*-subalgebra of $\mathcal{A}$ which contains the identity, then for every $b \in \mathcal{B}$ the equality*

$$\mathrm{sp}_{\mathcal{B}}\, b = \mathrm{sp}_{\mathcal{A}}\, b$$

holds. Shortly: unital C^*-algebras are always inverse closed.

Homomorphisms and isomorphisms. A map $\varphi : \mathcal{A} \to \mathcal{B}$ of a C^*-algebra $\mathcal{A}$ to a C^*-algebra $\mathcal{B}$ is referred to as a *C^*-algebra homomorphism* if φ is a Banach algebra homomorphism and $\varphi(a)^* = \varphi(a^*)$ for all $a \in \mathcal{A}$. Bijective C^*-algebra homomorphisms are called *C^*-algebra isomorphisms.*

One can show that if $\varphi : \mathcal{A} \to \mathcal{B}$ is a C^*-algebra homomorphism, then $\varphi(\mathcal{A})$ is a C^*-subalgebra of $\mathcal{B}$ (which includes that $\varphi(\mathcal{A})$ is always a closed subset of $\mathcal{B}$).

A C^*-algebra homomorphism $\varphi : \mathcal{A} \to \mathcal{B}$ of unital C^*-algebras $\mathcal{A}$ and $\mathcal{B}$ is said to *preserve spectra* if

$$\mathrm{sp}\, \varphi(a) = \mathrm{sp}\, a \quad \text{for every } a \in \mathcal{A}$$

and is referred to as an *isometry* if

$$\|\varphi(a)\| = \|a\| \quad \text{for every } a \in \mathcal{A}.$$

The following simple result will prove to be very useful in what follows.

Proposition 3.2. *Let $\mathcal{A}$ and $\mathcal{B}$ be two C^*-algebras with identities and let $\varphi : \mathcal{A} \to \mathcal{B}$ be a C^*-algebra homomorphism.*

(a) *If φ preserves spectra, then φ also preserves norms, i.e., φ is an isometry.*

(b) *If φ is injective, then φ preserves spectra.*

Proof. (a) Let $a \in \mathcal{A}$. Since both aa^* and $\varphi(a)\varphi(a)^*$ are selfadjoint and the norm of a selfadjoint element of a C^*-algebra coincides with its spectral radius, we obtain from (3.1) that

$$\begin{aligned} \|a\|^2 &= \|aa^*\| = \max\{|\lambda| : \lambda \in \operatorname{sp}(aa^*)\} \\ &= \max\{|\lambda| : \lambda \in \operatorname{sp}\varphi(aa^*)\} \\ &= \max\{|\lambda| : \lambda \in \operatorname{sp}(\varphi(a)\varphi(a)^*)\} \\ &= \|\varphi(a)\varphi(a)^*\| = \|\varphi(a)\|^2. \end{aligned}$$

(b) Clearly, $\operatorname{sp}\varphi(a) \subset \operatorname{sp} a$ for every $a \in \mathcal{A}$. Conversely, suppose $\varphi(a-\lambda e)$ is invertible. Since $\varphi(a-\lambda e)$ lies in $\varphi(\mathcal{A})$ and $\varphi(\mathcal{A})$ is a C^*-subalgebra of $\mathcal{B}$, it follows from Proposition 3.1 that the inverse of $\varphi(a-\lambda e)$ belongs to $\varphi(\mathcal{A})$ and is therefore of the form $\varphi(c)$ with $c \in \mathcal{A}$. Then injectivity of φ implies that c is the inverse of $a - \lambda e$. ■

For C^*-algebras, Theorems 2.25 and 2.29 can be strengthened.

Theorem 3.3 (Gelfand-Naimark). *If $\mathcal{A}$ is a commutative C^*-algebra with identity element then the Gelfand map $\Gamma : \mathcal{A} \to C(M)$ is an isometric C^*-algebra isomorphism of $\mathcal{A}$ onto $C(M)$.*

Theorem 3.4 (Allan-Douglas). *Let the situation be as in Theorem 2.29. In addition, suppose $\mathcal{A}$ is a C^*-algebra and Z is a C^*-subalgebra of $\mathcal{A}$. Then for every $a \in \mathcal{A}$,*

$$\|a\| = \max_{\omega \in \Omega} \|a + J_\omega\|$$

and the maximum is attained for some $\omega_0 \in \Omega$.

One can show that under the hypothesis of Theorem 3.4 we have $J_\omega \neq \mathcal{A}$ for all $\omega \in \Omega$.

For proofs of Theorems 3.3 and 3.4 we refer to [59, Theorems 4.29 and 7.49] and [39, Theorem 1.34]. ■

3.2 Continuous Symbols

Let $a \in C$ and suppose $T(a)$ is invertible. Then $\{T_n(a)\}$ is stable (Theorem 2.11), hence $T_n^{-1}(a) \to T^{-1}(a)$ strongly (Propositions 2.2 and 2.4), and thus

$$\liminf_{n\to\infty} \|T_n^{-1}(a)\| \geq \|T^{-1}(a)\| \tag{3.2}$$

by the Banach-Steinhaus theorem (Theorem 2.1). The stability of $\{T_n(a)\}$ is equivalent to the estimate

$$\limsup_{n\to\infty} \|T_n^{-1}(a)\| < \infty.$$

The purpose of this section is to show the stronger inequality

$$\limsup_{n\to\infty} \|T_n^{-1}(a)\| \le \|T^{-1}(a)\|. \tag{3.3}$$

Obviously, (3.2) and (3.3) imply that

$$\lim_{n\to\infty} \|T_n^{-1}(a)\| = \|T^{-1}(a)\|. \tag{3.4}$$

C^*-algebras in action. To prove (3.3), we will employ Proposition 3.2 and are thus led to the study of C^*-algebras generated by sequences of Toeplitz matrices.

Consider the algebra $\mathbf{S}(C)$, whose structure is described by Proposition 2.33. The direct sum $\mathcal{B}(l^2) \oplus \mathcal{B}(l^2)$ is a C^*-algebra with componentwise algebraic operations and the norm

$$\|(A, B)\| := \max\{\|A\|, \|B\|\}.$$

Define

$$\varphi : \mathbf{S}(C) \to \mathcal{B}(l^2) \oplus \mathcal{B}(l^2), \quad \{A_n\} \mapsto (A, \tilde{A}), \tag{3.5}$$

where

$$A := s\text{-}\lim_{n\to\infty} A_n, \qquad \tilde{A} := s\text{-}\lim_{n\to\infty} W_n A_n W_n. \tag{3.6}$$

Notice that if

$$A_n = T_n(a) + P_n K P_n + W_n L W_n + C_n \tag{3.7}$$

with $K, L \in \mathcal{K}(l^2)$ and $\{C_n\} \in \mathcal{N}$, then the two operators (3.6) are

$$A = T(a) + K, \qquad \tilde{A} = T(\tilde{a}) + L. \tag{3.8}$$

From (3.6) it is clear that φ is a C^*-algebra homomorphism. Taking into account (2.40) we see that the two operators A and $\tilde{A}$ are zero if and only if $a = 0$ and $K = L = 0$. Hence, $\operatorname{Ker}\varphi = \mathcal{N}$. This shows that the map

$$\mathrm{Sym} : \mathbf{S}(C)/\mathcal{N} \to \mathcal{B}(l^2) \oplus \mathcal{B}(l^2), \quad \{A_n\} + \mathcal{N} \mapsto (A, \tilde{A})$$

is a well-defined and injective C^*-algebra homomorphism. The Sym stands for "symbol". We now let Proposition 3.2 do its job.

Theorem 3.5. *A sequence $\{A_n\} \in \mathbf{S}(C)$ is stable if and only if the two operators A and $\tilde{A}$ are invertible.*

Proof. By Propositions 2.20 and 3.1, the stability of $\{A_n\}$ is equivalent to the invertibility of $\{A_n\}+\mathcal{N}$ in $\mathbf{S}(C)/\mathcal{N}$. As Sym is an injective C^*-algebra homomorphism, we deduce from Proposition 3.2(b) that $\{A_n\} + \mathcal{N}$ is invertible if and only if so are A and $\tilde{A}$. ■

Theorem 3.6. *If* $\{A_n\} \in \mathbf{S}(C)$, *then*

$$\lim_{n\to\infty} \|A_n\| = \max\{\|A\|,\, \|\tilde{A}\|\}. \tag{3.9}$$

Proof. Since Sym is injective, we can use Proposition 3.2 to conclude that

$$\limsup_{n\to\infty} \|A_n\| = \|\{A_n\} + \mathcal{N}\| = \max\{\|A\|,\, \|\tilde{A}\|\}. \tag{3.10}$$

Because

$$\|A\| \le \liminf_{n\to\infty} \|A_n\|, \qquad \|\tilde{A}\| \le \liminf_{n\to\infty} \|W_n A_n W_n\| = \liminf_{n\to\infty} \|A_n\|$$

by Theorem 2.1, we arrive at (3.9). ■

To be a little more explicit, we note that (3.7), (3.8), (3.9) give

$$\lim_{n\to\infty} \|T_n(a)+P_nKP_n+W_nLW_n\| = \max\{\|T(a)+K\|,\, \|T(\tilde{a})+L\|\} \tag{3.11}$$

whenever $a \in C$, $K \in \mathcal{K}(l^2)$, $L \in \mathcal{K}(l^2)$.

Theorem 3.6 will imply the desired equality (3.4). We first mention the following result for sequences $\{A_n\} \in \mathcal{F}_{cc}$ (recall Section 3.1). Notice that we put $\|R^{-1}\| = \infty$ in case R is a non-invertible operator.

Proposition 3.7. *If* $\{A_n\} \in \mathcal{F}_{cc}$ *and the strong limit of* A_n *is not invertible, then*

$$\lim_{n\to\infty} \|A_n^{-1}\| = \infty.$$

Proof. Since $\{A_n\} \in \mathcal{F}_{cc}$, we have an operator $A \in \mathcal{B}(l^2)$ such that $A_n \to A$ and $A_n^* \to A^*$ strongly. Assume there are $n_1 < n_2 < \dots$ and $M < \infty$ such that $\|A_{n_k}^{-1}\| \le M$. Then for every $x \in l^2$,

$$\|P_{n_k}x\| \le M\|A_{n_k}P_{n_k}x\|, \qquad \|P_{n_k}x\| \le M\|A_{n_k}^*P_{n_k}x\|,$$

and letting $n_k \to \infty$ we get

$$\|x\| \le M\|Ax\|, \qquad \|x\| \le M\|A^*x\|,$$

which is impossible if A is not invertible. ■

Corollary 3.8. *If* $a \in C$ *and* $K \in \mathcal{K}(l^2)$, *then*

$$\lim_{n\to\infty} \|(T_n(a) + P_nKP_n)^{-1}\| = \max\{\|(T(a) + K)^{-1}\|,\, \|T^{-1}(\tilde{a})\|\}.$$

Proof. If $T(a) + K$ is not invertible, the assertion follows from Proposition 3.7. So suppose $T(a) + K$ is invertible. Then $T(a)$ and thus also $T(\tilde{a})$ are

Fredholm of index zero and therefore invertible (Theorem 1.10 or Theorem 1.17).

Let $A_n := T_n(a) + P_n K P_n$. The two limits (3.6) are $A = T(a) + K$ and $\tilde{A} = T(\tilde{a})$. By Theorem 3.5, $\{A_n\}$ is stable. Hence, by Propositions 2.20 and 3.1, $\{A_n\} + \mathcal{N}$ is invertible in $\mathbf{S}(C)/\mathcal{N}$, i.e., there is a sequence $\{B_n\} \in \mathbf{S}(C)$ such that

$$\|A_n B_n - P_n\| \to 0, \qquad \|B_n A_n - P_n\| \to 0. \tag{3.12}$$

Denote the strong limits of B_n and $W_n B_n W_n$ by B and $\tilde{B}$, respectively. From (3.12) we get $AB = BA = I$ and $\tilde{A}\tilde{B} = \tilde{B}\tilde{A} = I$, whence $B = A^{-1}$ and $\tilde{B} = \tilde{A}^{-1}$. Theorem 3.6 gives

$$\lim_{n\to\infty} \|B_n\| = \max\{\|B\|, \|\tilde{B}\|\} = \max\{\|A^{-1}\|, \|\tilde{A}^{-1}\|\},$$

and because, again by (3.12), $\lim_{n\to\infty} \|A_n^{-1}\| = \lim_{n\to\infty} \|B_n\|$, we arrive at the assertion. ■

Corollary 3.9. *If $a \in C$, then*

$$\lim_{n\to\infty} \|T_n^{-1}(a)\| = \|T^{-1}(a)\|.$$

First proof. This follows from Corollary 3.8 for $K = 0$ along with the fact that $T(\tilde{a})$ is the transposed operator of $T(a)$, by virtue of which $\|T^{-1}(\tilde{a})\| = \|T^{-1}(a)\|$. ■

Second proof. If $T(a)$ is not invertible, we can make use of Proposition 3.7. In the case where $T(a)$ is invertible, Theorem 2.14 gives

$$T_n^{-1}(a) = T_n(a^{-1}) + P_n K(a) P_n + W_n K(\tilde{a}) W_n + C_n$$

with $K(a),\ K(\tilde{a}) \in \mathcal{K}(l^2)$ and $\|C_n\| \to 0$. Consequently, by (3.11),

$$\begin{aligned} \lim_{n\to\infty} \|T_n^{-1}(a)\| &= \max\left\{\|T(a^{-1}) + K(a)\|,\ \|T(\tilde{a}^{-1}) + K(\tilde{a})\|\right\} \\ &= \max\left\{\|T^{-1}(a)\|,\ \|T^{-1}(\tilde{a})\|\right\} = \|T^{-1}(a)\|. \ ■ \end{aligned}$$

Note. The method and the results of this section are from [155] and [23].

3.3 Piecewise Continuous Symbols

In this section we extend Theorems 3.5 and 3.6 to sequences $\{A_n\}$ in the smallest closed subalgebra $\mathbf{S}(PC)$ of $\mathcal{S}$ (or $\mathcal{F}$) containing $\{\{T_n(a)\} : a \in PC\}$. Let $\mathbf{A}(PC)$ stand for the closed subalgebra of $\mathcal{B}(l^2)$ generated by the operators $T(a)$ with $a \in PC$. Put

$$\mathbf{A}^\pi(PC) := \mathbf{A}(PC)/\mathcal{K}(l^2), \qquad \mathbf{S}^\pi(PC) := \mathbf{S}(PC)/\mathcal{J}$$

(note that $\mathcal{K}(l^2) \subset \mathbf{A}(PC)$ by Example 2.28 and $\mathcal{J} \subset \mathbf{S}(PC)$ due to Proposition 2.33).

Theorem 3.10. *The algebras* $\mathbf{A}^\pi(PC)$ *and* $\mathbf{S}^\pi(PC)$ *are commutative* C^*-*algebras, their maximal ideal spaces can be identified with the cylinder* $\mathbf{T} \times [0,1]$ (*with an exotic topology, see Figure* 15), *and the Gelfand maps are for* $a \in PC$ *given by*

$$\begin{aligned}\big(\Gamma T^\pi(a)\big)(t,\mu) &= (1-\mu)\,a(t-0) + \mu a(t+0),\\ \big(\Gamma\{T_n(a)\}^\pi\big)(t,\mu) &= (1-\mu)\,a(t-0) + \mu a(t+0),\end{aligned}$$

where $t \in \mathbf{T}$ *and* $\mu \in [0,1]$.

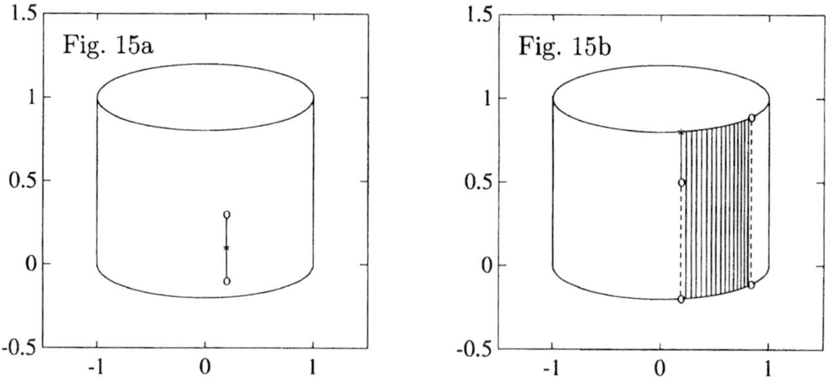

Neighborhood bases of a point $(t,\mu) \in \mathbf{T} \times [0,1]$ are formed by sets as in Figure 15a ($\mu \in (0,1)$) and Figure 15b ($\mu = 1$).

Proof. To show that $\mathbf{A}^\pi(PC)$ is commutative, we have to check whether $T^\pi(a)T^\pi(b) = T^\pi(b)T^\pi(a)$ whenever $a, b \in PC$. Obviously, it suffices to consider the case where a and b have only one jump, at α and β, say. Suppose first that $\alpha \neq \beta$. There are functions $\varphi, \psi \in C$ such that

$$\varphi^2 + \psi^2 = 1, \quad \varphi(\alpha) = 1, \quad \varphi(\beta) = 0, \quad \psi(\alpha) = 0, \quad \psi(\beta) = 1.$$

We have

$$T(ab) - T(a)T(b) = T(a\varphi^2 b) + T(a\psi^2 b) - T(a)\big(T(\varphi^2) + T(\psi^2)\big)T(b). \quad (3.13)$$

By Proposition 1.12,

$$\begin{aligned}&T^\pi(a\varphi^2 b) - T^\pi(a)T^\pi(\varphi^2)T^\pi(b)\\ &\quad = T^\pi(a\varphi\varphi b) - T^\pi(a\varphi)T^\pi(\varphi b) = H^\pi(a\varphi)H^\pi(\widetilde{\varphi b}),\end{aligned}$$

and since φb is continuous, $H^\pi(\tilde{\varphi}\tilde{b}) = 0$. Analogously one can show that

$$T^\pi(a\psi^2 b) - T^\pi(a)T^\pi(\psi^2)T^\pi(b) = 0.$$

This implies that (3.13) is compact and proves that $T^\pi(a)$ and $T^\pi(b)$ commute. If $\alpha = \beta$, there is a constant $\lambda \in \mathbf{C}$ and a function $c \in C$ such that $a = \lambda b + c$. It follows that

$$\begin{aligned} T(a)T(b) - T(b)T(a) &= T(\lambda b + c)T(b) - T(b)T(\lambda b + c) \\ &= T(c)T(b) - T(b)T(c), \end{aligned}$$

and as c is continuous, the latter commutator is compact. This completes the proof of the commutativity of $\mathbf{A}^\pi(PC)$.

Let $f : \mathbf{A}^\pi(PC) \to \mathbf{C}$ be a multiplicative linear functional. Then the restriction of f to $\mathbf{A}^\pi(C)$ is a multiplicative linear functional on $\mathbf{A}^\pi(C)$ and hence, by Example 2.28,

$$f\big(T^\pi(c)\big) = c(\tau) \quad \text{for all } \; c \in C$$

with some $\tau \in \mathbf{T}$. Let $\chi_\tau \in PC$ be the characteristic function of the half-circle $\{\tau e^{i\theta} : 0 < \theta < \pi\}$ and put

$$\mu := f\big(T^\pi(\chi_\tau)\big).$$

Then $\mu \in \operatorname{sp} T^\pi(\chi_\tau) = \operatorname{sp}_{\text{ess}} T(\chi_\tau) = [0,1]$ by Theorems 1.23 and 2.25. As every $a \in PC$ can be written in the form

$$a = a(t-0)(1-\chi_\tau) + a(t+0)\chi_\tau + c,$$

where $c \in PC$ is continuous at τ and $c(\tau) = 0$, it follows that

$$f\big(T^\pi(a)\big) = a(t-0)(1-\mu) + a(t+0)\mu. \tag{3.14}$$

We have shown that every multiplicative linear functional $f : \mathbf{A}^\pi(PC) \to \mathbf{C}$ acts on elements of the form $T^\pi(a)$ $(a \in PC)$ by the formula (3.14) with some $(t,\mu) \in \mathbf{T} \times [0,1]$. From Theorems 1.23 and 2.25 we see that for every $(t,\mu) \in \mathbf{T} \times [0,1]$ there must exist a linear multiplicative functional f such that (3.14) holds. At this point we have proved all assertions concerning $\mathbf{A}^\pi(PC)$. The commutativity of $\mathbf{S}^\pi(PC)$ can be shown as the commutativity of $\mathbf{A}^\pi(PC)$ (simply use Proposition 2.12 instead of Proposition 1.12). If f is a multiplicative linear functional on $\mathbf{S}^\pi(PC)$, then there is a $\tau \in \mathbf{T}$ such that

$$f\big(\{T_n(c)\}^\pi\big) = c(\tau) \quad \text{for all } \; c \in C$$

by virtue of Proposition 2.34. Let χ_τ be as above and put

$$\mu := f\big(\{T_n(\chi_\tau)\}^\pi\big).$$

Again $\mu \in \mathrm{sp}\{T_n(\chi_\tau)\}^\pi$, and since $\{T_n(\chi_\tau - \lambda)\}$ is stable and therefore $\{T_n(\chi_\tau) - \lambda P_n\}^\pi$ is invertible whenever $\lambda \notin [0,1]$ (Proposition 2.17), it follows that $\mu \in [0,1]$. Consequently, if $a \in PC$, then

$$f\big(\{T_n(a)\}^\pi\big) = a(t-0)(1-\mu) + a(t+0)\mu. \tag{3.15}$$

Finally, since $T(a-\lambda)$ is Fredholm if $\{T_n(a-\lambda)\}^\pi$ is invertible (just pass to the strong limit $n \to \infty$), it follows that for every $(t,\mu) \in \mathbf{T} \times [0,1]$ there is a multiplicative linear functional f for which (3.15) is valid. ■

In analogy to (3.5), consider the map

$$\varphi : \mathbf{S}(PC) \to \mathcal{B}(l^2) \oplus \mathcal{B}(l^2), \quad \{A_n\} \mapsto (A, \tilde{A}),$$

where A and $\tilde{A}$ are given by (3.6). As $\mathbf{S}(PC)$ is of much more intricate structure than $\mathbf{S}(C)$, the simple injectivity argument used in Section 3.2 does not work in the present case. However, it is obvious that $\mathcal{N} \subset \mathrm{Ker}\,\varphi$, hence

$$\mathrm{Sym} : \mathbf{S}(PC)/\mathcal{N} \to \mathcal{B}(l^2) \oplus \mathcal{B}(l^2), \quad \{A_n\} + \mathcal{N} \mapsto (A, \tilde{A})$$

is a well-defined C^*-algebra homomorphism, and we will show that Sym is isometric by proving that Sym preserves spectra.

Theorem 3.11. *A sequence $\{A_n\} \in \mathbf{S}(PC)$ is stable if and only if both A and $\tilde{A}$ are invertible.*

Proof. By Theorem 2.23, $\{A_n\}$ is stable if and only if A and $\tilde{A}$ are invertible and $\{A_n\}^\pi$ is invertible in $\mathcal{S}/\mathcal{J}$. We will prove that the invertibility of $\{A_n\}^\pi$ in $\mathcal{S}/\mathcal{J}$ is equivalent to the Fredholmness of A. This clearly gives the assertion.

By Proposition 3.1, $\{A_n\}^\pi$ is invertible in $\mathcal{S}/\mathcal{J}$ if and only if $\{A_n\}^\pi$ is invertible in $\mathbf{S}^\pi(PC) = \mathbf{S}(PC)/\mathcal{J}$. Combining Theorems 3.3 and 3.10 we see that this is equivalent to the invertibility of A^π in $\mathbf{A}^\pi(PC) = \mathbf{A}(PC)/\mathcal{K}(l^2)$, which, again by Proposition 3.1, is equivalent to the invertibility in $\mathcal{B}(l^2)/\mathcal{K}(l^2)$ and thus to the property of being Fredholm. ■

Theorem 3.12. *If $\{A_n\} \in \mathbf{S}(PC)$ then*

$$\lim_{n\to\infty} \|A_n\| = \max\big\{\|A\|,\ \|\tilde{A}\|\big\}.$$

Proof. Since $\{A_n\} \in \mathbf{S}(PC)$ is stable if and only if $\{A_n\} + \mathcal{N}$ is invertible in $\mathbf{S}(PC)/\mathcal{N}$ (Propositions 2.20 and 3.1), Theorem 3.11 says that Sym preserves spectra. Proposition 3.2(a) therefore gives (3.10), and the rest is as in the proof of Theorem 3.6. ■

Here are a few concrete examples.

Corollary 3.13. (a) *Let a_{jk} be a finite collection of functions in PC and put*

$$A_n = \sum_j \prod_k T_n(a_{jk}), \quad A = \sum_j \prod_k T(a_{jk}), \quad \tilde{A} = \sum_j \prod_k T(\tilde{a}_{jk}).$$

Then

$$\lim_{n\to\infty} \|A_n^{-1}\| = \max\{\|A^{-1}\|, \|\tilde{A}^{-1}\|\}.$$

(b) *If $a \in PC$ and $K \in \mathcal{K}(l^2)$, then*

$$\lim_{n\to\infty} \|(T_n(a) + P_n K P_n)^{-1}\| = \max\{\|(T(a)+K)^{-1}\|, \|T^{-1}(\tilde{a})\|\}.$$

(c) *If $a \in PC$, then*

$$\lim_{n\to\infty} \|T_n^{-1}(a)\| = \|T^{-1}(a)\|. \tag{3.16}$$

Proof. (a) If both A and $\tilde{A}$ are invertible, the assertion follows as in the proof of Corollary 3.8. In case A is not invertible, we deduce from Proposition 3.7 that $\|A_n^{-1}\| \to \infty$. Finally, if $\tilde{A}$ is not invertible, we obtain analogously that $\|\tilde{A}_n^{-1}\| \to \infty$ where $\tilde{A}_n := W_n A_n W_n$. Since

$$\|\tilde{A}_n^{-1}\| = \|W_n A_n^{-1} W_n\| = \|A_n^{-1}\|,$$

it results that $\|A_n^{-1}\| \to \infty$.

(b) Proceed as in the proof of Corollary 3.8.

(c) Since $T(\tilde{a})$ is the transposed operator of $T(a)$, this follows immediately from part (b). ■

Locally normal symbols. We do not know how to prove equality (3.16) under the sole assumption that a be locally sectorial on $\mathbf{T}$. However, Theorems 3.11, 3.12 and Corollary 3.13 remain valid for so-called locally normal symbols.

As first observed by Brown and Halmos [43], a Toeplitz operator $T(a)$ ($a \in L^\infty$) is normal, i.e., $T(a)T(\overline{a}) = T(\overline{a})T(a)$, if and only if $\operatorname{conv} \mathcal{R}(a)$ is a line segment. We therefore call a function $a \in L^\infty$ *locally normal on* $\mathbf{T}$ if $\operatorname{conv} \mathcal{R}_\tau(a)$ is a line segment for every $\tau \in \mathbf{T}$ (of course, this line segment may change with $\tau \in \mathbf{T}$). Obviously, functions in PC or functions for which $\mathcal{R}_\tau(a)$ consists of at most two points for every $\tau \in \mathbf{T}$ are locally normal.

Given a locally normal function $a \in L^\infty$, we define $\mathbf{A}(a, \overline{a}, C)$ and $\mathbf{S}(a, \overline{a}, C)$ as the smallest closed subalgebra of $\mathcal{B}(l^2)$ and $\mathcal{F}$ which contains

$$\{T(b) : b \in \{a, \overline{a}\} \cup C\} \quad \text{and} \quad \{\{T_n(b)\} : b \in \{a, \overline{a}\} \cup C\},$$

respectively, and consider the quotient algebras

$$\mathbf{A}^\pi(a, \overline{a}, C) := \mathbf{A}(a, \overline{a}, C)/\mathcal{K}(l^2), \quad \mathbf{S}^\pi(a, \overline{a}, C) := \mathbf{S}(a, \overline{a}, C)/\mathcal{J}.$$

One can show that these two algebras are commutative C^*-algebras. Each of them contains a copy of C (Example 2.28 and Proposition 2.33). We can therefore localize over the points $\tau \in \mathbf{T}$. In analogy to Theorem 3.10, we obtain that the local algebras

$$\mathbf{A}^\pi(a, \overline{a}, C)/J_\tau \quad \text{and} \quad \mathbf{S}^\pi(a, \overline{a}, C)/J_\tau$$

are both isomorphic to the C^*-algebra $C(\operatorname{conv} \mathcal{R}_\tau(a))$. Using this, we can show that Theorems 3.11 and 3.12 as well as Corollary 3.13(a) remain literally true with PC replaced by $\{a, \overline{a}\} \cup C$, where a is a (fixed but arbitrary) function, and that parts (b) and (c) of Corollary 3.13 hold for every locally normal function.

Notes. For $\mathbf{A}^\pi(PC)$, Theorem 3.10 is Gohberg and Krupnik's [88]. The part of Theorem 3.10 concerning $\mathbf{S}^\pi(PC)$ was established in our paper [36]. The latter paper is probably the first work in which C^*-algebra techniques were employed in the context of a concrete problem of numerical analysis. The proof of Theorem 3.10 given here is essentially from [35, pp. 35–38]. All other results of this section are taken from our papers [155] and [23].

3.4 Norm of the Resolvent

We will use the results of the previous two sections in order to determine the limiting sets of the pseudospectra of Toeplitz matrices. This will be done in the following section, and the theorem of the present section will be a main ingredient of the proof.

It is well known that nonconstant complex-valued analytic functions cannot have locally constant modulus. This is no longer true for operator-valued analytic functions: for example, if

$$A : \mathbf{C} \to \mathcal{B}(\mathbf{C}^2), \quad \lambda \mapsto \begin{pmatrix} \lambda & 0 \\ 0 & 1 \end{pmatrix},$$

then $\|A(\lambda)\| = \max\{|\lambda|, 1\}$, which is constant on the unit disk. The following theorem shows that such a phenomenon does not occur for the resolvent. We learned both this theorem and its proof from Andrzej Daniluk of Cracow (private communication; see [23] and [41]).

Theorem 3.14. *Let H be a Hilbert space and let $A \in \mathcal{B}(H)$. Suppose that $A - \lambda I$ is invertible for all λ in some open subset U of $\mathbf{C}$ and assume that $\|(A - \lambda I)^{-1}\| \le M$ for all $\lambda \in U$. Then $\|(A - \lambda I)^{-1}\| < M$ for all $\lambda \in U$.*

Proof. A little thought reveals that what we must show is the following: if U is an open subset of $\mathbf{C}$ containing the origin and $\|(A - \lambda I)^{-1}\| \le M$ for all $\lambda \in U$, then $\|A^{-1}\| < M$. To prove this, assume the contrary, i.e., let

$\|A^{-1}\| = M$. We have

$$(A - \lambda I)^{-1} = \sum_{j=0}^{\infty} \lambda^j A^{-j-1}$$

for all λ in some sufficiently small disk $|\lambda| \le r$. Given $x \in H$, we therefore get

$$\left\|(A - \lambda I)^{-1}x\right\|^2 = \sum_{j,k\ge 0} \lambda^j \overline{\lambda}^k (A^{-j-1}x,\, A^{-k-1}x)$$

whenever $|\lambda| \le r$. Integrating the latter equality along the circle $|\lambda| = r$, we obtain

$$\frac{1}{2\pi}\int_0^{2\pi} \left\|(A - re^{i\theta}I)^{-1}x\right\|^2 d\theta = \sum_{j=0}^{\infty} r^{2j}\|A^{-j-1}x\|^2,$$

and since $\|(A - re^{i\theta}I)^{-1}x\| \le M\,\|x\|$, we arrive at the inequality

$$\|A^{-1}x\|^2 + r^2\|A^{-2}x\|^2 \le M^2\|x\|^2.$$

Now pick an arbitrary $\varepsilon > 0$. Because $\|A^{-1}\| = M$, there is an $x_\varepsilon \in H$ such that $\|x_\varepsilon\| = 1$ and $\|A^{-1}x_\varepsilon\|^2 > M^2 - \varepsilon$. It follows that

$$M^2 - \varepsilon + r^2\|A^{-2}x_\varepsilon\|^2 < M^2,$$

i.e., $\|A^{-2}x_\varepsilon\|^2 < \varepsilon r^{-2}$, and consequently,

$$1 = \|x_\varepsilon\|^2 \le \|A^2\|^2\,\|A^{-2}x_\varepsilon\|^2 < \varepsilon r^{-2}\|A^2\|^2,$$

which is impossible if $\varepsilon > 0$ is small enough. ■

3.5 Limits of Pseudospectra

As will be seen in Chapter 5, the spectrum $\operatorname{sp} T_n(a)$ need not mimic $\operatorname{sp} T(a)$ as n goes to infinity (also recall Figure 13). In contrast to this, pseudospectra behave as nicely as we could ever expect.

Pseudospectra. For $\varepsilon > 0$, the *ε-pseudospectrum* $\operatorname{sp}_\varepsilon A$ of a bounded linear Banach space operator A is defined as the set

$$\operatorname{sp}_\varepsilon A := \left\{\lambda \in \mathbf{C} : \|(A - \lambda I)^{-1}\| \ge 1/\varepsilon\right\}. \qquad (3.17)$$

Here we put $\|(A - \lambda I)^{-1}\| = \infty$ if $A - \lambda I$ is not invertible. Thus, $\operatorname{sp} A \subset \operatorname{sp}_\varepsilon A$ for every $\varepsilon > 0$.

In the same way the question "Is A invertible?" is in numerical analysis better replaced by the question "What is $\|A^{-1}\|$?", the pseudospectra of matrices and operators are, in a sense, of even greater import than their usual spectra.

The following theorem provides an alternative description of pseudospectra.

Theorem 3.15. *Let H be a Hilbert space and let $A \in \mathcal{B}(H)$. Then for every $\varepsilon > 0$,*

$$\mathrm{sp}_\varepsilon A = \bigcup_{\|E\| \le \varepsilon} \mathrm{sp}(A + E), \tag{3.18}$$

the union over all $E \in \mathcal{B}(H)$ of norm at most ε.

Proof. Let S_1 and S_2 be the sets on the right of (3.17) and (3.18), respectively.

To prove that $S_2 \subset S_1$, let $\lambda \in S_2$ and choose an $E \in \mathcal{B}(H)$ such that $\|E\| \le \varepsilon$ and $A + E - \lambda I$ is not invertible. If $\lambda \in \mathrm{sp}\, A$, then clearly $\lambda \in S_1$. So assume $A - \lambda I$ is invertible. From the identities

$$\begin{aligned} A + E - \lambda I &= (A - \lambda I)\big(I + (A - \lambda I)^{-1} E\big) \\ &= \big(I + (A - \lambda I)^{-1} E\big)(A - \lambda I) \end{aligned}$$

we see that $I + (A - \lambda I)^{-1}E$ cannot be invertible. Hence

$$\|(A - \lambda I)^{-1}\| \, \|E\| \ge \|(A - \lambda I)^{-1} E\| \ge 1$$

and, consequently,

$$\|(A - \lambda I)^{-1}\| \ge 1/\|E\| \ge 1/\varepsilon,$$

which shows that $\lambda \in S_1$.

We now prove that $S_1 \subset S_2$. Contrary to what we want, let us assume that there is a $\lambda \in S_1 \backslash S_2$. Then $A + E - \lambda I$ is invertible whenever $\|E\| \le \varepsilon$. Letting $E = 0$, we obtain that $A - \lambda I$ and thus also $A^* - \overline{\lambda} I$ is invertible. Choosing $E = \mu(A^* - \overline{\lambda} I)^{-1}$ with an arbitrary $\mu \in \mathbf{C}$ satisfying

$$0 < |\mu| \le \varepsilon / \|(A^* - \overline{\lambda} I)^{-1}\| \tag{3.19}$$

we arrive at the conclusion that

$$\begin{aligned} A - \lambda I + \mu E &= A - \lambda I + \mu (A^* - \overline{\lambda} I)^{-1} \\ &= \mu (A - \lambda I)\big(\mu^{-1} I + (A - \lambda I)^{-1} (A^* - \overline{\lambda} I)^{-1}\big) \end{aligned}$$

is invertible. Thus,

$$\mu^{-1} I + (A - \lambda I)^{-1} (A^* - \overline{\lambda} I)^{-1}$$

is invertible for all μ subject to (3.19), which implies that the spectral radius of $(A - \lambda I)^{-1}(A^* - \overline{\lambda} I)^{-1}$ is less than $\|(A^* - \overline{\lambda} I)^{-1}\|/\varepsilon$. As the operator $(A - \lambda I)^{-1}(A^* - \overline{\lambda} I)^{-1}$ is selfadjoint, it results that

$$\|(A - \lambda I)^{-1}(A^* - \overline{\lambda} I)^{-1}\| < \|(A^* - \overline{\lambda} I)^{-1}\|/\varepsilon.$$

Hence (recall (3.1))

$$\|(A-\lambda I)^{-1}\|^2 < \|(A^* - \overline{\lambda} I)^{-1}\|/\varepsilon = \|(A-\lambda I)^{-1}\|/\varepsilon$$

and thus $\|(A-\lambda I)^{-1}\| < 1/\varepsilon$. This contradicts our assumption that λ be in S_1. ■

Example 3.16. The previous theorem can be used to get a good idea of the ε-pseudospectrum of a matrix A. Namely, we can randomly perturb A by matrices E satisfying $\|E\| \le \varepsilon$ and look at the superposition of the plots of the spectra (= eigenvalues) of $A+E$. For example, consider the symbol

$$a(t) = -t^4 - (3+2i)t^{-3} + it^2 + t^{-1} + 10t + (3+i)t^2 + 4t^3 + it^4,$$

where $t \in \mathbf{T}$. The range $a(\mathbf{T})$ (a "fish", or better, a "whale") and the eigenvalues of $T_{100}(a)$ are plotted in Figure 16, while Figure 17 shows the superposition of the (usual) spectra of $T_{50}(a) + E$ for 50 randomly chosen matrices E subject to the constraint $\|E\| = 10^{-2}$.

And what do we see on the screen of the computer when numerically computing the eigenvalues of $T_n(a)$ for large n? Using matlab, we computed the eigenvalues of $T_n(a)$ for $n = 200$, $n = 400$, $n = 500$, $n = 700$. The results are shown in Figures 18, 19, 20, and 21. Thus, if we blindly trusted in the computer, we could arrive at the conclusion that the eigenvalues of $T_n(a)$ eventually mimic the range of $a(\mathbf{T})$, i.e., $\mathrm{sp}_{\mathrm{ess}}\, T(a)$. In Chapter 5 we will show that this guess is wrong!

In Figure 22 we see the eigenvalues of $T_n(a)$ for $n = 300$ as they appear on the computer's screen. As already said, this picture is wrong. Clearly, this picture is the result of rounding errors. However, as the result of rounding errors we would rather expect a picture like Figure 23. So why do the rounding errors produce an eigenvalue distribution which, apart from a few outliers, mimics the essential range? A partial answer to this question will be given by Corollary 3.18, which tells us that the pseudospectra of $T_n(a)$ converge to the pseudospectrum of $T(a)$ as $n \to \infty$. ■

Limiting sets. Let $M_1, M_2, \ldots$ be a sequence of nonempty subsets of $\mathbf{C}$. The *uniform limiting set* of these sets,

$$\underset{n\to\infty}{u\text{-lim}}\, M_n, \tag{3.20}$$

is defined as the set of all $\lambda \in \mathbf{C}$ which are limits of some sequence $\{\lambda_n\}$ with $\lambda_n \in M_n$. In other words, λ belongs to the set (3.20) if and only if there are $\lambda_n \in M_n$ such that $\lambda_n \to \lambda$. We let

$$\underset{n\to\infty}{p\text{-lim}}\, M_n \tag{3.21}$$

stand for the set of all $\lambda \in \mathbf{C}$ which are partial limits of some sequence $\{\lambda_n\}$ with $\lambda_n \in M_n$, and we refer to (3.21) as the *partial limiting set* of the sets M_n.

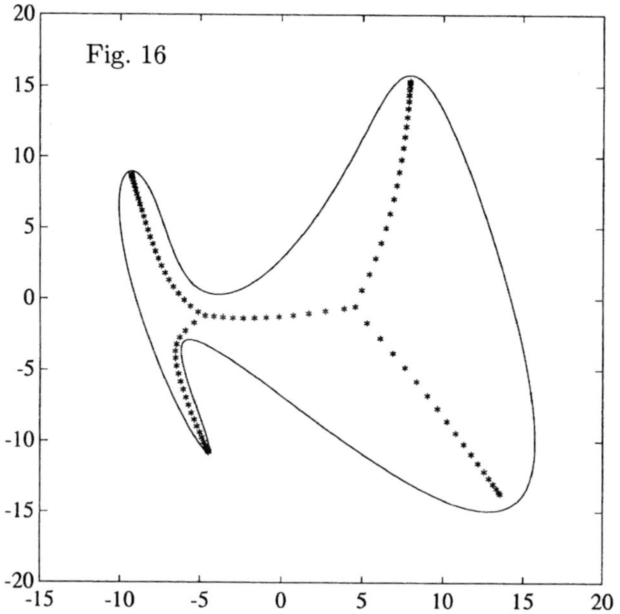

The essential range $\mathcal{R}(a)$ and the (actual) 100 eigenvalues of the matrix $T_{100}(a)$.

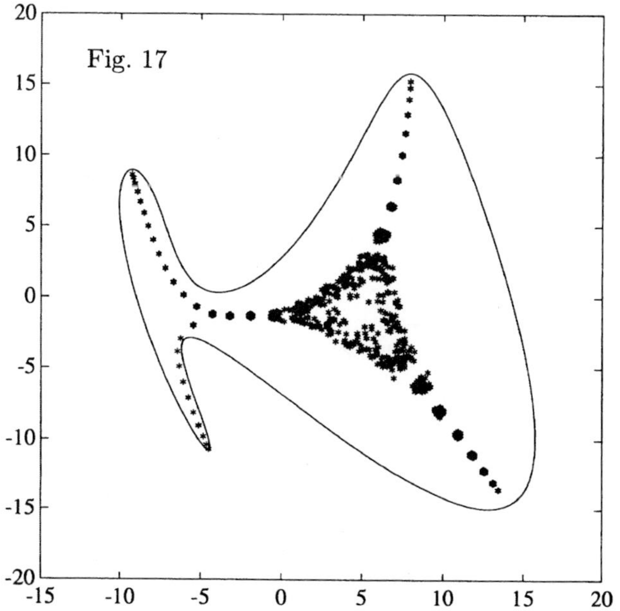

The superposition of the eigenvalues of 50 matrices $T_{50}(a) + E$ with randomly chosen matrices E for which $\|E\| = 10^{-2}$.

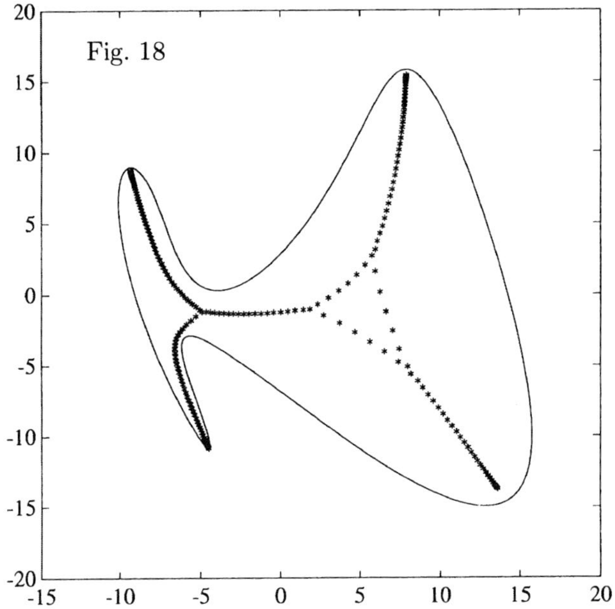

The (erroneous) 200 eigenvalues of the matrix $T_{200}(a)$ as they appear on the computer's screen.

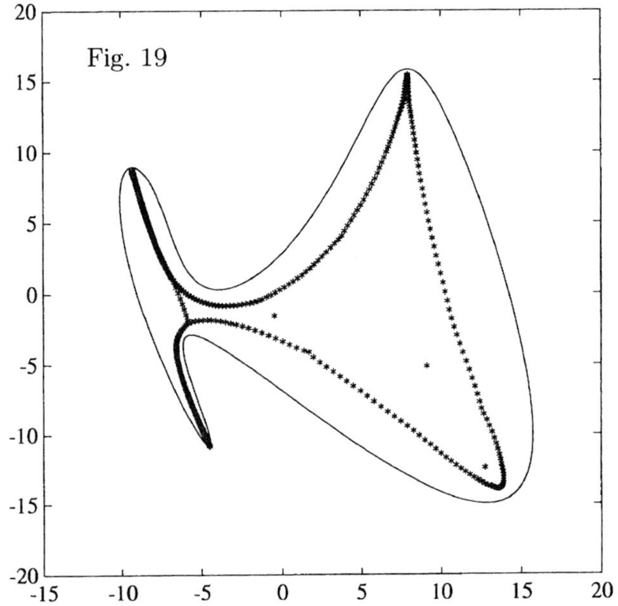

The (erroneous) 400 eigenvalues of the matrix $T_{400}(a)$ as they appear on the computer's screen.

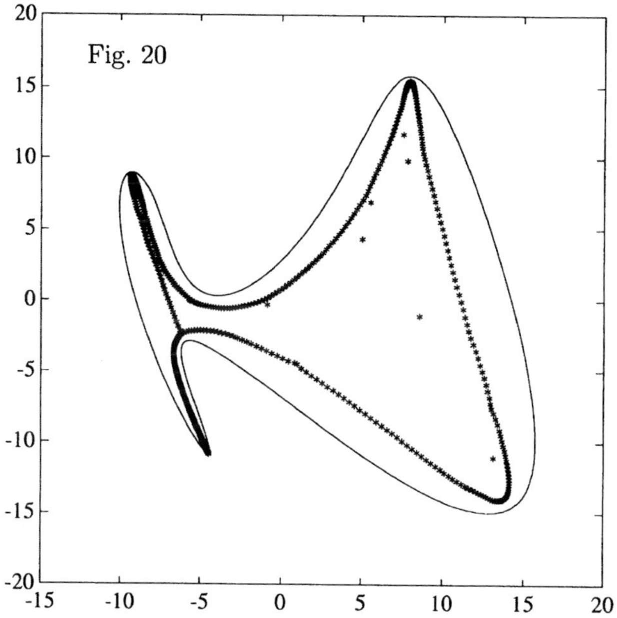

The (erroneous) 500 eigenvalues of the matrix $T_{500}(a)$ as they appear on the computer's screen.

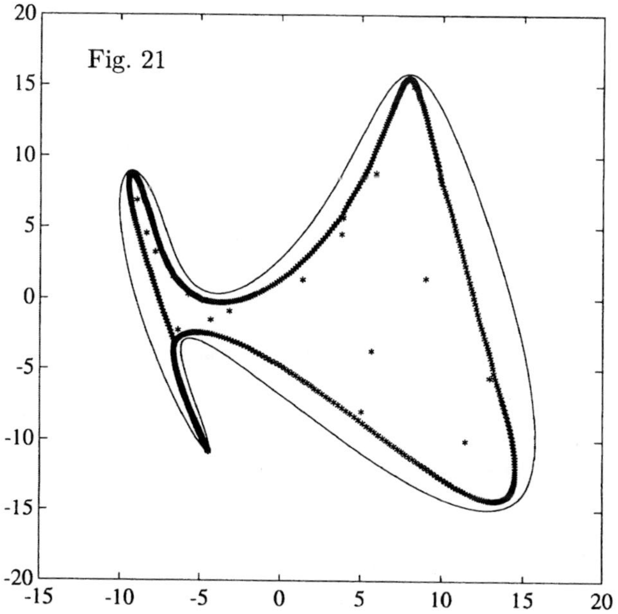

The (erroneous) 700 eigenvalues of the matrix $T_{700}(a)$ as they appear on the computer's screen.

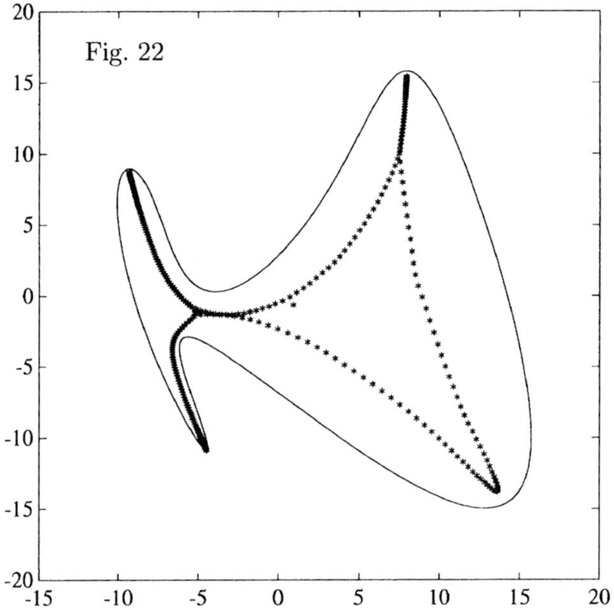

This is what we see on the screen as the result of rounding errors ($n = 300$).

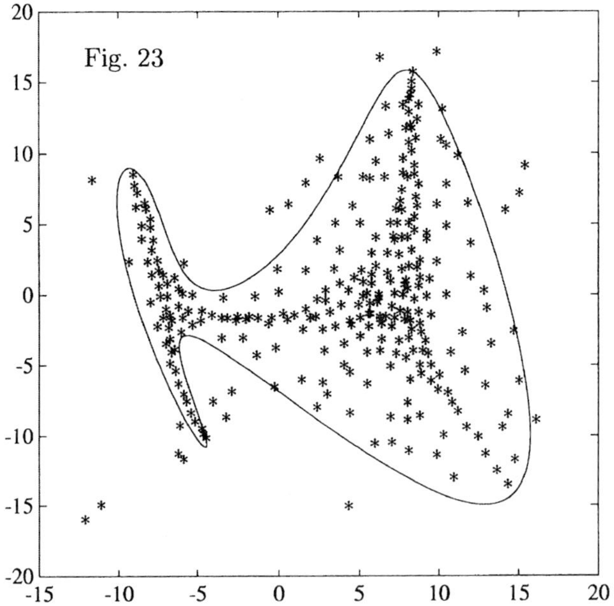

This is what we would expect on the screen as the result of rounding errors.

Equivalently, λ is in the set (3.21) if and only if there are $n_1 < n_2 < \dots$ and $\lambda_{n_k} \in M_{n_k}$ such that $\lambda_{n_k} \to \lambda$. Clearly, we always have

$$u\text{-}\lim_{n\to\infty} M_n \subset p\text{-}\lim_{n\to\infty} M_n.$$

If $M_n = \{1\}$ for odd n and $M_n = \{1,2\}$ for even n, then $u\text{-}\lim M_n$ equals $\{1\}$ while $p\text{-}\lim M_n$ is the doubleton $\{1,2\}$. In case all the sets M_n are contained in some disk $\{\lambda \in \mathbf{C} : |\lambda| \le R\}$, the partial limiting set $p\text{-}\lim M_n$ is never empty; however, $u\text{-}\lim M_n$ may be empty in this case (example: $M_n = \{1\}$ for odd n and $M_n = \{2\}$ for even n).

Here is the result of this section.

Theorem 3.17. *If* $\{A_n\} \in \mathbf{S}(PC)$ *and*

$$A_n \to A \ \textit{strongly}, \qquad W_n A_n W_n \to \tilde{A} \ \textit{strongly},$$

then for each $\varepsilon > 0$,

$$u\text{-}\lim_{n\to\infty} \operatorname{sp}_\varepsilon A_n = p\text{-}\lim_{n\to\infty} \operatorname{sp}_\varepsilon A_n = \operatorname{sp}_\varepsilon A \cup \operatorname{sp}_\varepsilon \tilde{A}.$$

Proof. We first show that $\operatorname{sp}_\varepsilon A \subset u\text{-}\lim \operatorname{sp}_\varepsilon A_n$. If $\lambda \in \operatorname{sp} A$, then $\|(A_n - \lambda I)^{-1}\| \to \infty$ by virtue of Proposition 3.7, which implies that λ belongs to $u\text{-}\lim \operatorname{sp}_\varepsilon A_n$. So suppose $\lambda \in \operatorname{sp}_\varepsilon A \backslash \operatorname{sp} A$. Then $\|(A - \lambda I)^{-1}\| \ge 1/\varepsilon$. Let $U \in \mathbf{C}$ be any open neighborhood of λ. From Theorem 3.14 we deduce that there is a point $\mu \in U$ such that $\|(A - \mu I)^{-1}\| > 1/\varepsilon$. Hence, we can find a k_0 such that

$$\|(A - \mu I)^{-1}\| \ge \frac{1}{\varepsilon - 1/k} \quad \text{for all } k \ge k_0.$$

As U was arbitrary, it follows that there exists a sequence $\lambda_1, \lambda_2, \dots$ such that $\lambda_k \in \operatorname{sp}_{\varepsilon - 1/k} A$ and $\lambda_k \to \lambda$. By Theorem 3.12,

$$\lim_{n\to\infty} \|(A_n - \lambda_k I)^{-1}\| \ge \frac{1}{\varepsilon - 1/k}.$$

Consequently, $\|(A_n - \lambda_k I)^{-1}\| \ge 1/\varepsilon$ and thus $\lambda_k \in \operatorname{sp}_\varepsilon A_n$ for all $n \ge n(k)$. This shows that $\lambda = \lim \lambda_k$ belongs to $u\text{-}\lim \operatorname{sp}_\varepsilon A_n$.

Repeating the above reasoning with $W_n A_n W_n$ and $\tilde{A}$ in place of A_n and A, respectively, we obtain

$$\operatorname{sp}_\varepsilon \tilde{A} \subset u\text{-}\lim_{n\to\infty} \operatorname{sp}_\varepsilon W_n A_n W_n.$$

Because W_n is an isometry and $W_n^2 = I$, we have $\operatorname{sp}_\varepsilon W_n A_n W_n = \operatorname{sp}_\varepsilon A_n$. In summary, we have proved that

$$\operatorname{sp}_\varepsilon A \cup \operatorname{sp}_\varepsilon \tilde{A} \subset u\text{-}\lim_{n\to\infty} \operatorname{sp}_\varepsilon A_n.$$

In order to prove the inclusion

$$\underset{n\to\infty}{p\text{-lim}}\ \mathrm{sp}_\varepsilon A_n \subset \mathrm{sp}_\varepsilon A \cup \mathrm{sp}_\varepsilon \tilde{A},$$

suppose $\lambda \notin \mathrm{sp}_\varepsilon A \cup \mathrm{sp}_\varepsilon \tilde{A}$. Then $\|(A-\lambda I)^{-1}\|$ and $\|(\tilde{A}-\lambda I)^{-1}\|$ are less than $1/\varepsilon$, whence

$$\|(A_n - \lambda I)^{-1}\| < 1/\varepsilon - \delta < 1/\varepsilon \quad \text{for all } n \geq n_0$$

with some $\delta > 0$ due to Theorem 3.12. If $n \geq n_0$ and $|\mu - \lambda| < \varepsilon\delta(1/\varepsilon - \delta)^{-1}$ then

$$\begin{aligned}\|(A_n - \mu I)^{-1}\| &\leq \frac{\|(A_n - \lambda I)^{-1}\|}{1 - |\mu - \lambda|\,\|(A_n - \lambda I)^{-1}\|} \\ &< \frac{1/\varepsilon - \delta}{1 - \varepsilon\delta(1/\varepsilon - \delta)^{-1}(1/\varepsilon - \delta)} = \frac{1}{\varepsilon},\end{aligned}$$

and thus $\mu \notin \mathrm{sp}_\varepsilon A_n$. This shows that λ cannot belong to $p\text{-lim}\,\mathrm{sp}_\varepsilon A_n$. ■

The following corollary is immediate from Theorem 3.17. For brevity, we put here

$$\lim_{n\to\infty} = \underset{n\to\infty}{u\text{-lim}} = \underset{n\to\infty}{p\text{-lim}}\ .$$

Corollary 3.18. (a) *If a_{jk} is a finite collection of functions in PC, then for each $\varepsilon > 0$,*

$$\lim_{n\to\infty} \mathrm{sp}_\varepsilon \sum_j \prod_k T_n(a_{jk}) = \mathrm{sp}_\varepsilon \sum_j \sum_k T(a_{jk}) \cup \mathrm{sp}_\varepsilon \sum_j \prod_k T(\tilde{a}_{jk}).$$

(b) *If $a \in PC$ and $K \in \mathcal{K}(l^2)$, then for each $\varepsilon > 0$,*

$$\lim_{n\to\infty} \mathrm{sp}_\varepsilon\big(T_n(a) + P_n K P_n\big) = \mathrm{sp}_\varepsilon\big(T(a) + K\big) \cup \mathrm{sp}_\varepsilon T(\tilde{a}).$$

(c) *If $a \in PC$, then for each $\varepsilon > 0$,*

$$\lim_{n\to\infty} \mathrm{sp}_\varepsilon T_n(a) = \mathrm{sp}_\varepsilon T(a). \ ■ \tag{3.22}$$

Once again Example 3.16. Comparing Figures 17 and 22 we arrive at the conclusion that the erroneous eigenvalues of $T_n(a)$ we see on the computer's screen are something like the pseudospectrum $\mathrm{sp}_\varepsilon T_n(a)$, with clear preference of points close to the boundary of $\mathrm{sp}_\varepsilon T_n(a)$. Corollary 3.18(c) says that the limiting set of $\mathrm{sp}_\varepsilon T_n(a)$ is $\mathrm{sp}_\varepsilon T(a)$ and hence, the erroneous eigenvalues must be asymptotically distributed close to the boundary of $\mathrm{sp}_\varepsilon T(a)$. In the case at hand, the boundary of $\mathrm{sp}_\varepsilon T(a)$ is a curve close to $a(\mathbf{T}) = \mathcal{R}(a)$, which indicates that the erroneous eigenvalues approach $a(\mathbf{T})$ as n goes to infinity. This is at least a rough explanation of the phenomenon we observe in Figures 18 to 21.

We remark, however, that the argument of the preceding paragraph does neither show why the erroneous eigenvalues are located near the boundary of $\mathrm{sp}_\varepsilon T_n(a)$ nor why they almost exactly approach $a(\mathbf{T})$ (note that the boundary of $\mathrm{sp}_\varepsilon T(a)$ is a curve lying outside the fish). These questions can only be answered by checking the concrete algorithm used to compute the eigenvalues of $T_n(a)$. ■

Notes. Henry Landau [114], [115], [116] was the first to study ε-pseudospectra of Toeplitz matrices and equality (3.22) (for smooth symbols) is in principle already in his papers. Independently, equality (3.22) (for symbols in the Wiener algebra W) was discovered by Reichel and Trefethen [140]. These three authors derived (3.22) with the help of different methods. The approach presented here and equality (3.22) for symbols in PC and even for locally normal symbols is from [23]. Theorem 3.17 appeared explicitly in the paper [145] by Roch and one of the authors for the first time.

For matrices (= operators on $\mathbf{C}^n$), Theorem 3.15 is a simple fact. For arbitrary Hilbert space operators (or, more generally, for elements of arbitrary unital C^*-algebras), this was first proved by T. Finck and T. Ehrhardt (see [145]). Example 3.16 is taken from [26] and is based on computations done by H. Heidler and P. Santos. Beautiful plots of pseudospectra of several (not necessarily Toeplitz) matrices can be found in [17], [140], [170], and [171].

Examples like Example 3.16 were discussed by Reichel and Trefethen [140] as well as by Beam and Warming [17]. The latter paper also pays attention to algorithms for computing Toeplitz eigenvalues. We will return to this problem in Section 5.8.

Finally, we note that with each trigonometric polynomial

$$a(t) = \sum_{k=-p}^{q} a_k t^k \quad (t \in \mathbf{T})$$

we may associate not only the sequence $\{T_n(a)\}$ of truncated Toeplitz matrices but also the sequence $\{C_n(a)\}$ of its so-called Toeplitz circulant cousins [17]; the matrix $C_n(a)$ is the circulant $n \times n$ matrix whose first row is

$$(a_0 \; a_1 \; \dots \; a_q \; 0 \; \dots \; 0 \; a_{-p} \; \dots \; a_{-1})$$

in case $n \geq p + q + 1$. The eigenvalues of the circulant $C_n(a)$ are

$$a(\omega_1), \dots, a(\omega_n),$$

where $\omega_1, \dots, \omega_n$ are the n solutions of $\omega^n = 1$. Clearly, $\mathrm{sp}\, C_n(a)$ asymptotically fills in $a(\mathbf{T})$. Accordingly, one of the conclusions of [17] is that "the numerical Toeplitz spectrum approaches the spectrum of its Toeplitz circulant cousin as $n \to \infty$."

3.6 Pseudospectra of Infinite Toeplitz Matrices

Corollary 3.18(c) does not relate the pseudospectrum of a large Toeplitz matrix $T_n(a)$ to the spectrum of $T(a)$ but to the pseudospectrum of $T(a)$. Thus, it is desirable to know more about the pseudospectra of infinite Toeplitz matrices.

Let $\Delta_\varepsilon := \{\lambda \in \mathbf{C} : |\lambda| \le \varepsilon\}$. If X is a Banach space and $A \in \mathcal{B}(X)$, we always have

$$\mathrm{sp}\, A + \Delta_\varepsilon \subset \mathrm{sp}_\varepsilon\, A, \tag{3.23}$$

where $\mathrm{sp}\, A + \Delta_\varepsilon := \{\mu + \nu : \mu \in \mathrm{sp}\, A : \nu \in \Delta_\varepsilon\}$. Indeed, if $\lambda \notin \mathrm{sp}_\varepsilon\, A$, then $\varepsilon < \|(A - \lambda I)^{-1}\|^{-1}$, which implies that $A - \lambda I - \delta I$ is invertible whenever $|\delta| \le \varepsilon$.

The following two theorems provide additional information about the pseudospectra of Toeplitz operators on l^2. We let $\mathrm{sp}_0\, A := \mathrm{sp}\, A$.

Theorem 3.19. *If $\varepsilon \ge 0$ and $a \in L^\infty$, then*

$$\mathrm{sp}\, T(a) + \Delta_\varepsilon \subset \mathrm{sp}_\varepsilon\, T(a) \subset \mathrm{conv}\, \mathcal{R}(a) + \Delta_\varepsilon\,.$$

Proof. The left inclusion is a special case of (3.23). For $\varepsilon = 0$, the right inclusion is the Brown-Halmos theorem (Theorem 1.18). In the case $\varepsilon > 0$, the right inclusion can be shown by arguments similar to those of the proof of Theorem 1.18. ∎

If $\mathrm{sp}\, T(a) = \mathrm{conv}\, \mathcal{R}(a)$, which happens, for instance, if $a \in C$ and $\mathcal{R}(a)$ is the boundary of some convex set or if a is any function in L^∞ for which $\mathrm{conv}\, \mathcal{R}(a)$ is a line segment, then Theorem 3.19 tells us that

$$\mathrm{sp}_\varepsilon\, T(a) = \mathrm{conv}\, \mathcal{R}(a) + \Delta_\varepsilon\,.$$

However, in general both inclusions of Theorem 3.19 may be proper.

Theorem 3.20. *Given $\varepsilon > 0$, there exist a and b in the Wiener algebra W such that*

$$\mathrm{sp}\, T(a) + \Delta_\varepsilon \ne \mathrm{sp}_\varepsilon\, T(a), \qquad \mathrm{sp}_\varepsilon\, T(b) \ne \mathrm{conv}\, \mathcal{R}(b) + \Delta_\varepsilon\,.$$

A full proof is in [30]. We here confine ourselves to the following. Suppose $a \in W$ and $\mathcal{R}(a) = a(\mathbf{T})$ is the half-circle $\{z \in \mathbf{T} : \mathrm{Im}\, z \ge 0\}$. Then $T(a)$ is invertible due to Theorem 1.15. Hence, if $\varepsilon > 0$ is small enough, then $\|T^{-1}(a)\| < 1/\varepsilon$, which implies that $0 \notin \mathrm{sp}_\varepsilon\, T(a)$ although $0 \in \mathrm{conv}\, \mathcal{R}(a) + \Delta_\varepsilon$. Second, put

$$b(e^{i\theta}) = \begin{cases} e^{2i\theta} & \text{for} \quad 0 \le \theta < \pi, \\ e^{-2i\theta} & \text{for} \quad \pi \le \theta < 2\pi. \end{cases} \tag{3.24}$$

If $e^{i\theta}$ traverses $\mathbf{T}$, then $b(e^{i\theta})$ twice traces out the unit circle, once in the positive and once in the negative direction. The Fourier coefficients of b are

$$b_0 = 0, \;\; b_2 = b_{-2} = 1/2, \;\; b_n = 0 \;\text{ if }\; n \neq \pm 2 \;\text{ is even,}$$
$$b_n = 4/\big(\pi i(n^2 - 4)\big) \;\text{ if }\; n \;\text{ is odd.}$$

Therefore $b \in W$. Theorem 1.15 implies that $\operatorname{sp} T(b) = \mathbf{T}$, but we can show that $\operatorname{sp}_{3/4} T(b) = \Delta_{7/4}$, which is larger than $\mathbf{T} + \Delta_{3/4}$.

Note. The above two theorems were established in our paper [30] with Grudsky.

4

Moore-Penrose Inverses and Singular Values

4.1 Singular Values of Matrices

Let H be a Hilbert space and let A be a bounded linear operator on H. Then $\operatorname{sp} A^*A \subset [0, \infty)$, and the non-negative square roots of the numbers in $\operatorname{sp} A^*A$ are called the *singular values* of A. The set of the singular values of A will be denoted by $\Sigma(A)$,

$$\Sigma(A) := \{s \in [0, \infty) : s^2 \in \operatorname{sp} A^*A\}.$$

It is well known (see, e.g., [132, p. 296]) that

$$\operatorname{sp} A^*A \cup \{0\} = \operatorname{sp} AA^* \cup \{0\}, \tag{4.1}$$

whence $\Sigma(A^*) \cup \{0\} = \Sigma(A) \cup \{0\}$.

Singular value decomposition. We think of $n \times n$ matrices as operators on $\mathbf{C}^n$, where $\mathbf{C}^n$ is equipped with the l^2 norm. If A_n is an $n \times n$ matrix, then $A_n^* A_n$ has n eigenvalues $\lambda_k(A_n^* A_n)$ and we can order them so that

$$0 \le \lambda_1(A_n^* A_n) \le \lambda_2(A_n^* A_n) \le \ldots \le \lambda_n(A_n^* A_n) \le \|A_n\|^2.$$

The singular values of A_n are $s_k(A_n) := (\lambda_k(A_n^* A_n))^{1/2}$. Thus,

$$0 \le s_1(A_n) \le s_2(A_n) \le \ldots \le s_n(A_n) \le \|A_n\|.$$

For the sake of convenience, let us also put $s_0(A_n) = 0$.

Theorem 4.1. *If $A_n \in \mathcal{B}(\mathbf{C}^n)$, then there exist unitary matrices $U_n, V_n \in \mathcal{B}(\mathbf{C}^n)$ such that*

$$A_n = U_n S_n V_n, \tag{4.2}$$

where $S_n := \operatorname{diag}(s_1(A_n), s_2(A_n), \dots, s_n(A_n))$.

This theorem is well known and proved in the majority of texts on linear algebra. The representation (4.2) is called the *singular value decomposition* of A.

Interpretation as approximation numbers. For $j \in \{0, 1, \dots, n\}$, let $\mathcal{F}_j^{(n)}$ denote the collection of all $n \times n$ matrices of rank at most j,

$$\mathcal{F}_j^{(n)} := \{F \in \mathcal{B}(\mathbf{C}^n) : \dim \operatorname{Im} F \le j\},$$

and define the jth *approximation number* of a matrix $A_n \in \mathcal{B}(\mathbf{C}^n)$ as

$$\operatorname{dist}(A_n, \mathcal{F}_j^{(n)}) := \inf\{\|A_n - F_j\| : F_j \in \mathcal{F}_j^{(n)}\}. \tag{4.3}$$

Since $\mathcal{F}_j^{(n)}$ is a closed subset of $\mathcal{B}(\mathbf{C}^n)$, the infimum in (4.3) is actually attained, i.e., we can also write

$$\operatorname{dist}(A_n, \mathcal{F}_j^{(n)}) := \min\{\|A_n - F_j\| : F_j \in \mathcal{F}_j^{(n)}\}.$$

Clearly, $\operatorname{dist}(A_n, \mathcal{F}_0^{(n)}) = \|A_n\|$ and $\operatorname{dist}(A_n, \mathcal{F}_n^{(n)}) = 0$.

Theorem 4.2. *If $A_n \in \mathcal{B}(\mathbf{C}^n)$, then*

$$s_k(A_n) = \operatorname{dist}(A_n, \mathcal{F}_{n-k}^{(n)}) \tag{4.4}$$

for every $k \in \{0, 1, \dots, n\}$.

This is again a well-known result. It was established by Dz.E. Allakhverdiev (1957) and M. Fiedler and V. Pták (1962) for general compact Hilbert space operators and led A. Pietsch to the introduction of the approximation numbers for compact Banach space operators; see [132, p. 293]. For a proof we refer to [87, Theorem II.2.1]. Note that by virtue of Theorem 4.1 the equality (4.4) is equivalent to saying that if $0 \le a_1 \le a_2 \le \dots \le a_n$, then

$$\operatorname{dist}(\operatorname{diag}(a_1, a_2, \dots, a_n), \mathcal{F}_{n-k}^{(n)}) = a_k.$$

The latter equality even holds if $\mathbf{C}^n$ is equipped with *any* norm (see [131, Theorem 11.11.3]).

4.2 The Lowest Singular Value

Since the norm of a diagonal matrix is the maximum of the moduli of the diagonal entries, we obtain from Theorem 4.1 that if $A_n \in \mathcal{B}(\mathbf{C}^n)$, then

$$s_1(A_n) = \begin{cases} 1/\|A_n^{-1}\| & \text{if } A_n \text{ is invertible,} \\ 0 & \text{if } A_n \text{ is not invertible.} \end{cases} \tag{4.5}$$

Thus, if $\{A_n\} = \{A_n\}_{n=1}^{\infty}$ is a sequence of $n \times n$ matrices A_n, we have

$$s_1(A_n) \to 0 \iff \|A_n^{-1}\| \to \infty. \tag{4.6}$$

This shows that the question of whether the lowest singular value converges to zero is closely connected with the stability of the sequence. More precisely,

$$\liminf_{n\to\infty} s_1(A_n) > 0 \iff \{A_n\} \text{ is stable.} \tag{4.7}$$

In the case where $\{A_n\} = \{T_n(a)\}$ is the sequence of the truncations of some Toeplitz matrix, we can have recourse to the results of Chapter 2.

Theorem 4.3. *Suppose $a \in L^\infty$ is locally sectorial on $\mathbf{T}$ or $a \in PC$. Then the following are equivalent:*

(i) $s_1\big(T_n(a)\big) \to 0$;

(ii) $\liminf_{n\to\infty} s_1\big(T_n(a)\big) = 0$;

(iii) $\big\{T_n(a)\big\}$ *is not stable;*

(iv) $T(a)$ *is not invertible.*

If $a \in L^\infty$ is locally normal on $\mathbf{T}$, then

$$\lim_{n\to\infty} s_1\big(T_n(a)\big) = 1/\|T^{-1}(a)\|, \tag{4.8}$$

the limit being zero if $T(a)$ is not invertible.

Proof. (i) $\Rightarrow$ (ii). Trivial.

(ii) $\Rightarrow$ (iii). Immediate from (4.7).

(iii) $\Rightarrow$ (iv). Theorem 2.18 and Corollary 2.19.

(iv) $\Rightarrow$ (i). If $\limsup s_1(T_n(a)) > 0$, then, by (4.5), there is a sequence $n_k \to \infty$ such that $\|T_{n_k}^{-1}(a)\| \le M < \infty$ with some M independent of n_k. Since Proposition 2.3 remains true with $\{T_{n_k}(a)\}$ in place of $\{A_n\}$, it follows that $T(a)$ must be invertible.

Finally, (4.8) results from (4.5) and (3.16) (and the discussion after Corollary 3.13). ■

For some recent results on the asymptotics of $s_1(T_n(a))$ for noninvertible operators $T(a)$ see also [151], [152], and [29].

Example 4.4: Cauchy-Toeplitz matrices. Let $\psi_\gamma \in PC$ be as in Examples 1.7 and 1.24, i.e., suppose

$$T_n(\psi_\gamma) = \left(\frac{1}{j-k+\gamma}\right)_{j,k=1}^{n} \qquad (\gamma \in \mathbf{C}\backslash\mathbf{Z}).$$

From what was said in Example 1.24 and from Theorem 4.3 we conclude that

$$\lim_{n\to\infty} s_1\big(T_n(\psi_\gamma)\big) = 0 \iff \liminf_{n\to\infty} s_1\big(T_n(\psi_\gamma)\big) = 0 \iff |\operatorname{Re}\gamma| \geq 1/2. \ \blacksquare$$

4.3 The Splitting Phenomenon

We say that the singular values of a sequence $\{A_n\}$ of $n \times n$ matrices A_n have the *splitting property* if there are $c_n \to 0$ and $d > 0$ such that

$$\Sigma(A_n) \subset [0, c_n] \cup [d, \infty) \quad \text{for all } n \geq 1, \tag{4.9}$$

and the singular values of $\{A_n\}$ are said to enjoy the *k-splitting property* if there exist $c_n \to 0$ and $d > 0$ such that (4.9) holds and, for all sufficiently large n, exactly k singular values of A_n lie in $[0, c_n]$ while the remaining $n-k$ singular values of A_n belong to $[d, \infty)$. Equivalently, the singular values of $\{A_n\}$ possess the k-splitting property if and only if

$$\lim_{n\to\infty} s_k(A_n) = 0 \quad \text{and} \quad \liminf_{n\to\infty} s_{k+1}(A_n) > 0.$$

The purpose of this section is to prove the following result.

Theorem 4.5. *Let $a \in PC$. If $T(a)$ is Fredholm of index $k \in \mathbf{Z}$, then the singular values of $\{T_n(a)\}$ have the $|k|$-splitting property, i.e.,*

$$\lim_{n\to\infty} s_{|k|}\big(T_n(a)\big) = 0 \quad \text{and} \quad \liminf_{n\to\infty} s_{|k|+1}\big(T_n(a)\big) > 0. \tag{4.10}$$

If $T(a)$ is not Fredholm, then

$$\lim_{n\to\infty} s_k\big(T_n(a)\big) = 0 \quad \text{for each } k \geq 1. \tag{4.11}$$

Note that $s_0(T_n(a)) = 0$ by definition. Figures 24 to 29 illustrate Theorem 4.5.

The proof is divided into several steps.

Lemma 4.6. *If A_n, B_n, C_n are $n \times n$ matrices, then*

$$s_k(A_n B_n C_n) \leq \|A_n\|\, s_k(B_n)\, \|C_n\|$$

for every $k \in \{1, \ldots, n\}$.

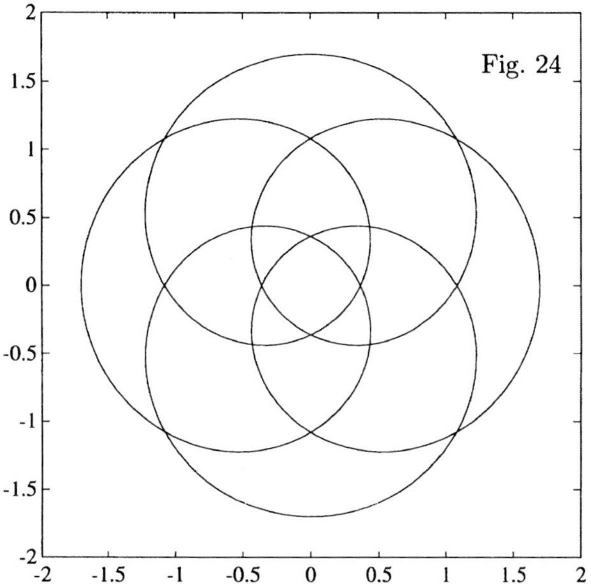

Figure 24 shows the essential range of the symbol $a(t) = 0.7t + t^5$ $(t \in \mathbf{T})$. We have $\text{wind}(a, 0.01) = \text{wind}(a, 0.1) = 5$.

Proof. If A_n and C_n are invertible, this follows easily from Theorem 4.2. The case of singular A_n and B_n can be reduced to the situation in which A_n and C_n are invertible by a perturbation argument. ■

Proposition 4.7. *If $a \in PC$ and $T(a)$ is Fredholm of index k, then*

$$\liminf_{n\to\infty} s_{|k|+1}\big(T_n(a)\big) > 0.$$

Proof. For the sake of convenience, we replace k by $-k$. We can then write $a = b\chi_k$ where $\chi_k(t) := t^k$ $(t \in \mathbf{T})$, $b \in PC$, and $T(b)$ is invertible (Theorems 1.10 and 1.23). Without loss of generality assume $k \geq 0$; otherwise consider adjoints. Because $\|T_n(\chi_{-k})\| = 1$, we obtain from Lemma 4.6 that

$$\begin{aligned} s_{k+1}\big(T_n(b\chi_k)\big) &= s_{k+1}\big(T_n(b\chi_k)\big)\|T_n(\chi_{-k})\| \\ &\geq s_{k+1}\big(T_n(b\chi_k)T_n(\chi_{-k})\big) = s_{k+1}\big(T_n(b) - P_nH(b\chi_k)H(\chi_k)P_n\big), \end{aligned}$$

the latter equality resulting from (2.13) and the identities

$$H(\tilde{\chi}_{-k}) = H(\chi_k), \qquad H(\chi_{-k}) = 0.$$

Since $\dim \operatorname{Im} H(\chi_k) = k$, we see that

$$F_k := P_nH(b\chi_k)H(\chi_k)P_n \in \mathcal{F}_k^{(n)},$$

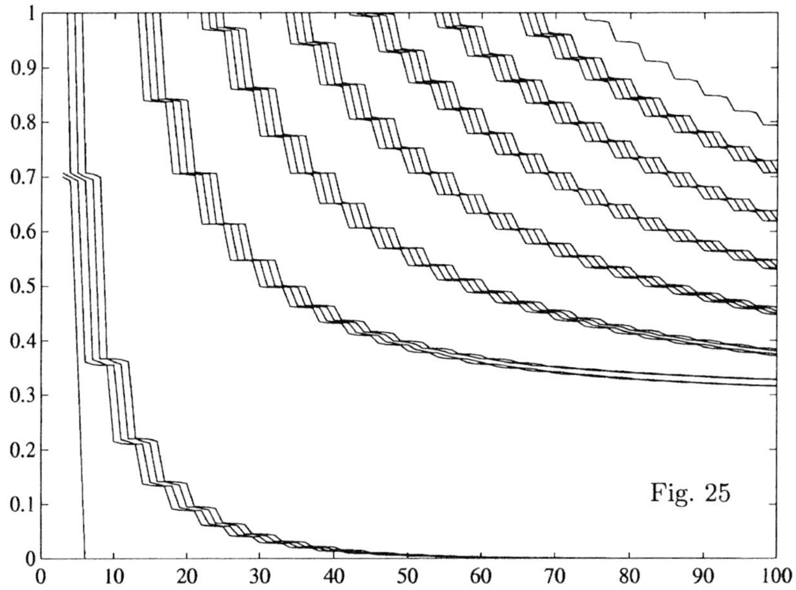

In Figure 25 we plotted the singular values $s_j(T_n(a - 0.01))$ for $3 \le n \le 100$ and $1 \le j \le \min\{n, 30\}$ in case a is as in Figure 24. In accordance with Theorem 4.5, the five lowest singular values go to zero, while the remaining singular values stay away from zero. The figure shows that, for example, the 6th singular value is waiting for the 7th, 8th, 9th singular values before making the next step downward. This is certainly a phenomenon caused by the high symmetry in Figure 24.

whence

$$\begin{aligned} s_{k+1}\big(T_n(b) - F_k\big) &= \inf\big\{\big\|T_n(b) - F_k - G_{n-k-1}\big\| : G_{n-k-1} \in \mathcal{F}^{(n)}_{n-k-1}\big\} \\ &\ge \inf\big\{\big\|T_n(b) - H_{n-1}\big\| : H_{n-1} \in \mathcal{F}^{(n)}_{n-1}\big\} = s_1\big(T_n(b)\big). \end{aligned}$$

As $T(b)$ is invertible, Theorem 4.3 shows that $s_1(T_n(b))$ is bounded away from zero. ■

Proposition 4.8. *If $a \in PC$ and $T(a)$ is Fredholm of index k, then*

$$\lim_{n \to \infty} s_{|k|}\big(T_n(a)\big) = 0.$$

Proof. Replace again k by $-k$ and assume $k > 0$ for the sake of definiteness. We write $a = \chi_k b$ as in the preceding proof. Using Proposition 2.12 and

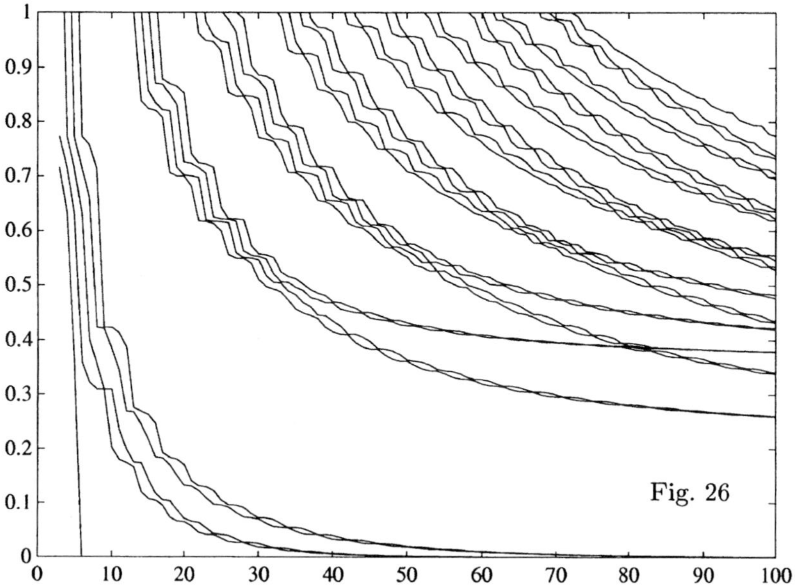

Fig. 26

Let a be as in Figure 24. In Figure 26 we plotted the singular values $s_j(T_n(a-0.1))$ for $1 \le j \le \min\{n, 30\}$ versus $3 \le n \le 100$. Again Theorem 4.5 is convincingly confirmed, which says that the five lowest singular values must tend to zero, while the remaing singular values stay away from zero. However, the way the sigular values decay differs from the pattern of Figure 25.

Lemma 4.6 we get

$$\begin{aligned} s_k\big(T_n(\chi_k b)\big) &= s_k\big(T_n(\chi_k)T_n(b) + P_n H(\chi_k)H(\tilde{b})P_n\big) \\ &\le \|T_n(b)\| s_k\big(T_n(\chi_k) + P_n H(\chi_k)H(\tilde{b})P_n T_n^{-1}(b)\big) \end{aligned}$$

for all sufficiently large n. Put

$$A_n := T_n(\chi_k) + P_n H(\chi_k)H(\tilde{b})P_n T_n^{-1}(b)$$

and write $A_n = A_n Q_{n-k} + A_n P_{n-k}$ with $Q_{n-k} := I - P_{n-k}$. Since

$$\operatorname{rank}(A_n P_{n-k}) \le \operatorname{rank} P_{n-k} = n-k,$$

we obtain from Theorem 4.2 that

$$s_n(A_n) \le \|A_n - A_n P_{n-k}\| = \|A_n Q_{n-k}\|$$

and hence, we are left with showing that $\|A_n Q_{n-k}\| \to 0$ as $n \to \infty$. As $T(\chi_k)Q_{n-k} = 0$, we have

$$A_n Q_{n-k} = P_n H(\chi_k)H(\tilde{b})P_n t_n^{-1}(b)Q_{n-k}. \tag{4.12}$$

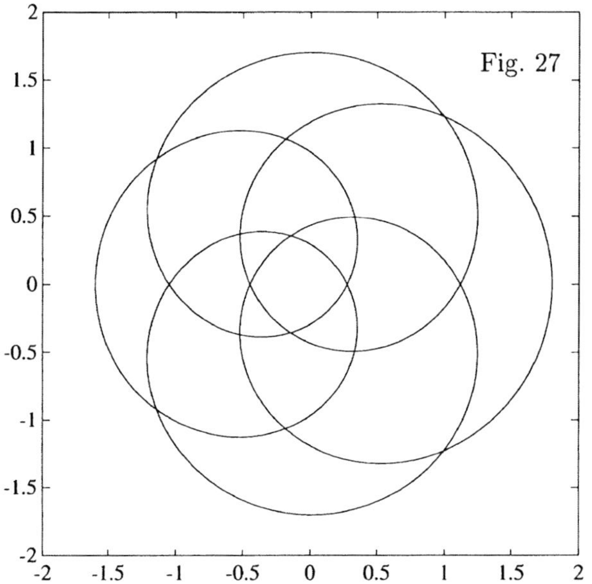

In Figure 27 we see the essential range of the symbol $a(t) = 0.7t + 0.1t^4 + t^5$ $(t \in \mathbf{T})$. Clearly, $\mathrm{wind}(a, 0.1) = 5$.

Because $H(\chi_k)$ is compact, $P_n \to I$ strongly, and

$$(P_n T_n^{-1}(b) Q_{n-k})^* = Q_{n-k} T_n^{-1}(\bar{b}) P_n \to 0 \cdot T^{-1}(\bar{b}) = 0 \text{ strongly,}$$

we deduce from Lemma 2.8 that $\|A_n Q_{n-k}\| \to 0$, as desired. ∎

Obviously, the first part of Theorem 4.5 is simply the union of Propositions 4.7 and 4.8.

More general symbols. We remark that the proofs given above actually yield more than part of Theorem 4.5. Namely, let Π^0 denote the set of all symbols $b \in L^\infty$ for which $\{T_n(b)\}$ is stable and let Π be the set of all symbols $a \in L^\infty$ such that $a\chi_k \in \Pi^0$ for some $k \in \mathbf{Z}$. For instance, we know that locally sectorial symbols belong to Π and we also have

$$G(C + H^\infty) \cup G(C + \overline{H^\infty}) \cup G(PQC) \subset \Pi,$$

where $G(B)$ stands for the invertible elements of a unital Banach algebra B, $C + H^\infty$ and $C + \overline{H^\infty}$ are as in Section 1.6, and PQC is the algebra of all piecewise quasicontinuous functions (see, e.g., [39, Section 3.35 and Chapter 7]). By Theorem 2.7, Π is a proper subset of L^∞. Repeating the proofs of Propositions 4.7 and 4.8 we arrive at the conclusion that (4.10) holds whenever $a \in \Pi$ and $T(a)$ is Fredholm of index k.

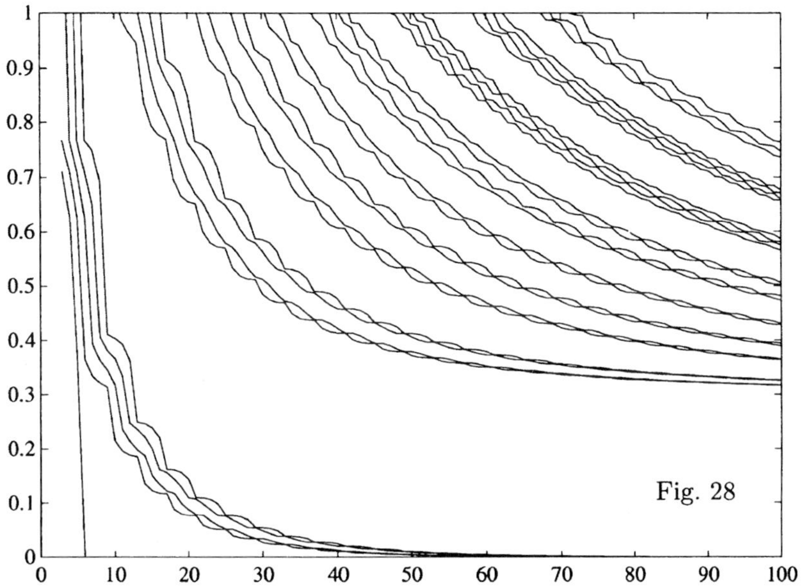

Fig. 28

In Figure 28, the symbol a is as in Figure 27 and we plotted the singular values $s_j(T_n(a-0.1))$ for $1 \le j \le \min\{n, 30\}$ versus $3 \le n \le 100$. As predicted by Theorem 4.5, it is precisely the five lower singular values which approach zero.

Speed of convergence. The proof of Proposition 4.8 also gives estimates for the speed of convergence of $s_{|k|}(T_n(a))$ to zero. Here is a sample result.

Corollary 4.9. *If*

$$\sum_{n \in \mathbf{Z}} |n|^\alpha |a_n| < \infty$$

for some $\alpha > 0$ *and* $T(a)$ *is Fredholm of index* k, *then*

$$s_{|k|}\big(T_n(a)\big) = O(1/n^\alpha) \quad \text{as } n \to \infty. \ \blacksquare$$

Proof. Let $l^{2,-\alpha}$ be the Hilbert space of all sequences $x = \{x_n\}_{n=1}^\infty$ such that

$$\|x\|_{2,-\alpha} := \left(\sum_{n=1}^\infty n^{-2\alpha}|x_n|^2\right)^{1/2} < \infty.$$

Given two Hilbert spaces X and Y, we denote by $\mathcal{B}(X,Y)$ the Banach space of all bounded linear operators from X to Y.

To prove the assertion, it suffices to show that

$$\|A_n Q_{n-k}\|_{\mathcal{B}(l^2,l^2)} = O(1/n^\alpha)$$

where A_nQ_{n-k} is given by (4.12). From (4.12) we get

$$\|A_nQ_{n-k}\|_{\mathcal{B}(l^2,l^2)} \le \|H(\chi_k)H(\tilde{b})\|_{\mathcal{B}(l^{2,-\alpha},l^2)} \times \|T_n^{-1}(b)P_n\|_{\mathcal{B}(l^{2,-\alpha},l^{2,-\alpha})}\|Q_{n-k}\|_{\mathcal{B}(l^2,l^{2,-\alpha})}. \tag{4.13}$$

Using the fact that $H(\chi_k)$ is of finite rank, it is easily seen that the first factor on the right of (4.13) is finite. The smoothness imposed on a and thus on b implies that the finite section method is applicable to $T(b)$ on $l^{2,-\alpha}$ (see [136, pp. 106–107] or [39, Theorem 7.25]), which gives the uniform boundedness of the second factor on the right of (4.13). Finally, it can be verified straightforwardly that $\|Q_{n-k}\|_{\mathcal{B}(l^2,l^{2,-\alpha})} = O(1/n^\alpha)$ as $n \to \infty$. ∎

The following proposition proves the second part of Theorem 4.5.

Proposition 4.10. *If $a \in PC$ and $T(a)$ is not Fredholm, then $s_k(T_n(a)) \to 0$ as $n \to \infty$ for each $k \ge 1$.*

Proof. The assertion is trivial if a vanishes identically. So let us suppose that a is not the zero function.

Assume there is a $k \ge 1$ such that $s_k(T_n(a))$ does not converge to zero. Let k_0 be the smallest k with this property. Then there are $n_j \to \infty$ and $d > 0$ such that

$$s_{k_0}\big(T_{n_j}(a)\big) \ge d \quad \text{and} \quad s_k\big(T_{n_j}(a)\big) \to 0 \ \text{ for } \ k < k_0. \tag{4.14}$$

To simplify notation, let us assume that $n_j = j$ for all j.

Write $T(a) = U_nS_nV_n$ as in Theorem 4.1. If $0 < \lambda < d^2$, then (4.14) implies that $S_n^2 - \lambda I_n$ is invertible for all sufficiently large n, say for $n \ge n_0$, and that

$$\|(S_n^2 - \lambda I_n)^{-1}\| \le M(\lambda)$$

with some $M(\lambda) < \infty$ independent of n. Because $T_n^*(a)T_n(a) - \lambda I_n = V_n^*(S_n^2 - \lambda I_n)V_n$, it follows that $T_n^*(a)T_n(a) - \lambda I_n$ is invertible for $n \ge n_0$ and that

$$\|\big(T_n^*(a)T_n(a) - \lambda I_n\big)^{-1}\| \le M(\lambda).$$

Thus, $T^*(a)T(a) - \lambda I$ is invertible due to Proposition 2.3 and, consequently,

$$\operatorname{sp}_{\text{ess}} T^*(a)T(a) \subset \operatorname{sp} T^*(a)T(a) \subset \{0\} \cup [d^2, \infty). \tag{4.15}$$

From Theorem 3.10 and the inverse closedness of $\mathbf{A}^\pi(PC)$ in $\mathcal{B}(l^2)/\mathcal{K}(l^2)$ (Proposition 3.1) we infer that $\operatorname{sp}_{\text{ess}} T^*(a)T(a)$ equals

$$\Big\{\big((1-\mu)\,\overline{a(t-0)} + \mu\,\overline{a(t+0)}\big)\big((1-\mu)\,a(t-0) + \mu\,a(t+0)\big) : (t,\mu) \in \mathbf{T}\times[0,1]\Big\}.$$

It is easily seen that this is a connected set. Since a does not vanish identically, we therefore obtain from (4.15) that

$$\operatorname{sp}_{\text{ess}} T^*(a)T(a) \subset [d^2, \infty),$$

implying that $T^*(a)T(a)$ is Fredholm. Analogously one can show that the operator $T(a)T^*(a)$ is Fredholm. This gives the Fredholmness of $T(a)$, which contradicts our hypothesis. ∎

At the end of Section 4.7 we will prove a result which essentially generalizes Propositions 4.8 and 4.10.

Example 4.11: Cauchy-Toeplitz matrices. Let $\psi_\gamma \in PC$ be the function of Examples 1.7, 1.24, 4.4. From Example 1.24 and Theorem 4.5 we see that if

$$k - 1/2 < \operatorname{Re}\gamma < k + 1/2,$$

then $\sum(T_n(\psi_\gamma))$ has the $|k|$-splitting property, while if

$$\operatorname{Re}\gamma - 1/2 \in \mathbf{Z},$$

we have $s_j(T_n(\psi_\gamma)) \to 0$ for each $j \geq 1$. ∎

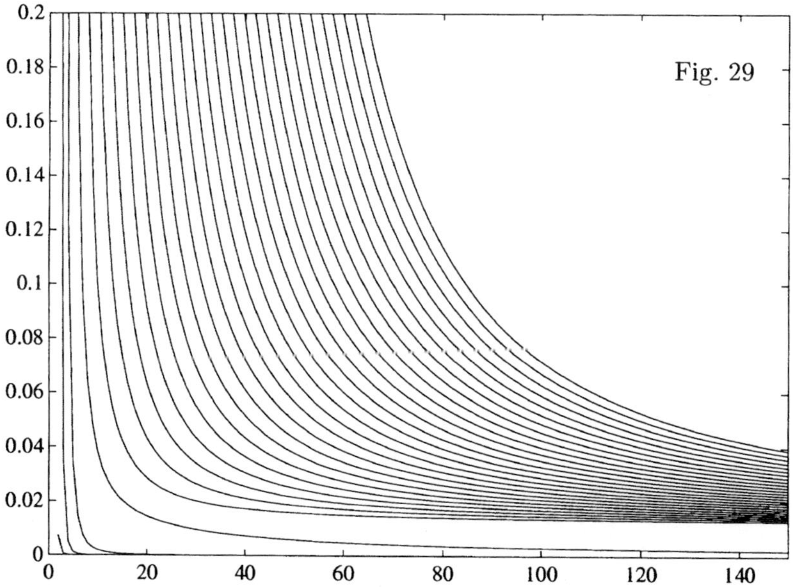

Let $\gamma = 4 + i$. In Figure 29, we plotted the singular values $s_j(T_n(\psi_\gamma))$ for $1 \leq j \leq \min\{30, n\}$ versus $2 \leq n \leq 150$. In accordance with Example 4.11, four singular values go to zero and the remaining singular values stay away from zero.

Note. Theorem 4.5 was established by Roch and Silbermann only in [143], [144], the proof given here follows Böttcher [27], with some improvements by T. Ehrhardt.

4.4 Upper Singular Values

The main result of this section is Theorem 4.13. This theorem was obtained in [27] and shows that the upper singular values $s_{n-k}(T_n(a))$ always approach $\|T(a)\| = \|a\|_\infty$ as $n \to \infty$, independently of whether $T(a)$ is Fredholm or not. We start with an auxiliary result.

Lemma 4.12. *Fix an integer $k \ge 0$ and let $\{F_n\}_{n=1}^\infty$ be a uniformly bounded sequence of operators on l^2 such that* $\operatorname{rank} F_n \le k$ *for every $n \ge 1$. Then there exists an operator $F \in \mathcal{B}(l^2)$ with* $\operatorname{rank} F \le k$ *which enjoys the following property: for each $x, y \in l^2$, the number (y, Fx) is a partial limit of the sequence $\{(y, F_n x)\}_{n=1}^\infty$.*

Proof. By assumption, there are orthonormal sets $\{e_j^{(n)}\}_{n=1}^k$ and $\{f_j^{(n)}\}_{n=1}^k$ in l^2 as well as numbers $\{\gamma_j^{(n)}\}_{n=1}^k$ such that

$$F_n x = \sum_{j=1}^k \gamma_j^{(n)}\big(x, f_j^{(n)}\big)e_j^{(n)} \quad \text{for } x \in l^2)$$

and $|\gamma_j^{(n)}| \le \|F_n\|$ (this is essentially Theorem 4.1; see also [139, Theorem VI.17], for example). Let $B_1 := \{x \in l^2 : \|x\| \le 1\}$ be the unit ball of l^2. The Banach-Alaoglu theorem says that B_1 is compact in the $*$-weak topology. Put $M := \sup_{n \ge 1} \|F_n\|$. Since $D_M := \{\gamma \in \mathbf{C} : |\gamma| \le M\}$ is also compact, we see that the set

$$A := D_M \times \ldots \times D_M \times B_1 \times \ldots \times B_1$$

(D_M occurring k times and B_1 occurring $2k$ times) is compact. The sequence

$$\{(\gamma_1^{(n)}, \ldots, \gamma_k^{(n)}, e_1^{(n)}, \ldots, e_k^{(n)}, f_1^{(n)}, \ldots, f_k^{(n)})\}_{n=1}^\infty$$

is contained in A, and hence it has an accumulation point in A, say the point

$$(\gamma_1, \ldots, \gamma_k, e_1, \ldots, e_k, f_1, \ldots, f_k).$$

Put

$$Fx = \sum_{j=1}^k \gamma_j (x, f_j)\, e_j \quad \text{for } x \in l^2.$$

By the definition of the $*$-weak topology, for each $x, y \in l^2$ there is a sequence $\{n_l\}_{l=1}^\infty$ such that

$$(y, e_j^{(n_l)}) \to (y, e_j), \qquad (x, f_j^{(n_l)}) \to (x, f_j)$$

as $n_l \to \infty$ and, at the same time, $\gamma_j^{(n_l)} \to \gamma_j$ as $n_l \to \infty$. This shows that $(y, F_{n_l}) \to (y, Fx)$ as $n_l \to \infty$. ∎

Theorem 4.13. *If $a \in L^\infty$, then for each $k \geq 0$,*

$$\lim_{n\to\infty} s_{n-k}\big(T_n(a)\big) = \|T(a)\| = \|a\|_\infty.$$

Proof. Contrary to what we want, let us assume that there is a $c < \|T(a)\|$ such that $s_{n-k}(T_n(a)) \leq c$ for all n in some infinite set $\mathcal{N}$. Since $s_{n-k}(T_n(a)) = \text{dist}(T_n(a), \mathcal{F}_k^{(n)})$ by Theorem 4.2, we can find $F_n \in \mathcal{F}_k^{(n)}$ $(n \in \mathcal{N})$ so that $\|T_n(a) - F_n\| \leq c$. Hence,

$$\|F_n\| \leq \|T_n(a)\| + \|T_n(a) - F_n\| \leq \|T(a)\| + c.$$

Using Lemma 4.12 we get an operator $F \in \mathcal{B}(l^2)$ with $\text{rank}\, F \leq k$ such that for each $x, y \in l^2$ the number (y, Fx) is a partial limit of the sequence $\{(y, F_n P_n x)\}_{n=1}^\infty$. Now assume that $\|x\| = \|y\| = 1$. Then

$$|(y, T_n(a)P_n x) - (y, F_n P_n x)| \leq \|T_n(a) - F_n\| \leq c,$$

which, by the construction of F, implies that $|(y, T(a)x) - (y, Fx)| \leq c$ and thus $\|T(a) - f\| \leq c$. Consequently,

$$\|T(a)\|_{\text{ess}} := \text{dist}\big(T(a), \mathcal{K}(l^2)\big) \leq c < \|T(a)\|.$$

However, Theorem 1.25 shows that

$$\|T(a)\|_{\text{ess}} \geq \max\big\{|\lambda| : \lambda \in \mathcal{R}(a)\big\} = \|a\|_\infty,$$

whence $\|T(a)\|_{\text{ess}} = \|T(a)\| = \|a\|_\infty$ for every $a \in L^\infty$. This contradiction completes the proof. ■

Steffen Roch observed that the preceding proof also yields the following more general result: if $A \in \mathcal{B}(l^2)$ and $\|A\| = \|A\|_{\text{ess}}$, then for each $k \geq 0$,

$$\lim_{n\to\infty} s_{n-k}(P_n A P_n) = \|A\|.$$

4.5 Moler's Phenomenon

In 1985, Cleve Moler discovered the remarkable fact that about 20 singular values of

$$T_{30}(\psi_{1/2}) = \left(\frac{1}{j-k+1/2}\right)_{j,k=1}^{30}$$

lie in the segment $[\pi-\varepsilon, \pi]$ where ε is *very* small. As pointed out in [174], appropriate computations show that the number of singular values of $T_n(\psi_{1/2})$ located in $(0, \pi - 10^{-4})$ is $6, 7, 8$ for $n = 40, 60, 100$, respectively. Figure 30 nicely illustrates this observation. Note that nevertheless $s_j(T_n(\psi_{1/2})) \to 0$ as $n \to \infty$ for each fixed j due to Example 4.11. This phenomenon has motivated much recent work on the singular values of Toeplitz matrices. A rigorous explanation of the Moler phenomenon was first given by Parter [128]. His arguments were simplified by Tyrtyshnikov [174], [175]. In this section we present an explanation of Moler's phenomenon which uses the ideas of [174] and [175].

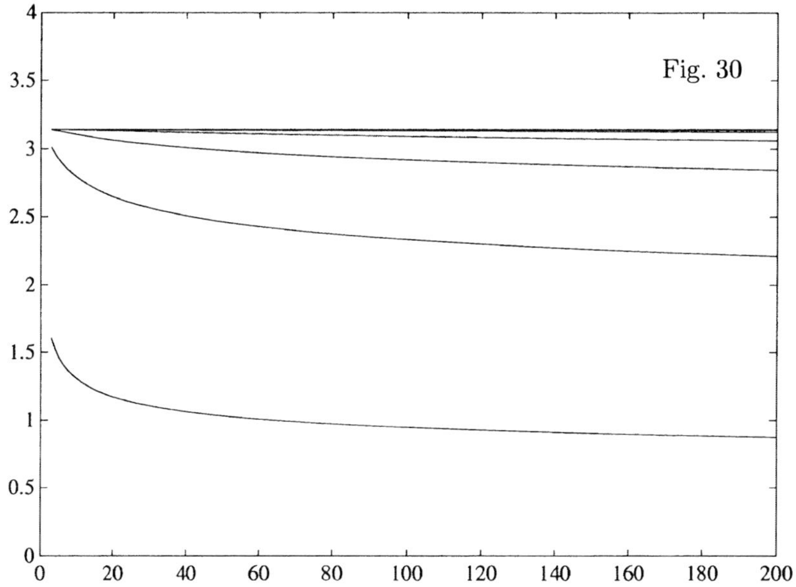

Figure 30 shows the singular values $s_j(T_n(\psi_{1/2}))$ for $1 \le j \le \min\{n, 20\}$ versus $3 \le n \le 200$. Although, $s_j(T_n(\psi_{1/2})) \to 0$ as $n \to \infty$ for each fixed j, this is hardly seen in the figure. In [175], it was proved that there are constants c_1 and c_2 such that $c_1/\log n \le s_1(T_n(\psi_{1/2})) \le c_2/\log n$.

Given a function $a \in L^2 := L^2(\mathbf{T})$ with Fourier coefficients a_l $(l \in \mathbf{Z})$, we put $T_n(a) := (a_{j-k})_{j,k=1}^n$. The *Frobenius* (or *Hilbert-Schmidt*) norm of an $n \times n$ matrix $A_n = (a_{jk})_{j,k=1}^n$ is defined as

$$\|A_n\|_F := \left(\sum_{j,k=1}^{n} |a_{jk}|^2 \right)^{1/2}.$$

Since multiplication by unitary matrices does not change the Frobenius norm, we see from Theorem 4.1 that

$$\|A_n\|_F^2 = s_1^2(A_n) + \ldots + s_n^2(A_n), \tag{4.16}$$

where $s_k^2(A_n) := (s_k(A_n))^2$.

Proposition 4.14. *If $a \in L^2$, then*

$$\lim_{n\to\infty} \frac{\|T_n(a)\|_F^2}{n} = \|a\|_2^2 := \frac{1}{2\pi}\int_0^{2\pi} |a(e^{i\theta})|^2\, d\theta.$$

Proof. Obviously,

$$\frac{\|T_n(a)\|_F^2}{n} = |a_0|^2 + \frac{n-1}{n}\big(|a_1|^2 + |a_{-1}|^2\big) + \ldots + \frac{1}{n}\big(|a_{n-1}|^2 + |a_{-n+1}|^2\big),$$

whence $\|T_n(a)\|_F^2/n \le \|a\|_2^2$ and $\|T_n(P_k a)\|_F^2/n \to \|P_k a\|_2^2$, where $P_k a$ is the kth partial sum of the Fourier series of a:

$$(P_k a)(t) = \sum_{|l|\le k} a_l t^l \quad (t \in \mathbf{T}).$$

Given $\varepsilon > 0$, pick k so that $\|a - P_k a\|_2 < \varepsilon/3$ and then choose n_0 so that

$$\|P_k a\|_2 - \|T_n(P_k a)\|_F/\sqrt{n} < \varepsilon/3 \quad \text{for all } n \ge n_0.$$

For $n \ge n_0$, we have $0 \le \|a\|_2 - \|T_n(a)\|_F/\sqrt{n}$ and

$$\begin{aligned}
&\|a\|_2 - \|T_n(a)\|_F/\sqrt{n} \\
&\quad \le \|a - P_k a\|_2 + \|P_k a\|_2 - \|T_n(a - P_k a) + T_n(P_k a)\|_F/\sqrt{n} \\
&\quad \le \|a - P_k a\|_2 + \big(\|P_k a\|_2 - \|T_n(P_k a)\|_F/\sqrt{n}\big) + \|T_n(a - P_k a)\|_F/\sqrt{n},
\end{aligned}$$

the first and second terms are less than $\varepsilon/3$, and the third term is at most $\|a - P_k a\|_2 < \varepsilon/3$. ∎

Corollary 4.15 (Parter). *Let $a \in L^\infty \setminus \{0\}$ and suppose $|a(t)| = \|a\|_\infty$ for almost all $t \in \mathbf{T}$. Fix a number $\varepsilon \in (0, \|a\|_\infty)$ and denote by γ_n the number of singular values of $T_n(a)$ (counted up to multiplicity) which are located in $[0, \|a\|_\infty - \varepsilon)$. Then*

$$\gamma_n/n \to 0 \quad \textit{as } n \to \infty.$$

Proof. Since $s_k(T_n(a)) \le \|T_n(a)\| \le \|a\|_\infty$, we get

$$\frac{1}{n}\sum_{k=1}^{n} \big(s_k(T_n(a))\big)^2 \le \frac{\gamma_n(\|a\|_\infty - \varepsilon)^2 + (n - \gamma_n)\|a\|_\infty^2}{n}. \tag{4.17}$$

Let c be any partial limit of the sequence $\{\gamma_n/n\}$ and suppose $\gamma_{n_j}/n_j \to c$. Combining (4.17), (4.16), and Proposition 4.14 we obtain

$$\begin{aligned}
&c(\|a\|_\infty - \varepsilon)^2 + (1-c)\|a\|_\infty^2 \\
&\quad \ge \lim_{j\to\infty} \frac{\|T_{n_j}(a)\|_F^2}{n_j} = \frac{1}{2\pi}\int_0^{2\pi} |a(e^{i\theta})|^2 d\theta = \|a\|_\infty^2.
\end{aligned}$$

Hence $c\,(\|a\|_\infty - \varepsilon)^2 \geq c\,\|a\|_\infty^2$, which is only possible if $c = 0$. Consequently, $\gamma_n/n \to 0$. ■

For $\gamma \in \mathbf{R}\setminus\mathbf{Z}$, we know from Example 1.7 that

$$|\psi_\gamma(t)| = \pi/|\sin \pi\gamma| = \|\psi_\gamma\|_\infty.$$

Thus, Corollary 4.15 tells us that if ε is any number in $(0, \pi/|\sin\pi\gamma|)$, then the percentage of the singular values of $T_n(\psi_\gamma)$ in $[0, \pi/|\sin\pi\gamma| - \varepsilon)$ goes to zero as n increases. In the case $\gamma = 1/2$, this is Moler's observation.

Proposition 4.14 states that

$$\frac{1}{n}\sum_{k=1}^{n} f\big(s_k(T_n(a))\big) \to \frac{1}{2\pi}\int_0^{2\pi} f\big(|a(e^{i\theta})|\big)\,d\theta \tag{4.18}$$

in case $a \in L^2$ and $f(\lambda) = \lambda^2$. Parter [128] and Avram [5] proved that (4.18) is true for every $a \in L^\infty$ and every continuous function f. Tyrtyshnikov [176], [177] extended (4.18) to arbitrary $a \in L^2$ and arbitrary continuous functions f with compact support. As (4.18) is basically a problem about the eigenvalue distribution of $T_n(|a|^2)$, we defer the matter to Chapter 5.

4.6 Limiting Sets of Singular Values

In this section we determine the uniform and partial limiting sets of

$$\Sigma(A_n) = \big\{s_1(A_n), \ldots, s_n(A_n)\big\}$$

provided $\{A_n\} \subset \mathbf{S}(PC)$ (recall (3.20) and (3.21) for the limiting sets and Section 3.3 for the definition of $\mathbf{S}(PC)$).

Theorem 4.16. *Let $\{B_n\} \in \mathbf{S}(PC)$ and denote the strong limits of B_n and $W_nB_nW_n$ by B and $\tilde{B}$, respectively. If $B_n = B_n^*$ for every n, then*

$$u\text{-}\lim_{n\to\infty} \operatorname{sp} B_n = p\text{-}\lim_{n\to\infty} \operatorname{sp} B_n = \operatorname{sp} B \cup \operatorname{sp} \tilde{B}.$$

Proof. The operators $B_n, B, \tilde{B}$ are all selfadjoint and hence all occurring spectra are real. So let $\lambda \in \mathbf{R}$.

If $\lambda \notin u\text{-}\lim \operatorname{sp} B_n$, then there are $n_j \to \infty$ and $\delta > 0$ such that $|\lambda - \mu| > \delta$ for all $\mu \in \bigcup_{j=1}^\infty \operatorname{sp} B_{n_j}$. It follows that $B_{n_j} - \lambda I_{n_j}$ is invertible for all j and that

$$\sup_{j\geq 1} \big\|(B_{n_j} - \lambda I_{n_j})^{-1}\big\| = \sup_{j\geq 1} \max_{\mu\in \operatorname{sp} B_{n_j}} \frac{1}{|\lambda - \mu|} \leq \frac{1}{\delta}$$

(note that $(B_{n_j} - \lambda I_{n_j})^{-1}$ is selfadjoint). Hence, the sequence $\{B_{n_j} - \lambda I_{n_j}\}$ is stable, which, by Proposition 2.3, implies that $B - \lambda I$ is invertible. Since

$\{W_{n_j} B_{n_j} W_{n_j} - \lambda I_{n_j}\}$ is stable together with $\{B_{n_j} - \lambda I_{n_j}\}$, we also obtain the invertibility of $\tilde{B} - \lambda I$. Consequently,

$$\operatorname{sp} B \cup \operatorname{sp} \tilde{B} \subset u\text{-}\lim_{n\to\infty} \operatorname{sp} B_n \subset p\text{-}\lim_{n\to\infty} \operatorname{sp} B_n.$$

On the other hand, if $\lambda \notin \operatorname{sp} B \cup \operatorname{sp} \tilde{B}$, we deduce from Theorem 3.11 that

$$M := \limsup_{n\to\infty} \|(B_n - \lambda I)^{-1}\| < \infty.$$

Therefore

$$\max_{\mu \in \operatorname{sp} B_n} \frac{1}{|\lambda - \mu|} < 2M$$

for all sufficiently large n, say for $n \geq n_0$. Thus, $|\lambda - \mu| > 1/(2M)$ whenever $\mu \in \bigcup_{n \geq n_0} \operatorname{sp} B_n$, which shows that $\lambda \notin p\text{-}\lim \operatorname{sp} B_n$. We so have proved that

$$p\text{-}\lim_{n\to\infty} \operatorname{sp} B_n \subset \operatorname{sp} B \cup \operatorname{sp} \tilde{B}. \blacksquare$$

Corollary 4.17. *If $\{A_n\} \in \mathbf{S}(PC)$ and A and $\tilde{A}$ are the strong limits of A_n and $W_n A_n W_n$, respectively, then*

$$u\text{-}\lim_{n\to\infty} \Sigma(A_n) = p\text{-}\lim_{n\to\infty} \Sigma(A_n) = \Sigma(A) \cup \Sigma(\tilde{A}).$$

Proof. This is Theorem 4.16 with $B_n := A_n^* A_n$. ■

Corollary 4.18. *If $a \in PC$ and $f : \mathbf{R} \to \mathbf{R}$ is continuous, then*

$$\begin{aligned} u\text{-}\lim_{n\to\infty} \operatorname{sp} f\big(T_n^*(a) T_n(a)\big) &= p\text{-}\lim_{n\to\infty} \operatorname{sp} f\big(T_n^*(a) T_n(a)\big) \\ &= \operatorname{sp} f\big(T^*(a) T(a)\big) \cup \operatorname{sp} f\big(T(a) T^*(a)\big). \end{aligned}$$

Proof. Theorem 4.16 gives the assertion in the case where f is a polynomial. Since

$$\operatorname{sp} T_n^*(a) T_n(a) \subset [0, \|T_n(a)\|^2] \subset [0, \|a\|_\infty^2]$$

and (recall (4.1))

$$\operatorname{sp} T^*(a) T(a) \cup \{0\} = \operatorname{sp} T(a) T(a^*) \cup \{0\} \subset [0, \|T(a)\|^2] = [0, \|a\|_\infty^2],$$

the case of general continuous f can be reduced to the case of polynomials by the approximation theorem of Weierstrass. ■

Corollary 4.19. *If $a \in PC$, then*

$$u\text{-}\lim_{n\to\infty} \Sigma\big(T_n(a)\big) = p\text{-}\lim_{n\to\infty} \Sigma\big(T_n(a)\big) = \Sigma\big(T(a)\big). \tag{4.19}$$

Proof. Corollary 4.17 with $A_n = T(a)$ or Corollary 4.18 with $f(\lambda) = \lambda$ imply that the two limiting sets equal

$$\Sigma(T(a)) \cup \Sigma(T(\overline{a}))$$

(notice that $\Sigma(T(\tilde{a})) = \Sigma(T(\overline{a}))$). From (4.1) we obtain

$$\Sigma\big(T(a)\big) \cup \{0\} = \Sigma\big(T(\overline{a})\big) \cup \{0\}.$$

If $T(a)$ is invertible, then $0 \notin \Sigma(T(a)) \cup \Sigma(T(\overline{a}))$ and Theorem 4.3 shows that 0 does not belong to the limit sets. In case $T(a)$ is not invertible, we have $0 \in \Sigma(T(a)) \cap \Sigma(T(\overline{a}))$, and again using Theorem 4.3 we see that 0 is in the two limiting sets. ■

Notes. Corollary 4.19 was established in Widom's paper [188]. He also proved that (4.19) holds if a is locally selfadjoint, i.e., if for each $\tau \in \mathbf{T}$ there exists a number $\gamma(\tau) \in \mathbf{C}$ such that $\mathcal{R}_\tau(a) \subset \gamma(\tau)\mathbf{R}$. In [155] it was shown that (4.19) holds for every locally normal symbol a. Theorem 4.16 was established in the paper [142] by Roch and one of the authors.

4.7 The Moore-Penrose Inverse

Throughout this section we let $\mathcal{B}$ stand for a C^*-algebra with identity element e. An element $a \in \mathcal{B}$ is said to be *Moore-Penrose invertible* if there is an element $b \in \mathcal{B}$ such that

$$aba = a, \qquad bab = b, \qquad (ab)^* = ab, \qquad (ba)^* = ba. \tag{4.20}$$

Proposition 4.20. *If $a \in \mathcal{B}$ is Moore-Penrose invertible, then there is only one element $b \in \mathcal{B}$ satisfying* (4.20).

Proof. If (4.20) holds for some b and is also true with b replaced by c, then

$$\begin{aligned} b &= bab = b(ab)^* = bb^*a^* = bb^*(aca)^* = bb^*a^*c^*a^* \\ &= bb^*a^*(ac)^* = bb^*a^*ac = bb^*a^*(aca)c \\ &= bb^*a^*a(ca)^*c = bb^*a^*aa^*c^*c = b(ab)^*aa^*c^*c \\ &= (bab)aa^*c^*c = baa^*c^*c = (ba)^*a^*c^*c \\ &= a^*b^*a^*c^*c = (aba)^*c^*c = a^*c^*c = (ca)^*c = cac = c. \ ■ \end{aligned}$$

If it exists, the (uniquely determined) element $b \in \mathcal{B}$ satisfying (4.20) is denoted by a^+ and called the *Moore-Penrose inverse* of a. The next theorem provides a criterion for the existence of the Moore-Penrose inverse.

Theorem 4.21. *An element $a \in \mathcal{B}$ is Moore-Penrose invertible if and only if there exists a number $d > 0$ such that*

$$\operatorname{sp}(a^*a) \subset \{0\} \cup [d^2, \infty). \tag{4.21}$$

Proof. Suppose (4.21) is valid. If a^*a is invertible, then $b := (a^*a)^{-1}a^*$ obviously satisfies (4.20). So assume a^*a is not invertible. Let $\mathcal{A}$ be the smallest closed subalgebra of $\mathcal{B}$ containing e and a^*a. Clearly, $\mathcal{A}$ is a commutative C^*-algebra. The maximal ideal space of $\mathcal{A}$ can be identified with $\text{sp}(a^*a)$ and the Gelfand map Γ, given by $(\Gamma e)(\lambda) = 1$ and $(\Gamma a^*a)(\lambda) = \lambda$, is a C^*-algebra isomorphism of $\mathcal{A}$ onto $C(\text{sp}(a^*a))$ (Example 2.26 and Theorem 3.3). Let $p \in \mathcal{A}$ be the element for which

$$(\Gamma p)(\lambda) = \begin{cases} 1 & \text{if} \quad \lambda = 0 \\ 0 & \text{if} \quad \lambda \in \text{sp}(a^*a) \setminus \{0\} \end{cases} \tag{4.22}$$

(note that the right-hand side of (4.22) is a continuous function by virtue of (4.21)). Considering Gelfand transforms we see that

$$p^2 = p = p^*, \qquad a^*a + p \in G\mathcal{A}, \qquad pa^*a = a^*ap = 0. \tag{4.23}$$

From (3.1) and (4.23) we obtain that

$$\|ap\|^2 = \|(ap)^*ap\| = \|p^*a^*ap\| = \|0\| = 0,$$

whence $ap = 0$. By (4.23), we may define $b := (a^*a + p)^{-1}a^*$. Because $\mathcal{A}$ is commutative and $ap = 0$, we get

$$\begin{aligned} aba &= a(a^*a + p)^{-1}a^*a \\ &= a(a^*a + p)^{-1}(a^*a + p) - a(a^*a + p)^{-1}p \\ &= a - ap(a^*a + p)^{-1} = a \end{aligned}$$

and similarly we can verify that $bab = b$. Taking into account that the element $(a^*a + p)^{-1}$ is selfadjoint, we see that $(ab)^* = ab$ and $(ba)^* = ba$. Thus, b is a Moore-Penrose inverse of a.

Conversely, suppose a has a Moore-Penrose inverse b. If $b = 0$, then $a = 0$ and (4.21) follows. Assume $b \neq 0$. If $0 < \lambda < \|bb^*\|^{-1}$, then $e - \lambda bb^*$ is invertible. From (4.20) we deduce that

$$\begin{aligned} (e - \lambda bb^*)ba &= ba - \lambda bb^*ba = baba - \lambda bb^*a^*b^* \\ &= b(ab)^*a - \lambda b(bab)^* = bb^*a^*a - \lambda bb^* = bb^*(a^*a - \lambda e), \end{aligned}$$

whence

$$(e - \lambda bb^*)^{-1}bb^*(a^*a - \lambda e) = ba. \tag{4.24}$$

Again using (4.20) we obtain

$$\begin{aligned} -\frac{1}{\lambda}(e - ba)(a^*a - \lambda e) &= -\frac{1}{\lambda}a^*a + e + \frac{1}{\lambda}baa^*a - ba \\ &= -\frac{1}{\lambda}a^*a + e + \frac{1}{\lambda}a^*b^*a^*a - ba = e - ba. \end{aligned} \tag{4.25}$$

Adding (4.24) and (4.25) we arrive at the conclusion that

$$(e - \lambda bb^*)^{-1}bb^* - \frac{1}{\lambda}(e - ba)$$

is a left inverse of $a^*a - \lambda e$. Analogously we can show that $a^*a - \lambda e$ is invertible from the right. This gives (4.21) with $d^2 = \|bb^*\|^{-1} = \|b\|^{-2}$. ■

For further referencing, we record the following consequence of the previous two results.

Corollary 4.22. *Let $\mathcal{B}$ be a C^*-algebra with identity e and let $\mathcal{A}$ be a C^*-subalgebra of $\mathcal{B}$ which contains e. If $a \in \mathcal{A}$ is Moore-Penrose invertible in $\mathcal{B}$ then $a^+ \in \mathcal{A}$.*

Proof. This is immediate from Propositions 3.1 and 4.20 and from Theorem 4.21. ■

The Kato number. Suppose (4.21) holds for some $d > 0$. Denote by $d_0 \in (0, \infty]$ the largest d for which (4.21) is true. The number $\kappa(a) \in [0, \infty)$ given by

$$\kappa(a) := \frac{1}{d_0} = \max\left\{\frac{1}{\sqrt{\lambda}} : \lambda \in \mathrm{sp}(a^*a)\setminus\{0\}\right\}$$

is called the *Kato number* of a.

Theorem 4.23. *If $a \in \mathcal{B}$ is Moore-Penrose invertible, then*

$$\|a^+\| = \kappa(a).$$

Proof. Let $d_0 \in (0, \infty]$ be the largest d for which (4.21) is valid. If a^*a is invertible, we know from the proof of Theorem 4.21 that $a^+ = (a^*a)^{-1}a^*$. Hence, by (3.1),

$$\begin{aligned} \|a^+\|^2 &= \|a^+(a^+)^*\| = \|(a^*a)^{-1}a^*a(a^*a)^{-1}\| \\ &= \|(a^*a)^{-1}\| = \max\{\lambda^{-1} : \lambda \in \mathrm{sp}(a^*a)\} = 1/d_0^2. \end{aligned}$$

In case a^*a is not invertible, we construct p as in the proof of Theorem 4.21 and have $a^+ = (a^*a + p)^{-1}a^*$. Again by (3.1),

$$\|a^+\|^2 = \|a^+(a^+)^*\| = \|(a^*a + p)^{-1}a^*a(a^*a + p)^{-1}\|.$$

Put $c := (a^*a+p)^{-1}a^*a(a^*a+p)^{-1}$. Since $(\Gamma a^*a)(\lambda) = \lambda$ for all $\lambda \in \mathrm{sp}(a^*a)$ and $(\Gamma p)(\lambda)$ is given by (4.22), it follows that

$$(\Gamma c)(\lambda) = \begin{cases} 0 & \text{if } \lambda = 0, \\ 1/\lambda & \text{if } \lambda \in \mathrm{sp}(a^*a)\setminus\{0\}, \end{cases}$$

whence $\|\Gamma c\|_\infty = 1/d_0^2$. From Theorem 3.3 (applied to the algebra $\mathcal{A}$ of the proof of Theorem 4.21) we therefore obtain

$$\|a^+\|^2 = \|c\|^2 = \|\Gamma c\|_\infty^2 = 1/d_0^2. \blacksquare$$

Now suppose H is a Hilbert space and consider the C^*-algebra $\mathcal{B}(H)$. In accordance with the above definitions, an operator $A \in \mathcal{B}(H)$ is said to be *Moore-Penrose invertible* if there exists an operator $B \in \mathcal{B}(H)$ such that

$$ABA = A, \quad BAB = B, \quad (AB)^* = AB, \quad (BA)^* = BA. \tag{4.26}$$

In case such an operator B exists, it is uniquely determined. It is then denoted by A^+ and referred to as the *Moore-Penrose inverse* of A. By Theorem 4.23,

$$\|A^+\| = \kappa(A) = \max\left\{\frac{1}{\sqrt{\lambda}} : \lambda \in \text{sp}(A^*A)\backslash\{0\}\right\}.$$

An operator $A \in \mathcal{B}(H)$ is called *normally solvable* if its range, $\text{Im}\, A$, is closed. It is well known that A is normally solvable if and only if A^* is.

Theorem 4.24. *An operator $A \in \mathcal{B}(H)$ is Moore-Penrose invertible if and only if it is normally solvable. In that case $A^*A + P$ is invertible and*

$$A^+ = (A^*A + P)^{-1}A^*,$$

where P is the orthogonal projection of H onto $\text{Ker}\, A$.

Proof. Suppose A has a Moore-Penrose inverse B. Let $x_n \in H$ and assume $Ax_n \to y$. Then $ABAx_n \to ABy$, and since $ABA = A$, it follows that $y = ABy$. Consequently, $\text{Im}\, A$ is closed.

Conversely, suppose A is normally solvable. Denote by P the orthogonal projection of H onto $\text{Ker}\, A$. Obviously,

$$AP = 0 \quad \text{and} \quad PA^* = 0. \tag{4.27}$$

We have the orthogonal decomposition $H = \text{Ker}\, P \oplus \text{Im}\, P$ and the operator given by

$$A_1 : \text{Ker}\, P \to \text{Im}\, A, \ x_1 \mapsto Ax$$

is obviously bijective. As $\text{Im}\, A$ is closed and therefore a Hilbert space, it follows that A_1 is invertible. Hence, there is a $C_1 \in (0, \infty)$ such that

$$\|x_1\| \le C_1\|A_1x_1\| \quad \text{for all } x_1 \in \text{Ker}\, P.$$

An arbitrary $x \in H$ may be written in the form $x = x_1 + x_2$ with $x_1 \in \text{Ker}\, P$ and $x_2 = Px \in \text{Im}\, P$, whence

$$\|x\| \le \|x_1\| + \|x_2\| \le C_1\|A_1x_1\| + \|Px\| = C_1\|Ax\| + \|Px\|.$$

Consequently, letting $\delta := 1/\max\{C_1, 1\}$ we get

$$\delta\|x\| \le \|Ax\| + \|Px\| \quad \text{for all } x \in H.$$

Squaring the latter inequality we obtain

$$\begin{aligned}\delta^2\|x\|^2 &\leq \left(\|Ax\|+\|Px\|\right)^2 \leq 2\|Ax\|^2+2\|Px\|^2\\ &= 2\big(x,(A^*A+p)x\big) \leq 2\|x\|\,\|(A^*A+P)x\|\end{aligned}$$

and thus $\delta^2\|x\| \leq 2\|(A^*A+P)x\|$ for all $x \in H$. This shows that the (selfadjoint) operator A^*A+P is invertible.

Finally, since P and $(A^*A+P)^{-1}$ commute by virtue of (4.27), it can be shown as in the proof of Theorem 4.21 that $B=(A^*A+P)^{-1}A^*$ satisfies (4.26). ■

Example 4.25: finite matrices. Operators on finite-dimensional spaces are always normally solvable and thus Moore-Penrose invertible. Of course, as their spectrum is finite, it is also obvious that (4.21) is automatically satisfied.

Although this will not be needed in the following, we recall the connection between the Moore-Penrose inverse and the minimal least-squares solution. Let $A \in \mathcal{B}(\mathbf{C}^n)$ be given by an $n\times n$ matrix. For $y \in \mathbf{C}^n$, denote by $\mathcal{X}_A(y)$ the set of all $x \in \mathbf{C}^n$ at which $\|Ax-y\|$ attains its minimum. One can show that $\mathcal{X}_A(y)$ contains a unique element of minimal norm, i.e., there is a unique $x_0 \in \mathcal{X}_A(y)$ such that $\|x_0\| \leq \|x\|$ for all $x \in \mathcal{X}_A(y)$. This x_0 is given by $x_0 = A^+y$ and is called the minimal least-squares solution of the equation $Ax=y$. Equivalently, the jth column of the matrix A^+ is the minimal least-squares solution of the system $Ax=e_j$ where e_j is the jth column of the identity matrix.

Now suppose $A \in \mathcal{B}(\mathbf{C}^n)$ is a nonzero $n\times n$ matrix and let $A=USV$ be the singular value decomposition (Theorem 4.1). We have

$$S=\operatorname{diag}(s_1,s_2,\ldots,s_n)$$

where $0 \leq s_1 \leq s_2 \leq \ldots \leq s_n = \|A\|$ are the singular values of A. Denote by s_{j+1} the smallest nonzero singular value:

$$0=s_1=\ldots=s_j<s_{j+1}\leq\ldots\leq s_n=\|A\|.$$

It is easy to check that

$$S^+=\operatorname{diag}(0,\ldots,0,s_{j+1}^{-1},\ldots,s_n^{-1}) \tag{4.28}$$

is the Moore-Penrose inverse of S. This implies that $A^+=V^*S^+U^*$ is the Moore-Penrose inverse of A. Indeed,

$$\begin{aligned}AA^+A &= USVV^*S^+U^*USV = USS^+SV = USV = A,\\ (AA^+)^* &= (A^+)^*A^* = US^+VV^*SU^* = US^+SU^*\\ &= USS^+U^* = USVV^*S^+U^* = AA^+,\end{aligned}$$

and the remaining two equalities can be verified analogously.

In particular, we immediately see that

$$\|A^+\| = \|S^+\| = s_{j+1}^{-1}, \tag{4.29}$$

i.e., the norm of the Moore-Penrose inverse is the inverse of the smallest nonzero singular value.

Let $D_j = \operatorname{diag}(1, \ldots, 1, 0, \ldots, 0)$ (j units and $n - j$ zeros) and put $P := V^* D_j V$. Clearly, P is a selfadjoint projection, and it is not difficult to show that $\operatorname{Im} P = \operatorname{Ker} A$. We have

$$(A^*A + P)^{-1}A^* = V^*(S^2 + D_j)^{-1}SU^*,$$

and, obviously, $(S^2 + D_j)^{-1}S$ is nothing but the diagonal matrix (4.28). ■

We are now in a position to establish the generalizations of Propositions 4.8 and 4.10 promised in Section 4.3.

Theorem 4.26. *Let $A \in \mathcal{B}(l^2)$ be an operator and let $\{A_n\}_{n=1}^{\infty}$ be a sequence of matrices $A_n \in \mathcal{B}(\mathbf{C}^n)$ such that $A_n \to A$ and $A_n^* \to A^*$ strongly as $n \to \infty$. If*

$$\limsup_{n\to\infty} s_{k+1}(A_n) > 0 \tag{4.30}$$

for some integer $k \geq 0$, then A is Fredholm and

$$\alpha(A) := \dim \operatorname{Ker} A \leq k, \qquad \beta(A) := \dim(l^2/\operatorname{Im} A) \leq k.$$

Proof. Let k_0 be the smallest k for which (4.30) holds. This means that there are a sequence $\{n_j\}$ and a number $d > 0$ such that

$$s_{k_0+1}(A_{n_j}) \geq d \text{ for all } j \quad \text{and} \quad \lim_{j\to\infty} s_{k_0}(A_{n_j}) = 0. \tag{4.31}$$

To simplify notation, assume that $n_j = j$.

By literally repeating the first part of the proof of Proposition 4.10 we arrive at the inclusion $\operatorname{sp} A^*A \subset \{0\} \cup [d, \infty)$. From Theorems 4.21 and 4.24 we infer that A is normally solvable, i.e., that $\operatorname{Im} A$ is closed.

Our next objective is to prove that $\dim \operatorname{Ker} A \leq k_0$. Let

$$A_n = U_n \operatorname{diag}(s_1, s_2, \ldots, s_n) V_n$$

be the singular value decomposition of A_n (Theorem 4.1 with $s_j := s_j(A_n)$), put

$$B_n = U_n \operatorname{diag}(s_{k_0+1}, \ldots, s_{k_0+1}, s_{k_0+2}, \ldots, s_n) V_n, \tag{4.32}$$

the number s_{k_0+1} occurring $k_0 + 1$ times, and set

$$F_n = V_n^* \operatorname{diag}(s_{k_0+1}^2 - s_1^2, \ldots, s_{k_0+1}^2 - s_{k_0}^2, 0, \ldots, 0) V_n.$$

Obviously, $B_n^*B_n = A_n^*A_n + F_n$ and $\operatorname{rank} F_n \le k_0$. Because $\|B_n\| = s_n = \|A_n\|$, we get

$$\|F_n\| \le \|B_n^*B_n\| + \|A_n^*A_n\| \le 2\|A_n\|^2.$$

Hence, $\|F_n\|$ is uniformly bounded. From Lemma 4.12 we therefore obtain the existence of an operator $F \in \mathcal{B}(l^2)$ with $\operatorname{rank} F \le k_0$ such that for each $x, y \in l^2$ the value (y, Fx) is a partial limit of the sequence $\{(y, F_n x)\}_{n=1}^{\infty}$. Taking into account (4.31) and (4.32), we see that

$$s_1(B_n) = s_{k_0+1}(A_n) \ge d > 0,$$

and hence (4.5) tells us that $\|B_n^{-1}\| \le 1/d$. Consequently,

$$\|P_n x\| = \|B_n^{-1}B_n P_n x\| \le (1/d)\|B_n P_n x\|$$

for $x \in l^2$. Taking squares yields

$$d^2\|P_n x\|^2 \le \|B_n P_n x\|^2 = (x, B_n^* B_n P_n x),$$

whence

$$d^2\|P_n x\|^2 \le (x, A_n^* A_n P_n x) + (x, F_n P_n x).$$

Passing to an appropriate partial limit in the last inequality we obtain

$$d^2\|x\|^2 \le (x, A^*Ax) + (x, Fx) \le \|x\|\big(\|A^*Ax\| + \|Fx\|\big)$$

and thus,

$$d^2\|x\| \le \|A^*Ax\| + \|Fx\|. \tag{4.33}$$

Now assume $\dim \operatorname{Ker} A^*A > k_0$. Let $x_1, \ldots, x_{k_0+1}$ be linearly independent elements of $\operatorname{Ker} A^*A$. As $\operatorname{rank} F \le k_0$, the elements $Fx_1, \ldots, Fx_{k_0+1}$ are linearly dependent. Suppose $\sum \alpha_j F x_j = 0$ is a nontrivial linear combination and put $x = \sum \alpha_j x_j$. Obviously, $x \in \operatorname{Ker} A^*A$ and $Fx = 0$. From (4.33) we therefore deduce that $x = 0$, contradicting our assumption that $x_1, \ldots, x_{k_0+1}$ be linearly independent. Thus, we have shown that $\operatorname{Ker} A^*A$ is at most k_0-dimensional. Since $\operatorname{Ker} A \subset \operatorname{Ker} A^*A$, it follows that $\dim \operatorname{Ker} A \le k_0$.

Analogously one can show that $\dim \operatorname{Ker} A^* \le k_0$. Because

$$\dim \operatorname{Ker} A^* = \dim(l^2/\operatorname{Im} A)$$

whenever A is normally solvable, this completes the proof. ∎

Corollary 4.27. *Let $A_n \in \mathcal{B}(\mathbf{C}^n)$ and $A \in \mathcal{B}(l^2)$. Suppose $A_n \to A$ strongly and $A_n^* \to A^*$ strongly as $n \to \infty$. If A is not Fredholm, then*

$$\lim_{n\to\infty} s_k(A_n) = 0 \quad \text{for each } k \ge 1.$$

Proof. Immediate from Theorem 4.26. ∎

The preceding corollary implies in particular that Proposition 4.10 is true under the sole hypothesis that $a \in L^\infty$.

Corollary 4.28. *If $a \in L^\infty$ and $T(a)$ is Fredholm of index k, then*

$$\lim_{n\to\infty} s_{|k|}(T_n(a)) = 0.$$

Proof. There is nothing to prove for $k = 0$. So let $|k| \geq 1$, assume $T(a)$ is Fredholm of index k but $\limsup_{n\to\infty} s_{|k|}(T_n(a)) > 0$. Then, by Theorem 4.26, $\alpha(T(a)) \leq |k| - 1$ and $\beta(T(a)) \leq |k| - 1$. Since one of the numbers $\alpha(T(a))$ and $\beta(T(a))$ is zero by virtue of Theorem 1.10, it follows that $|\operatorname{Ind} T(a)| \leq |k| - 1$. This contradiction proves the assertion. ∎

Of course, the last corollary extends Proposition 4.8 to symbols $a \in L^\infty$.

Notes. Proposition 4.20 and Theorem 4.24 are well known. Theorems 4.21 and 4.23 are implicit in Harte and Mbekhta's paper [96] and appeared explicitly in the papers [143], [145] by Roch and one of the authors. The proof of the "only if" portion of Theorem 4.21 given here is due to T. Ehrhardt. In [144], it was shown that if $a \in L^\infty$ and $T(a)$ is not normally solvable, then $s_k(T_n(a)) \to 0$ as $n \to \infty$ for each $k \geq 1$; note that for $a \in PC$ Fredholmness is equivalent to normal solvability (see, e.g., [87, Theorem 9.2.5] or [39, Remark on p. 85]).

Theorem 4.26 as well as its Corollaries 4.27 and 4.28 were communicated to us by T. Ehrhardt. We remark that this theorem gives the middle horizontal implication in the following scheme, which summarizes some relations between a sequence $\{A_n\} \in \mathcal{F}_{cc}$ and its strong limit $A = \text{s-lim}\, A_n$:

$\{A_n\}$ is stable	$\Longrightarrow$	A is invertible
$\Downarrow$		$\Downarrow$
$\{\Sigma(A_n)\}$ has the k-splitting property	$\Longrightarrow$	A is Fredholm and $\alpha(A) \leq k$, $\beta(A) \leq k$
$\Downarrow$		$\Downarrow$
$\{\Sigma(A_n)\}$ has the splitting property	$\Longrightarrow$	A is normally solvable

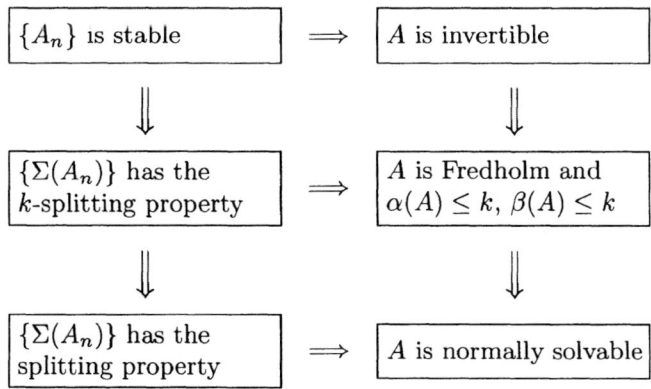

The vertical implications are all trivial; we note that the upper left vertical implication has to be interpreted as "if $\{A_n\}$ is stable, then $\{\Sigma(A_n)\}$ has the k-splitting property with $k = 0$." The upper horizontal implication is nothing but Proposition 2.3(b). The lower horizontal implication is not yet available at the present time, but it follows from the implication (i) $\Rightarrow$ (ii) of Corollary 4.31, which will be established in Section 4.9.

4.8 Asymptotic Moore-Penrose Inversion

Let $A \in \mathcal{B}(l^2)$ and let $\{A_n\} = \{A_n\}_{n=1}^{\infty}$ be an approximating sequence of $n \times n$ matrices A_n, i.e., suppose A_n $(:= A_n P_n)$ converges strongly to A. In Chapter 2 we studied the question whether $A_n^{-1} \to A^{-1}$ strongly provided A is invertible. Obviously, this question makes no sense in case A is not invertible. In that case we can pose other questions.

Question I. Suppose A is normally solvable. Then the Moore-Penrose inverse A^+ of A exists, and as finite matrices are always Moore-Penrose invertible, one can ask whether $A_n^+ \to A^+$ strongly. Clearly, this question is equivalent to the following: Is there a sequence $\{B_n\}$ of $n \times n$ matrices B_n such that

$$A_n B_n A_n - A_n = 0, \qquad B_n A_n B_n - B_n = 0, \tag{4.34}$$

$$(A_n B_n)^* - A_n B_n = 0, \qquad (B_n A_n)^* - B_n A_n = 0, \tag{4.35}$$

and $B_n \to A^+$ strongly?

Question II. Let A again be normally solvable. We now slightly modify the preceding question: is there a sequence $\{B_n\}$ of $n \times n$ matrices B_n such that

$$\|A_n B_n A_n - A_n\| \to 0, \qquad \|B_n A_n B_n - B_n\| \to 0, \tag{4.36}$$

$$\|(A_n B_n)^* - A_n B_n\| \to 0, \qquad \|(B_n A_n)^* - B_n A_n\| \to 0, \tag{4.37}$$

and $B_n \to A^+$ strongly?

Both questions address the problem of asymptotic Moore-Penrose inversion of the infinite matrix A. Although the differences in the formulations of Questions I and II are only minor ones, it turns out that the answers to these two questions differ significantly. In Sections 4.9 and 4.11 we study Question II, Section 4.10 is devoted to Question I.

4.9 Moore-Penrose Sequences

Recall that $\mathcal{F}$ stands for the C^*-algebra of all sequences $\{A_n\} = \{A_n\}_{n=1}^{\infty}$ of matrices (operators) $A_n \in \mathcal{B}(\mathbf{C}^n)$ for which

$$\|\{A_n\}\| := \sup_{n \geq 1} \|A_n\| < \infty$$

and that $\mathcal{N}$ is the closed two-sided ideal of $\mathcal{F}$ which consists of the sequences $\{C_n\}$ such that $\|C_n\| \to 0$ as $n \to \infty$. We know from Chapter 2 that the invertibility of the coset $\{A_n\} + \mathcal{N}$ in $\mathcal{F}/\mathcal{N}$ is equivalent to the stability of the sequence $\{A_n\}$. Note that stability is in turn equivalent to the existence of a number $d > 0$ such that

$$\Sigma(A_n) := (\operatorname{sp} A_n^* A_n)^{1/2} \subset [d, \infty)$$

for all sufficiently large n.

A sequence $\{A_n\} \in \mathcal{F}$ is called a *Moore-Penrose sequence* if $\{A_n\} + \mathcal{N}$ is Moore-Penrose invertible in $\mathcal{F}/\mathcal{N}$. In other words, $\{A_n\} \in \mathcal{F}$ is a Moore-Penrose sequence if and only if there exists a sequence $\{B_n\} \in \mathcal{F}$ such that (4.36) and (4.37) hold.

Theorem 4.30 will give a characterization of Moore-Penrose sequences. To prove this theorem, we need the following fact from the elementary theory of C^*-algebras.

Lemma 4.29. *Let $\mathcal{B}$ be a C^*-algebra, let $\mathcal{A}$ be a C^*-subalgebra of $\mathcal{B}$, and let J be a closed two-sided ideal of $\mathcal{B}$. Then $\mathcal{A} + J$ is a C^*-subalgebra of $\mathcal{B}$ and the map*

$$\varphi : \mathcal{A}/(\mathcal{A} \cap J) \to (\mathcal{A} + J)/J, \quad a + (\mathcal{A} \cap J) \mapsto a + J$$

is an (isometric) C^*-algebra *isomorphism.*

For a proof see, e.g., [56, Corollary 1.8.4] or [71, pp. 17–18]. We remark that the point of this lemma is the closedness of $\mathcal{A} + J$; the isomorphism of the two algebras is general ring theory.

Theorem 4.30. *A sequence $\{A_n\} \in \mathcal{F}$ is a Moore-Penrose sequence if and only if*

$$\Sigma(A_n) = \{s_1(A_n), \ldots, s_n(A_n)\}$$

has the splitting property, i.e., if and only if there exist $c_n \to 0$ and $d > 0$ such that

$$\Sigma(A_n) \subset [0, c_n] \cup [d, \infty)$$

for all $n \geq 1$.

Proof. Write $A_n = U_n S_n V_n$ as in Theorem 4.1. We have

$$\begin{aligned} \|A_n B_n A_n - A_n\| &= \|U_n S_n V_n B_n U_n S_n V_n - U_n S_n V_n\| \\ &= \|S_n \tilde{B}_n S_n - S_n\| \text{ where } \tilde{B}_n := V_n B_n U_n, \end{aligned}$$

and since analogous equalities hold for the remaining three norms in (4.36) and (4.37), we arrive at the conclusion that $\{A_n\}$ is a Moore-Penrose sequence if and only if $\{S_n\}$ is one, that is, if and only if $\{S_n\} + \mathcal{N}$ is Moore-Penrose invertible in $\mathcal{F}/\mathcal{N}$. By Theorem 4.21, this is equivalent to the existence of a number $d > 0$ such that

$$\mathrm{sp}_{\mathcal{F}/\mathcal{N}}\big(\{S_n^2\} + \mathcal{N}\big) \subset \{0\} \cup [d^2, \infty). \tag{4.38}$$

Let $\mathcal{D} \subset \mathcal{F}$ denote the sequences $\{A_n\}$ constituted by the diagonal matrices A_n. From Proposition 3.1 and Lemma 4.29 we infer that

$$\begin{aligned} \mathrm{sp}_{\mathcal{F}/\mathcal{N}}\big(\{S_n^2\} + \mathcal{N}\big) &= \mathrm{sp}_{(\mathcal{D}+\mathcal{N})/\mathcal{N}}\big(\{S_n^2\} + \mathcal{N}\big) \\ &= \mathrm{sp}_{\mathcal{D}/(\mathcal{D}\cap\mathcal{N})}\big(\{S_n^2\} + (\mathcal{D}\cap\mathcal{N})\big). \end{aligned} \tag{4.39}$$

Consider the infinite diagonal matrix

$$\mathrm{diag}\big(S_1^2, S_2^2, \ldots\big) = \mathrm{diag}(\varrho_1, \varrho_2, \varrho_3, \ldots)$$

where $\varrho_1 = s_1^2(A_1)$, $\varrho_2 = s_1^2(A_2)$, $\varrho_3 = s_2^2(A_2)$, ... (i.e., $S_m \in \mathcal{B}(\mathbf{C}^m)$ and $\varrho_m \in \mathbf{C}$). Obviously, the spectrum on the right of (4.39) coincides with the set $\mathcal{P}(\varrho_m)$ of the partial limits of the sequence $\{\varrho_m\}$. Consequently, (4.38) holds if and only if $\mathcal{P}(\varrho_m) \subset \{0\} \cup [d^2, \infty)$, which is easily seen to be equivalent to the splitting property of $\Sigma(A_n)$. ■

As in Section 3.1, let $\mathcal{F}_{cc}$ be the C^*-subalgebra of $\mathcal{F}$ which consists of all sequences $\{A_n\} \in \mathcal{F}$ such that

$$A_n \to A \text{ strongly} \quad \text{and} \quad A_n^* \to A^* \text{ strongly}$$

for some $A \in \mathcal{B}(l^2)$. Here is the answer to Question II of Section 4.8.

Corollary 4.31. *Let $A \in \mathcal{B}(l^2)$, let $\{A_n\} \in \mathcal{F}_{cc}$, and suppose $A_n \to A$ strongly. Then the following are equivalent:*

(i) *$\Sigma(A_n)$ has the splitting property;*

(ii) *the operator A is normally solvable and there exists a sequence $\{B_n\} \in \mathcal{F}_{cc}$ such that* (4.36) *and* (4.37) *hold.*

If (i) *and* (ii) *are satisfied, then $B_n \to A^+$ strongly and $B_n^* \to (A^+)^*$ strongly.*

Proof. (i) ⇒ (ii). Theorem 4.30 guarantees the existence of a sequence $\{B_n\}$ in $\mathcal{F}$ for which (4.36) and (4.37) hold. Since $\{B_n\}+\mathcal{N}$ is the Moore-Penrose inverse of $\{A_n\} + \mathcal{N} \in \mathcal{F}/\mathcal{N}$ and $\{A_n\} + \mathcal{N}$ belongs to $\mathcal{F}_{cc}/\mathcal{N}$, we deduce from Corollary 4.22 that $\{B_n\} \in \mathcal{F}_{cc}$. Hence, there is a $B \in \mathcal{B}(l^2)$ such that $B_n \to B$ and $B_n^* \to B^*$ strongly. Passing in (4.36) and (4.37) to the strong limit, we get (4.26). This shows that A is normally solvable (Theorem 4.24) and that $B = A^+$.

(ii) ⇒ (i). Immediate from Theorem 4.30. ■

We now turn to the case where $\{A_n\} = \{T_n(a)\}$ is the sequence of the finite sections of some infinite Toeplitz matrix.

Theorem 4.32. *For $a \in PC$, the following are equivalent:*

(i) *$T(a)$ is normally solvable;*

(ii) *$T(a)$ is Fredholm or $T(a) = 0$;*

(iii) *there exists a sequence $\{B_n\} \in \mathcal{F}$ such that*

$$\|T_n(a)B_nT_n(a) - T_n(a)\| \to 0, \quad \|B_nT_n(a)B_n - B_n\| \to 0, \tag{4.40}$$

$$\|(T_n(a)B_n)^* - T_n(a)B_n\| \to 0, \quad \|(B_nT_n(a))^* - B_nT_n(a)\| \to 0. \tag{4.41}$$

If these conditions are satisfied, then $B_n \to T^+(a)$ strongly and $B_n^* \to (T^+(a))^*$ strongly.

Proof. (i) $\Rightarrow$ (ii). This is a well-known fact (see, e.g., [87, Theorem 9.2.5] or [39, Remark on p. 85]).

(ii) $\Rightarrow$ (iii). Theorem 4.5 shows that $\Sigma(T_n(a))$ has the splitting property (even the k-splitting property for some $k \geq 0$). Hence, by Corollary 4.31, there exists a sequence $\{B_n\} \in \mathcal{F}_{cc}$ satisfying (4.40), (4.41) such that $B_n \to T^+(a)$ and $B_n^* \to (T^+(a))^*$ strongly.

(iii) $\Rightarrow$ (i). Clearly, $\{B_n\} + \mathcal{N}$ is the Moore-Penrose inverse of $\{T_n(a)\} + \mathcal{N}$ in $\mathcal{F}/\mathcal{N}$. Since $\{T_n(a)\} + \mathcal{N}$ lies in $\mathcal{F}_{cc}/\mathcal{N}$, it follows from Corollary 4.22 that $\{B_n\}$ belongs to $\mathcal{F}_{cc}$. Passing in (4.40) and (4.41) to the strong limit, we see that the strong limit of B is the Moore-Penrose inverse of $T(a)$. Theorem 4.24 completes the proof. ∎

Notes. Several versions of the two questions raised in Section 4.8 have been studied for a long time; see, for example, the paper [123] by Moore and Nashed and the article [125] by Nashed. For Toeplitz operators, Question I of Section 4.8 was probably first considered by Heinig and Hellinger [100], while Question II was taken up in this context by one of the authors only in [156]. In the latter paper, Theorem 4.32 (and an essential generalization of this theorem) was proved by different methods; see Section 6.5. Note that the splitting property does not occur in the statement of Theorem 4.32. Neither does it in [156]. The splitting property was introduced by Roch and Silbermann in [143], and this paper also contains Theorem 4.30 and Corollary 4.31. The proof of Theorem 4.30 given here is from Böttcher's paper [27].

4.10 Exact Moore-Penrose Sequences

A sequence $\{A_n\} \in \mathcal{F}$ is invertible in $\mathcal{F}$ if and only if A_n is invertible for all $n \geq 1$ and $\sup_{n\geq 1} \|A_n^{-1}\| < \infty$, or equivalently, if and only if there is a $d > 0$ such that

$$\Sigma(A_n) := \left(\mathrm{sp}(A_n^* A_n)\right)^{1/2} \subset [d, \infty) \qquad \text{for all } n \geq 1.$$

We call a sequence $\{A_n\} \in \mathcal{F}$ an *exact Moore-Penrose sequence* if $\{A_n\}$ is Moore-Penrose invertible in (the C^*-algebra) $\mathcal{F}$. Of course,

$$\begin{aligned} &\{A_n\} \text{ is Moore-Penrose invertible in } \mathcal{F} \\ &\quad \Longleftrightarrow (4.34),\ (4.35) \text{ hold with some } \{B_n\} \in \mathcal{F} \\ &\quad \Longleftrightarrow \{A_n^+\} \in \mathcal{F}. \end{aligned}$$

Here is the analogue of Theorem 4.30.

Theorem 4.33. *A sequence $\{A_n\} \in \mathcal{F}$ is an exact Moore-Penrose sequence if and only if there is a number $d > 0$ such that*

$$\Sigma(A_n) \subset \{0\} \cup [d, \infty) \quad \textit{for all } n \geq 1. \tag{4.42}$$

Proof. As in the proof of Theorem 4.30 we see that $\{A_n\}$ is an exact Moore-Penrose sequence if and only if $\{S_n\}$ enjoys this property. Write

$$\operatorname{diag}(S_1, S_2, \ldots) = \operatorname{diag}(\mu_1, \mu_2, \mu_3, \ldots)$$

($S_m \in \mathcal{B}(\mathbf{C}^m)$ and $\mu_m \in \mathbf{C}$) and define $f : [0, \infty) \to [0, \infty)$ by

$$f(x) := \begin{cases} 0 & \text{if } \ x = 0, \\ x^{-1} & \text{if } \ x > 0. \end{cases}$$

By virtue of (4.28),

$$\operatorname{diag}\big(S_1^+, S_2^+, \ldots\big) = \operatorname{diag}\big(f(\mu_1), f(\mu_2), f(\mu_3), \ldots\big).$$

Hence, $\{S_n^+\} \in \mathcal{F}$ if and only if $\{f(\mu_m)\}$ is a bounded sequence, which is equivalent to (4.42). ■

We say that $\Sigma(A_n)$ has the *exact splitting property* if (4.42) holds for some $d > 0$. Notice that (4.42) is equivalent to the existence of a number $\delta > 0$ such that

$$\Sigma(A_n) \subset \{0\} \cup [\delta, \infty) \quad \text{for all sufficiently large } n; \tag{4.43}$$

indeed, if (4.43) holds for $n \geq n_0$, then (4.42) is valid with $d := \min\{\delta, \sigma\}$ where σ is the minimal nonzero singular value of A_n for $n \in \{1, \ldots, n_0 - 1\}$.

Question I of Section 4.8 can now be answered as follows.

Corollary 4.34. *Let $A \in \mathcal{B}(l^2)$, let $\{A_n\} \in \mathcal{F}_{cc}$, and suppose $A_n \to A$ strongly. Then the following are equivalent:*

(i) *$\Sigma(A_n)$ has the exact splitting property;*

(ii) *$\{A_n\}$ is an exact Moore-Penrose sequence;*

(iii) *the sequence $\{A_n^+\}$ is strongly convergent;*

(iv) *A is normally solvable and $A_n^+ \to A^+$ strongly.*

Proof. (i) $\Rightarrow$ (ii). Immediate from Theorem 4.33.

(ii) $\Rightarrow$ (iii). If $\{A_n\}$ is Moore-Penrose invertible in $\mathcal{F}$, then $\{A_n\}$ is Moore-Penrose invertible in $\mathcal{F}_{cc}$ (Corollary 4.22). The only candidate for the Moore-Penrose inverse of $\{A_n\}$ is $\{A_n^+\}$, whence $\{A_n^+\} \in \mathcal{F}_{cc}$.

(iii) $\Rightarrow$ (iv). If A_n^+ is strongly convergent, then $\{A_n^+\} \in \mathcal{F}$ by Theorem 2.1. As $\{A_n^+\}$ is the Moore-Penrose inverse of $\{A_n\}$, we obtain from Corollary 4.22 that $\{A_n^+\} \subset \mathcal{F}_{cc}$. Denoting by B the strong limit of A_n^+, we get (4.26). From Theorem 4.24 we therefore conclude that A is normally solvable and Proposition 4.20 shows that $A^+ = B$.

(iv) $\Rightarrow$ (ii). The sequence $\{A_n^+\}$ belongs to $\mathcal{F}$ (Theorem 2.1) and is the Moore-Penrose inverse of $\{A_n\} \in \mathcal{F}$. ∎

Now let $A_n = T_n(a)$ with $a \in PC$. We know from Theorem 4.32 that $\{T_n(a)\}$ is a Moore-Penrose sequence if and only if $T(a)$ is Fredholm. Being an exact Moore-Penrose sequence is a much more delicate property.

Theorem 4.35. *Let $a \in PC$. Then $\{T_n(a)\}$ is an exact Moore-Penrose sequence if and only if $T(a)$ is Fredholm and*

$$\dim \operatorname{Ker} T_n(a) = |\operatorname{Ind} T(a)| \tag{4.44}$$

for all sufficiently large n.

Proof. If $\{T_n(a)\}$ is an exact Moore-Penrose sequence, then $T(a)$ is Fredholm due to Theorem 4.32. Suppose $T(a)$ is Fredholm of index k. From Theorem 4.5 we infer that there are $d > 0$ and $n_1 \in \mathbf{N}$ such that

$$s_{|k|}\big(T_n(a)\big) \to 0 \quad \text{and} \quad s_{|k|+1}\big(T_n(a)\big) \geq d \ \text{ for } n \geq n_1. \tag{4.45}$$

By Theorem 4.2, $\operatorname{dist}(T_n(a), \mathcal{F}^{(n)}_{n-|k|-1}) \geq d > 0$ and thus,

$$\operatorname{rank} T_n(a) \geq n - |k| \quad \text{for all } n \geq n_1. \tag{4.46}$$

From (4.45) and Theorem 4.33 we conclude that $\{T_n(a)\}$ is an exact Moore-Penrose sequence if and only if $s_{|k|}(T_n(a)) = 0$ for all $n \geq n_2 \geq n_1$. Because

$$s_{|k|}(T_n(a)) = \operatorname{dist}(T_n(a),\ \mathcal{F}^{(n)}_{n-|k|})$$

and $\mathcal{F}^{(n)}_{n-|k|}$ is a closed subset of $\mathcal{B}(\mathbf{C}^n)$, we have $s_{|k|}(T_n(a)) = 0$ for all $n \geq n_2$ if and only if

$$\operatorname{rank} T_n(a) \leq n - |k| \quad \text{for all } n \geq n_2. \tag{4.47}$$

Combining (4.46) and (4.47) we obtain that $\{T_n(a)\}$ is an exact Moore-Penrose sequence if and only if $T(a)$ is Fredholm of some index k and

$$\dim \operatorname{Ker} T_n(a) = n - \operatorname{rank} T_n(a) = |k| \quad \text{for all } n \geq n_2. \ ∎$$

If $a \in PC$ and $T(a)$ is Fredholm of index zero and thus invertible, then the sequence $\{T_n(a)\}$ is stable (Corollary 2.19). In this case

$$\Sigma\big(T_n(a)\big) \subset [d, \infty) \quad \text{and} \quad \dim \operatorname{Ker} T_n(a) = 0$$

with some $d > 0$ for all sufficiently large n and hence Theorem 4.33 yields that $\{T_n(a)\}$ is an exact Moore-Penrose sequence. However, we have $T_n^+(a) = T_n^{-1}(a)$ for all n large enough and therefore consideration of Moore-Penrose inverses is not at all necessary in this situation.

The really interesting case is the one in which $T(a)$ is Fredholm of nonzero index. The rest of this section is devoted to the proof of the following result.

Theorem 4.36 (Heinig-Hellinger). *Let $a \in PC$. Suppose $T(a)$ is Fredholm and* $\operatorname{Ind} T(a) \neq 0$. *If* $\operatorname{Ind} T(a) < 0$, *then the following are equivalent:*

(i) $\{T_n(a)\}$ *is an exact Moore-Penrose sequence;*

(ii) $\dim \operatorname{Ker} T_n(a) = |\operatorname{Ind} T(a)|$ *for all n large enough;*

(iii) $\operatorname{Ker} T(\tilde{a}) \subset \operatorname{Im} P_{n_0}$ *for some $n_0 \geq 1$;*

(iv) $(a^{-1})_{-m} = 0$ *for all sufficiently large m.*

(Here $(a^{-1})_j$ denotes the jth Fourier coefficient of a^{-1}.) If $\operatorname{Ind} T(a) > 0$, *then the following are equivalent:*

(i′) $\{T_n(a)\}$ *is an exact Moore-Penrose sequence;*

(ii′) $\dim \operatorname{Ker} T_n(a) = \operatorname{Ind} T(a)$ *for all n large enough;*

(iii′) $\operatorname{Ker} T(a) \subset \operatorname{Im} P_{n_0}$ *for some $n_0 \geq 1$;*

(iv′) $(a^{-1})_m = 0$ *for all sufficiently large m.*

For the sake of definiteness, let us assume that

$$\operatorname{Ind} T(a) = -k < 0.$$

The equivalence (i) $\Leftrightarrow$ (ii) is nothing but a repetition of Theorem 4.35. The proofs of the implications (iv) $\Rightarrow$ (iii) $\Rightarrow$ (ii) are easy. Here they are.

Proof of the implication (iv) $\Rightarrow$ (iii). Let $x \in \operatorname{Ker} T(\tilde{a})$. Then $\tilde{a}x =: \varphi_- \in H^2_-$, whence $x = \tilde{a}^{-1}\varphi_-$, which shows that $x_m = 0$ for all sufficiently large m. Another proof follows from the identity

$$T(\tilde{a}^{-1})T(\tilde{a}) = I - H(\tilde{a}^{-1})H(a),$$

which gives $x = H(\tilde{a}^{-1})H(a)x$, and since $H(\tilde{a}^{-1})$ has only a finite number of nonzero rows, it results that $x_m = 0$ for all sufficiently large m. ■

Proof of the implication (iii) $\Rightarrow$ (ii). If n is large enough, then

$$s_{k+1}(T_n(\tilde{a})) \geq d > 0$$

by Theorem 4.5, so rank $T_n(\tilde{a}) > n - k - 1$ (Theorem 4.2) and thus

$$\dim \operatorname{Ker} T_n(\tilde{a}) < k + 1. \tag{4.48}$$

If $x \in \operatorname{Ker} T(\tilde{a}) \subset \operatorname{Im} P_{n_0}$ and $n \geq n_0$, then

$$T_n(\tilde{a}) P_n x = P_n T(\tilde{a}) x = 0,$$

which shows that

$$\dim \operatorname{Ker} T_n(\tilde{a}) \geq \dim \operatorname{Ker} T(\tilde{a}) = k \tag{4.49}$$

(recall Theorem 1.10 for the last equality). Clearly, (ii) follows from (4.48) and (4.49). ■

The proof of the implication (ii) $\Rightarrow$ (iv) is less trivial and is based on the following deep theorem.

Theorem 4.37 (Heinig). *Let $a \in L^\infty$ and let $k > 0$ be an integer. Then*

$$\dim \operatorname{Ker} T_n(a) = k \quad \textit{for all sufficiently large } n$$

if and only if a or $\tilde{a}$ is of the form $\chi_{p+k}(r+h)$ where h is a function in H^∞, r is a rational function whose restriction to $\mathbf{T}$ is bounded and which has exactly p poles in the open unit disk (multiplicities taken into account) and no pole at the origin, $r(0)+h(0) \neq 0$, and χ_{p+k} is defined by $\chi_{p+k}(t) = t^{p+k}$ $(t \in \mathbf{T})$.

A proof is in [99, Satz 6.2 and formula (8.4)]. Also see the book [101, Theorem 8.6].

Proof of the implication (ii) $\Rightarrow$ (iv). Let $k = |\operatorname{Ind} T(a)|$. Furthermore, let $\chi_{p+k}(r+h)$ be the representation of a or $\tilde{a}$ ensured by Theorem 4.37. Put $\chi_{p+k}(r+h) =: b$ and denote by $\alpha_1, \ldots, \alpha_p$ and $\beta_1, \ldots, \beta_q$ the poles of r inside and outside $\mathbf{T}$, respectively. For $t \in \mathbf{T}$,

$$\begin{aligned} r(t) &= \frac{u_+(t)}{(t-\alpha_1)\ldots(t-\alpha_p)(t-\beta_1)\ldots(t-\beta_q)} \\ &= \frac{t^{-p} v_+(t)}{(1-\alpha_1/t)\ldots(1-\alpha_p/t)(1-t/\beta_1)\ldots(1-t/\beta_q)} \end{aligned}$$

with polynomials $u_+, v_+ \in H^\infty$. Clearly,

$$s_+(t) := (1-t/\beta_1)^{-1} \ldots (1-t/\beta_q)^{-1} \in H^\infty.$$

Letting

$$c_+(t) := t^k v_+(t) s_+(t) + t^{p+k}(1-\alpha_1/t)\ldots(1-\alpha_p/t) h(t),$$

we get

$$b(t) = (1-\alpha_1/t)^{-1} \ldots (1-\alpha_p/t)^{-1} c_+(t).$$

The function c_+ belongs to H^∞ and has zero of order at least k at the origin. Obviously, $(1-\alpha_1/t)\ldots(1-\alpha_p/t)$ is a function in $G\overline{H^\infty}$. If c_+

would have infinitely many zeros in $\mathbf{D}$, then $T(c_+)$ and thus also $T(b)$ were not Fredholm (see, e.g., [59, Theorem 7.36] or [39, Theorem 2.64]). Thus, c_+ has only a finite number $\varkappa \geq k$ of zeros in $\mathbf{D}$. It follows that $\operatorname{Ind} T(c_+) = -\varkappa$ (again see, e.g., [59, Theorem 7.36] or [39, Theorem 2.64]) and hence,

$$\operatorname{Ind} T(b) = \operatorname{Ind} T(c_+) = -\varkappa.$$

If $b = a$, then $\varkappa$ must equal k. Consequently, $c_+(z) = z^k \varphi_+(z)$ with φ_+ in GH^∞. This implies that

$$a^{-1}(t) = t^{-k}(1 - \alpha_1/t) \dots (1 - \alpha_p/t)\varphi_+^{-1}(t)$$

has only finitely many nonzero Fourier coefficients with negative index. If b would equal $\tilde{a}$, it would result that $\operatorname{Ind} T(\tilde{a})$ is negative, which is impossible due to the equality $\operatorname{Ind} T(\tilde{a}) = -\operatorname{Ind} T(a) = k$. ■

Corollary 4.38. *If $a \in PC \backslash C$, then $\{T_n(a)\}$ is an exact Moore-Penrose sequence if and only if $\{T_n(a)\}$ is stable.*

Proof. The "if" part is trivial. To prove the "only if" portion, suppose $\{T_n(a)\}$ is an exact Moore-Penrose sequence. Then $T(a)$ is Fredholm by Theorem 4.35 (or by Theorem 4.32). If $T(a)$ has index zero, then $\{T_n(a)\}$ is stable (Theorem 1.10 and Corollary 2.19). In case $\operatorname{Ind} T(a) \neq 0$, we obtain from Theorem 4.36 that a^{-1} is a polynomial times a function in H^∞ or $\overline{H^\infty}$. As functions in $H^\infty \cup \overline{H^\infty}$ cannot have jumps (Lindelöf's theorem; see, e.g., [135, p.60]), this contradicts our hypotheses that $a \in PC \setminus C$. ■

Thus, studying Question I of Section 4.8 for symbols $a \in PC \setminus C$ does actually not go beyond the finite section method.

Notes. Theorems 4.33 and 4.35 as well as Corollary 4.34 were established by Roch and Silbermann in [143], the simple proofs of Theorems 4.33 and 4.35 presented in the text are due to Böttcher [27]. Heinig and Hellinger [100] proved the equivalences (ii) ⇔ (iv) and (ii′) ⇔ (iv′) of Theorem 4.36 for symbols in the Wiener algebra. Conditions (iii) and (iii′) were introduced by Silbermann in [156]. The proof of Theorem 4.36 contained in [100] also makes use of Theorem 4.37 and is rather difficult. The above proof of the implication (ii) ⇒ (iv) is taken from [27]. Our proof of the implication (ii)⇒(iv) nicely illustrates the usefulness of Douglas' Fredholm theory of Toeplitz operators with $C + H^\infty$ symbols (see [58] and [59]) even when dealing with Wiener symbols.

4.11 Regularization and Kato Numbers

Question I versus Question II. Let $a \in PC$ and suppose $T(a)$ is Fredholm of index k. We know from Theorem 4.32 that then $\{T_n(a)\} + \mathcal{N}$ is always Moore-Penrose invertible in $\mathcal{F}/\mathcal{N}$, but Theorem 4.36 tells us that $\{T_n^+(a)\} + \mathcal{N}$ is the Moore-Penrose inverse of $\{T_n(a)\} + \mathcal{N}$ in rare cases only. For example, if $a \in PC\backslash C$ and $k \neq 0$, then $\{T_n^+(a)\} + \mathcal{N}$ is never the Moore-Penrose inverse of $\{T_n(a)\} + \mathcal{N}$ (Corollary 4.38). In other words, while the answer to Question I is in general "yes", the answer to Question II is "no" in the majority of interesting cases (recall Section 4.8).

Regularization. In Example 4.25 we observed that

$$T_n^+(a) = V_n^* S_n^+ U_n^*,$$

where

$$S_n^+ = \operatorname{diag}\big(0, \ldots, 0, s_{j+1}^{-1}(T_n(a)), \ldots, s_n^{-1}(T_n(a))\big)$$

and $s_{j+1}(T_n(a))$ is the smallest nonzero singular value of $T_n(a)$. The unboundedness of

$$\|T_n^+(a)\| = \|S_n^+\| = 1/s_{j+1}\big(T_n(a)\big)$$

as n goes to infinity is caused by the circumstance that $s_{j+1}(T_n(a))$, though nonzero, may approach zero.

Let again $a \in PC$ and let $T(a)$ be Fredholm of index k. Then, by Theorem 4.5, the first $|k|$ singular values of $T_n(a)$ tend to zero, while the remaining $n - |k|$ singular values of $T_n(a)$ stay away from zero. Thus, let us consider the "regularized" Moore-Penrose inverse

$$T_n^\#(a) := V_n^* S_n^\# U_n^*,$$

where

$$S_n^\# := \operatorname{diag}\big(0, \ldots, 0, s_{|k|+1}^{-1}(T_n(a)), \ldots, s_n^{-1}(T_n(a))\big)$$

($|k|$ zeros). We then have

$$\|T_n^\#(a)\| = \|S_n^\#\| = 1/s_{|k|+1}\big(T_n(a)\big), \tag{4.50}$$

and this remains bounded as $n \to \infty$. As the following theorem shows, $\{T_n^\#(a)\}$ is in fact always an asymptotic Moore-Penrose inverse of $\{T_n(a)\}$.

Theorem 4.39. *If $a \in PC$ and $T(a)$ is a Fredholm operator of index k, then $\{T_n^\#(a)\} + \mathcal{N}$ is the Moore-Penrose inverse of $\{T_n(a)\} + \mathcal{N}$ in $\mathcal{F}/\mathcal{N}$, i.e., $\{B_n\} := \{T_n^\#(a)\}$ belongs to $\mathcal{F}$ and satisfies* (4.40) *and* (4.41). *Moreover, $T_n^\#(a) \to T^+(a)$ strongly.*

Proof. From (4.50) and Theorem 4.5 we see that $\{T_n^\#(a)\} \in \mathcal{F}$. We have

$$\begin{aligned} T_n(a) T_n^\#(a) T_n(a) - T_n(a) &= U_n S_n S_n^\# S_n V_n - U_n S_n V_n \\ &= -U_n \operatorname{diag}\Big(s_1(T_n(a)), \ldots, s_{|k|}(T_n(a)), 0, \ldots, 0\Big) V_n, \end{aligned}$$

and thus

$$\big\|T_n(a) T_n^\#(a) T_n(a) - T_n(a)\big\| = s_{|k|}\big(T_n(a)\big) \to 0$$

by Theorem 4.5. An easy computation gives

$$\begin{aligned} &T_n^{\#}(a)\,T_n(a)\,T_n^{\#}(a) - T_n^{\#}(a) = 0, \\ &\big(T_n^{\#}(a)\,T_n(a)\big)^* - T_n^{\#}(a)\,T_n(a) = 0, \\ &\big(T_n(a)\,T_n^{\#}(a)\big)^* - T_n(a)\,T_n^{\#}(a) = 0. \end{aligned}$$

Finally, Theorem 4.32 shows that $T_n^{\#}(a) \to T^+(a)$ strongly. ■

Kato numbers. Once more suppose $a \in PC$ and that $T(a)$ is Fredholm of index k. Then

$$d_0 := \liminf_{n\to\infty} s_{|k|+1}\big(T_n(a)\big) > 0 \tag{4.51}$$

by virtue of Theorem 4.5. In Section 4.7 we defined the Kato number $\kappa(A)$ of a normally solvable operator as the inverse of the smallest nonzero singular value of A, i.e., as the inverse of the smallest number in the set

$$(\mathrm{sp}(A^*A)\setminus\{0\})^{1/2},$$

and Theorem 4.23 says that $\kappa(A) = \|A^+\|$. The following theorem identifies the d_0 in (4.51) as the inverse of the Kato number of $T(a)$.

Theorem 4.40. *If $a \in PC$ and $T(a)$ is Fredholm of index k, then*

$$\lim_{n\to\infty} s_{|k|+1}\big(T_n(a)\big) = \|T^+(a)\|^{-1} \tag{4.52}$$

or, equivalently,

$$\lim_{n\to\infty} \|T_n^{\#}(a)\| = \|T^+(a)\|. \tag{4.53}$$

Proof. The equivalence of (4.52) and (4.53) follows from (4.50). Put

$$d_n := s_{|k|+1}\big(T_n(a)\big) = \|T_n^{\#}(a)\|^{-1}, \qquad d := \|T^+(a)\|^{-1}.$$

Then $\Sigma(T(a)) \subset \{0\} \cup [d,\infty)$ by Theorem 4.23. Since $d_n \in \Sigma(T_n(a))$, we infer from Corollary 4.19 that

$$d_0 := \liminf_{n\to\infty} d_n \in \Sigma\big(T(a)\big) \subset \{0\} \cup [d,\infty).$$

As $d_0 > 0$ by Theorem 4.5, it follows that $d_0 \geq d$. Thus,

$$\limsup_{n\to\infty} \|T_n^{\#}(a)\| = \frac{1}{d_0} \leq \frac{1}{d} = \|T^+(a)\|. \tag{4.54}$$

From Theorems 4.39 and 2.1 we obtain

$$\|T^+(a)\| \leq \liminf_{n\to\infty} \|T_n^{\#}(a)\|. \tag{4.55}$$

Obviously, (4.54) and (4.55) imply (4.53). ■

If $k = 0$, then $T_n^{\#}(a) = T_n^{-1}(a)$ for all sufficiently large n and $T^+(a) = T^{-1}(a)$. Hence, in this case (4.53) goes over into (3.16). We emphasize again that the equality

$$\lim_{n\to\infty} \|T_n^+(a)\| = \|T^+(a)\|$$

is in general NOT TRUE. It is definitely not true if $a \in PC\setminus C$ and $T(a)$ is Fredholm with nonzero index, since in this case

$$\limsup_{n\to\infty} \|T_n^+(a)\| = +\infty$$

by virtue of Corollary 4.38.

Notes. Theorem 4.39 was established by Roch and one of the authors in [143]. Theorem 4.40 is perhaps new.

5

Determinants and Eigenvalues

5.1 The Strong Szegö Limit Theorem

We now study the behavior of the determinants

$$D_n(a) := \det T_n(a) := \det \begin{pmatrix} a_0 & a_{-1} & \dots & a_{-(n-1)} \\ a_1 & a_0 & \dots & a_{-(n-2)} \\ \vdots & \vdots & \ddots & \vdots \\ a_{n-1} & a_{n-2} & \dots & a_0 \end{pmatrix}$$

as n goes to infinity. The strong Szegö limit theorem says that, after appropriate normalization, the determinants $D_n(a)$ approach a nonzero limit provided a is sufficiently smooth and $T(a)$ is invertible. Before stating and proving this theorem, we need a few more auxiliary facts.

Hilbert-Schmidt and trace class operators. These are especially nice compact operators and they may be defined in various ways. One definition is as follows: an operator K on a separable Hilbert space is called a *Hilbert-Schmidt operator* if there is an orthonormal basis with respect to which A has a matrix representation (a_{jk}) such that $\sum_{j,k} |a_{jk}|^2 < \infty$, and a *trace class* (or *nuclear*) operator is an operator which can be represented as the product of two Hilbert-Schmidt operators.

One can show that if K is a Hilbert-Schmidt operator, then the (finite) number $\sum_{j,k} |a_{jk}|^2$ is independent of the particular choice of the orthonormal basis which gives the matrix representation.

Another definition has recourse to the approximation numbers. Given a separable Hilbert space H and an operator $K \in \mathcal{B}(H)$, the nth *approximation number* is defined as

$$\sigma_n(K) := \inf\{\|K - F\| : F \in \mathcal{K}(H),\ \dim \operatorname{Im} F \le n\}.$$

Note that if $H = \mathbf{C}^n$, then the numbers $\sigma_n(K)$ are just the singular values of K in reversed order: from Theorem 4.2 we see that $\sigma_j(K) = s_{n-j}(K)$ $(j = 0, \ldots, n-1)$. Clearly, K is compact if and only if $\sigma_n(K) \to 0$ as $n \to \infty$. For $1 \le p < \infty$, the set of all operators K for which

$$\|K\|_p := \left(\sum_{n=0}^{\infty} \left(\sigma_n(K)\right)^p\right)^{1/p} < \infty \tag{5.1}$$

is denoted by $\mathcal{K}_p(H)$ and called the pth *Schatten class* (or *Schatten ideal*) of H. The Hilbert-Schmidt operators are the operators in $\mathcal{K}_2(H)$, while the trace class operators are the operators in $\mathcal{K}_1(H)$. We remark that $\mathcal{K}_p(H)$ is a (non-closed) ideal of $\mathcal{B}(H)$. Obviously, $\mathcal{K}_1(H) \subset \mathcal{K}_2(H) \subset \mathcal{K}(H)$. We also note that $\mathcal{K}_p(H)$ is a Banach space with the norm (5.1).

Example 5.1: Hilbert-Schmidt Hankel operators. An operator A in $\mathcal{B}(l^2)$ given by an infinite matrix $(a_{jk})_{j,k=1}^{\infty}$ is Hilbert-Schmidt if and only if $\sum |a_{jk}|^2 < \infty$. Hence,

$$H(a) \in \mathcal{K}_2(l^2) \iff \sum_{n=1}^{\infty} n|a_n|^2 < \infty,$$

$$H(\tilde{a}) \in \mathcal{K}_2(l^2) \iff \sum_{n=1}^{\infty} n|a_{-n}|^2 < \infty,$$

$$\{H(a), H(\tilde{a})\} \subset \mathcal{K}_2(l^2) \iff \sum_{n=-\infty}^{\infty} |n|\,|a_n|^2 < \infty.$$

Clearly, $\sum |n|\,|a_n|^2 < \infty$ if and only if $\sum (|n|+1)|a_n|^2 < \infty$. The latter sum is usually preferred because it represents a norm (while the former vanishes for constant functions). The set of all functions $a \in L^2$ for which $\sum(|n|+1)|a_n|^2$ is finite is nowadays denoted by $B_2^{1/2}$ and is referred to as a Besov space. Taking into account Proposition 1.12, we see in particular that if $a, b \in L^\infty$ are sufficiently smooth, e.g., if

$$\sum_{n \in \mathbf{Z}} (|n|+1)|a_n|^2 < \infty \quad \text{and} \quad \sum_{n \in \mathbf{Z}} (|n|+1)|b_n|^2 < \infty,$$

then the semi-commutator $T(ab) - T(a)T(b) = H(a)H(\tilde{b})$ is a trace class operator. ■

Krein algebras. It was Mark Krein [111] who first discovered that the set

$$L^\infty \cap B_2^{1/2} := \left\{ a \in L^\infty : \sum_{n\in\mathbf{Z}} \big(|n|+1\big)|a_n|^2 < \infty \right\}$$

is a (non-closed) subalgebra of L^∞. Indeed, for $a, b \in L^\infty$ we have

$$H(ab) = H(a)T(\tilde{b}) + T(a)H(b); \tag{5.2}$$

with the operators P and Q of Section 1.4, this is nothing but the identity $PabQ = PaQbQ + PaPbQ$. Thus, if $H(a)$ and $H(b)$ are in $\mathcal{K}_2(l^2)$, then so also is $H(ab)$, which implies that $L^\infty \cap B_2^{1/2}$ is an algebra.

From what was said in the previous paragraph we see that

$$W \cap B_2^{1/2} := \left\{ a \in W : \sum_{n\in\mathbf{Z}} \big(|n|+1\big)|a_n|^2 < \infty \right\}$$

is a (non-closed) subalgebra of the Wiener algebra W. If $a \in W \cap B_2^{1/2}$ has no zeros on $\mathbf{T}$, then $a^{-1} \in W$ by virtue of Wiener's theorem. Actually, a^{-1} again belongs to $W \cap B_2^{1/2}$. To see this, notice first that we may without loss of generality assume that $\operatorname{wind}(a,0) = 0$. From (5.2) we obtain

$$0 = H(a^{-1}a) = H(a^{-1})T(\tilde{a}) + T(a^{-1})H(a)$$

and since $T(\tilde{a})$ is invertible, it follows that

$$H(a^{-1}) = -T(a^{-1})H(a)T^{-1}(\tilde{a}) \in \mathcal{K}_2(l^2).$$

Analogously we get $H(\tilde{a}^{-1}) \in \mathcal{K}_2(l^2)$.

One can show that the norm

$$\|a\| := \sum_{n\in\mathbf{Z}} |a_n| + \left(\sum_{n\in\mathbf{Z}} \big(|n|+1\big)|a_n|^2 \right)^{1/2}$$

makes $W \cap B_2^{1/2}$ into a Banach algebra whose maximal ideal space may be identified with $\mathbf{T}$. This implies that if $a \in W \cap B_2^{1/2}$ has no zeros on $\mathbf{T}$ and $\operatorname{wind}(a,0) = 0$, then $a = e^b$ with some $b \in W \cap B_2^{1/2}$. Equivalently, a has a logarithm $\log a$ in $W \cap B_2^{1/2}$, i.e.,

$$\sum_{n\in\mathbf{Z}} |(\log a)_n| + \sum_{n\in\mathbf{Z}} \big(|n|+1\big)|(\log a)_n|^2 < \infty.$$

This inequality tells us that the two terms

$$G(a) := \exp(\log a)_0, \tag{5.3}$$

$$E(a) := \exp \sum_{k=1}^{\infty} k(\log a)_k(\log a)_{-k} \tag{5.4}$$

are well-defined (and independent of the particular choice of $\log a$ in the algebra $W \cap B_2^{1/2}$).

Theorem 5.2 (Szegö's strong limit theorem). *If $a \in W \cap B_2^{1/2}$ has no zeros on $\mathbf{T}$ and $\text{wind}(a, 0) = 0$, then*

$$\lim_{n\to\infty} \frac{D_n(a)}{G(a)^n} = E(a). \tag{5.5}$$

Since, obviously, $D_n(\gamma a) = \gamma^n D_n(a)$ for every $\gamma \in \mathbf{C}$, formula (5.5) says that after replacing a by $(1/G(a))a$ ("normalization") the determinants converge to $E(a)$.

The rest of this section is devoted to the main steps of the proof of Theorem 5.2.

Operator determinants. Let $K \in \mathcal{K}_1(H)$ be a trace class operator and let $\{\lambda_j(K)\}$ denote the sequence of its eigenvalues counted up to algebraic multiplicity, which is the dimension of the linear space

$$\{x \in H : (K - \lambda_j(K))^n x = 0 \text{ for some } n \geq 0\}.$$

Weyl's inequality states that

$$\sum_j |\lambda_j(K)| \leq \sum_n |\sigma_n(K)| =: \|K\|_1 < \infty,$$

and hence the *determinant*

$$\det(I + K) := \prod_j \big(1 + \lambda_j(K)\big) \tag{5.6}$$

is well-defined. If $H = \mathbf{C}^n$, this is the usual determinant of $I + K$. For general separable Hilbert spaces H, the determinant (5.6) has a lot of properties we know for the determinant of finite matrices. In what follows we need the fact that the map

$$\mathcal{K}_1(H) \to \mathbf{C}, \quad K \mapsto \det(I + K)$$

is continuous (provided $\mathcal{K}_1(H)$ is equipped with the trace norm $\|\cdot\|_1$ given by (5.1)). We will also employ the trace class version of Lemma 2.8, which says that if $K \in \mathcal{K}_1(H)$, $B_n \to B$ strongly, $C_n^* \to C^*$ strongly, then $\|B_n K C_n - BKC\|_1 \to 0$.

The *trace* of an operator $K \in \mathcal{K}_1(H)$ is defined as

$$\text{tr} K := \sum_j \lambda_j(K).$$

One can show that if $K \in \mathcal{K}_1(H)$ and $\{e_i\}$ is any orthonormal basis of H, then

$$\operatorname{tr} K = \sum_i (Ke_i, e_i), \tag{5.7}$$

the latter series always being absolutely convergent.

In Section 2.4 we observed that it is useful to decompose a given operator into an operator for which the finite section method converges and a compact operator. The following simple result fits in with this philosophy.

Lemma 5.3. *Let $A \in \mathcal{B}(l^2)$ be invertible, suppose $\{P_nAP_n\}$ is a stable sequence, and let $K \in \mathcal{K}_1(l^2)$ be a trace class operator. Then*

$$\lim_{n\to\infty} \frac{\det P_n(A+K)P_n}{\det P_nAP_n} = \det(I + A^{-1}K).$$

Here we have the usual determinant of $n \times n$ matrices on the left and the determinant of operators in $I + \mathcal{K}_1(l^2)$ defined above on the right.

Proof. Put $A_n = P_nAP_n$ and $K_n = P_nKP_n$. Since A_n is invertible for all sufficiently large n and A_n^{-1} ($:= A_n^{-1}P_n$) converges strongly to A^{-1} (Propositions 2.2 and 2.4), it follows that $A_n^{-1}K_n = A_n^{-1}P_nKP_n$ converges to $A^{-1}K$ in the trace norm, and consequently,

$$\frac{\det(A_n + K_n)}{\det A_n} = \det(I + A_n^{-1}K_n) \to \det(I + A^{-1}K). \blacksquare$$

Thus, if we know the asymptotic behavior of $\det P_nAP_n$, then the preceding lemma says something about the asymptotic behavior of the determinants $\det P_n(A+K)P_n$. Here is the way Lemma 5.3 can be used to prove Szegö's theorem: since

$$I = T(aa^{-1}) = T(a)T(a^{-1}) + H(a)H(\tilde{a}^{-1})$$

and thus

$$\begin{aligned} T(a) &= T^{-1}(a^{-1}) - H(a)H(\tilde{a}^{-1})T^{-1}(a^{-1}) \\ &= T^{-1}(a^{-1}) + \text{trace class operator}, \end{aligned}$$

we may try our luck with applying Lemma 5.3 to $A = T^{-1}(a^{-1})$ (recall Theorem 2.10). This works indeed.

Proposition 5.4. *Let $a \in W$ and suppose $T(a)$ is invertible. Then*

$$\det P_nT^{-1}(a^{-1})P_n = G(a)^n \quad \textit{for all } n \geq 1.$$

Proof. Let $a = a_-a_+$ be a Wiener-Hopf factorization of a (Theorem 1.14). Then $T^{-1}(a^{-1}) = T(a_+)T(a_-)$, so

$$P_nT^{-1}(a^{-1})P_n = P_nT(a_+)T(a_-)P_n = P_nT(a_+)P_nT(a_-)P_n,$$

whence

$$\det P_n T^{-1}(a^{-1}) P_n = \det T_n(a_+) \det T_n(a_-).$$

Since $T_n(a_+)$ and $T_n(a_-)$ are triangular, we see that $\det P_n T^{-1}(a^{-1}) P_n$ is $C(a)^n$ with $C(a) := (a_+)_0 (a_-)_0$. Writing $a_+ = e^{b_+}$ and $a_- = e^{b_-}$ with $b_+ \in W_+$ and $b_- \in W_-$, we get

$$C(a) = e^{(b_+)_0} e^{(b_-)_0} = e^{(b_+ + b_-)_0} = e^{(\log a)_0} = G(a). \blacksquare$$

Notice that the previous proposition gives us the determinants of *all* finite sections of the inverse of a Toeplitz matrix with a Wiener symbol.

Proof of Theorem 5.2. Put

$$A = T^{-1}(a^{-1}) \quad \text{and} \quad K = -H(a)H(\tilde{a}^{-1})T^{-1}(a^{-1})$$

and apply Theorem 2.10, Lemma 5.3, and Proposition 5.4. What results is that

$$\begin{aligned} \lim_{n \to \infty} \frac{D_n(a)}{G(a)^n} &= \det\big(I - T(a^{-1})H(a)H(\tilde{a}^{-1})T^{-1}(a^{-1})\big) \\ &= \det\big(I - T(a^{-1})\big(I - T(a)T(a^{-1})\big)T^{-1}(a^{-1})\big) \\ &= \det T(a^{-1})T(a). \end{aligned}$$

We remark that $T(a^{-1})T(a) - I = H(a^{-1})H(\tilde{a})$ is of trace class due to the assumption that a be in $W \cap B_2^{1/2}$. We are left with showing that $\det T(a^{-1})T(a)$ equals the expression $E(a)$ given by (5.4). As first observed by Widom, this follows from writing

$$\begin{aligned} T(a^{-1})T(a) &= T(a_-^{-1})T(a_+^{-1})T(a_-)T(a_+) \\ &= e^{-T(\log a_-)} e^{-T(\log a_+)} e^{T(\log a_-)} e^{T(\log a_+)} \end{aligned}$$

and invoking the remarkable identity

$$\det e^A e^B e^{-A} e^{-B} = \exp \operatorname{tr}(AB - BA), \tag{5.8}$$

which was shown by Helton, Howe, and Pincus to be valid whenever A and B are bounded Hilbert space operators for which $AB - BA$ is of trace class. Notice that

$$\begin{aligned} &\operatorname{tr}\big(T(\log a_-)T(\log a_+) - T(\log a_+)T(\log a_-)\big) \\ &\quad = \operatorname{tr}\big(H(\log a_+)H(\log a_-)^{\sim}\big) = \sum_{k=1}^{\infty} k (\log a_+)_k (\log a_-)_{-k} \\ &\quad = \sum_{k=1}^{\infty} k (\log a)_k (\log a)_{-k}. \blacksquare \end{aligned}$$

Notes. For Hilbert-Schmidt and trace class operators and for operator determinants we refer the reader to the books [85], [131], [132], [139] and to the articles [82], [157].

The history of Theorem 5.2 is long. In 1952, Szegö [166] proved the theorem for positive symbols which have a derivate satisfying a Hölder condition. This smoothness condition was step by step relaxed by many mathematicians, including M. Kac (1954), G. Baxter (1961), I.I. Hirschman (1966), M.G. Krein (1966), Ya.L. Geronimus (1967), A. Devinatz (1967) (see [39] for precise references). In 1971, Golinski and Ibragimov [89] finally proved that (5.5) is true whenever it makes sense (still requiring that a be positive!).

Baxter [16] and Hirschman [102] were the first to replace the positivity of a by the condition that $a(t) \neq 0$ for $t \in \mathbf{T}$ and $\text{wind}(a, 0) = 0$. Theorem 5.2 as it is stated here was established by Hirschman [102] ($a \in W \cap B_2^{1/2}$) and Devinatz [55] ($a \in C \cap B_2^{1/2}$) using different methods. The development culminated with Widom's paper [185], in which (the block case version of) Theorem 5.2 was proved under the sole assumption that a be in $L^\infty \cap B_2^{1/2}$. The proof given here is also Widom's. Basor [12] writes: "The proofs of the various Szegö theorems were for the most part difficult, indirect, and worst of all gave no 'natural' indication why the terms in the expansion, especially the $E(a)$, occurred. Fortunately, this state of affairs was considerably altered in 1976 by Widom [185], whose elegant application of ideas from operator theory extended Szegö's theorem to the block case and gave easy proofs of the results."

A short and self-contained proof of the formula (5.8) is in T. Ehrhardt's dissertation [69, pp. 45–47].

For further development of the matter after Widom the reader is referred to [35], [39] and the references there. For example, in [35] we showed that if $a \in L^\infty$, $T(a)$ is invertible, and

$$\sum_{n=1}^{\infty} n^{2\alpha} |a_{-n}|^2 + \sum_{n=1}^{\infty} n^{2\beta} |a_n|^2 < \infty \qquad (\alpha > 0,\ \beta > 0,\ \alpha + \beta = 1),$$

then (5.5) holds. Lemma 5.3 appeared explicitly in [21] for the first time, but it is implicit already in [185] and [11]. Finally, papers [106] and [107] by Johansson provide one more approach to Szegö's strong limit theorem and shed new light on the topic.

5.2 Ising Model and Onsager Formula

In this section we present a few pieces of the story behind Szegö's limit theorem and, at the same time, a concrete example to Theorem 5.2.

It turns out that if a ferromagnet is heated, its magnetization deceases. This is no surprise, since we may think of the magnet as consisting of millions and millions of adjusted elementary magnets that are thrown into disorder in the process of heating. However, there is a temperature, the so-called *Curie point* T_c (which equals 1043 Kelvin degrees for iron), at which the magnetization disappears completely and above which it remains zero. If, conversely, a hot ferromagnet is cooled down, magnetization appears at the Curie point and increases as the temperature decreases. In other words, we would understand a behavior as in Figure 31a, but in fact we observe a behavior as in Figure 31b.

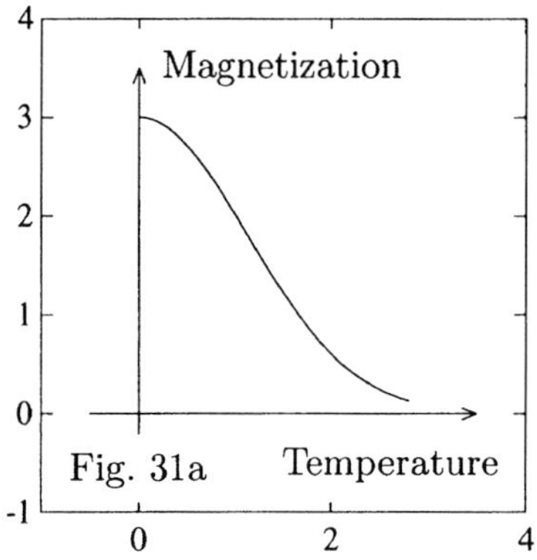

Fig. 31a

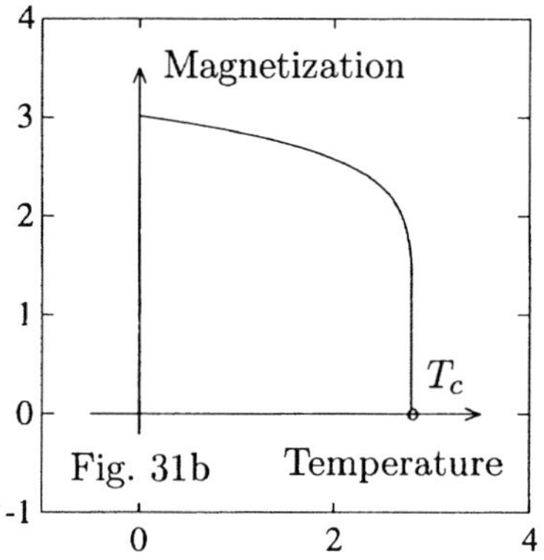

Fig. 31b

What we encounter is a so-called phase transition. Statistical physics provides the apparatus for explaining such macroscopic phenomena by microscopic interactions. In the case at hand, one models a magnet by a large number of elementary magnets located at the sites of the lattice $\mathbf{Z}^d$. Of course, in the real-world case d equals 3. Minimization of energy causes the elementary magnets to point in the same direction, while maximization of entropy effects the opposite tendency. The so-called *Ising model* includes the theory of computing the correlation between two elementary magnets at a large distance ($n = 10^{23}$, say) via the interaction of neighboring elementary magnets. The presence of such a correlation ("long-range order") means that we observe a global magnetic field, whereas absence of a correlation is interpreted as lack of macroscopic magnetization. The temperature T enters the formula for the correlation as a parameter, and hence the problem is to show that the correlation is positive for $T < T_c$ and zero for $T \geq T_c$.

The disappointing but pioneering 1925 result of Ernst Ising was that there is no phase transition, i.e., no Curie point, for a "one-dimensional magnet" ($d = 1$). As far as we know, until now no rigorous *and* satisfactory results are available in three (or more) dimensions. However, in 1948, Lars Onsager (who received the Nobel Prize for Chemistry in 1968) surprised the physicists with a beautiful formula for the magnetization of a two-dimensional ferromagnet ($d = 2$). This formula implied that a two-dimensional ferromagnet has a Curie point T_c and that its magnetization decays as $(T_c - T)^{1/8}$ as T approaches T_c from below.

Onsager has never published a proof of his formula, but from his correspondence with Bruria Kaufman it is known that he derived the formula by considering Toeplitz determinants.

So what has a hot magnet to do with Toeplitz determinants? Throughout what follows we consider the two-dimensional Ising model, i.e., we let $d = 2$. The interaction between two neighboring elementary magnets at the sites (j, k) and $(j + 1, k)$ (or (j, k) and $(j, k + 1)$) is expressed by a quantity J_x (and J_y, respectively); see Figures 32 and 33. The correlation $\langle \sigma_{0,0}\, \sigma_{0,n} \rangle$

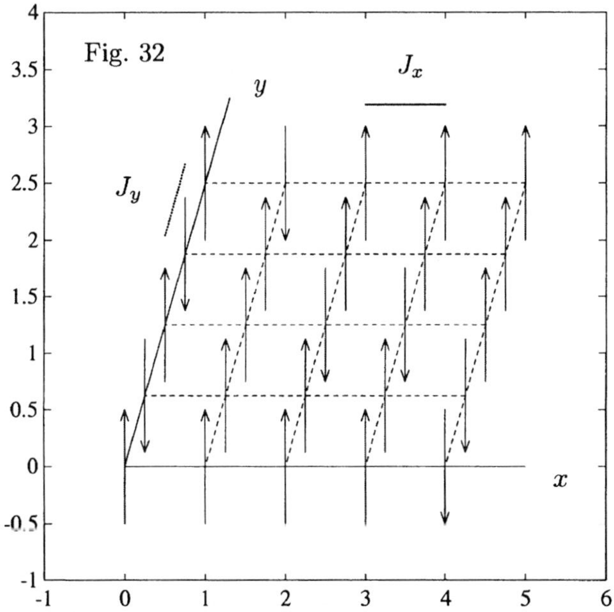

between elementary magnets at the sites $(0, 0)$ and $(0, n)$ (n large) is a huge sum indexed by the possible positions ("spin up" or "spin down") of the elementary magnets. Each term of this sum is a product and involves the parameters J_x, J_y and the temperature T. And what is the determinant of a large matrix? It is also a huge sum of products. The reader's guess is right: one can indeed show that the correlation $\langle \sigma_{0,0} \sigma_{0,n} \rangle$ may be expressed as the determinant of some Toeplitz (!) matrix,

$$\langle \sigma_{0,0}\, \sigma_{0,n} \rangle = D_n(a). \tag{5.9}$$

As to our knowledge, it was Montroll, Potts, and Ward [122] who first rigorously proved such an equality. A physically relevant quantity, the so-called *spontaneous magnetization* M, is given by

$$M^2 = \lim_{n\to\infty} \langle \sigma_{0,0}\, \sigma_{0,n} \rangle. \tag{5.10}$$

Now Szegö enters the scene. In 1915, he had published his "first" theorem (Theorem 5.8). At the turn of the 1950s, he was told by S. Kakutani that

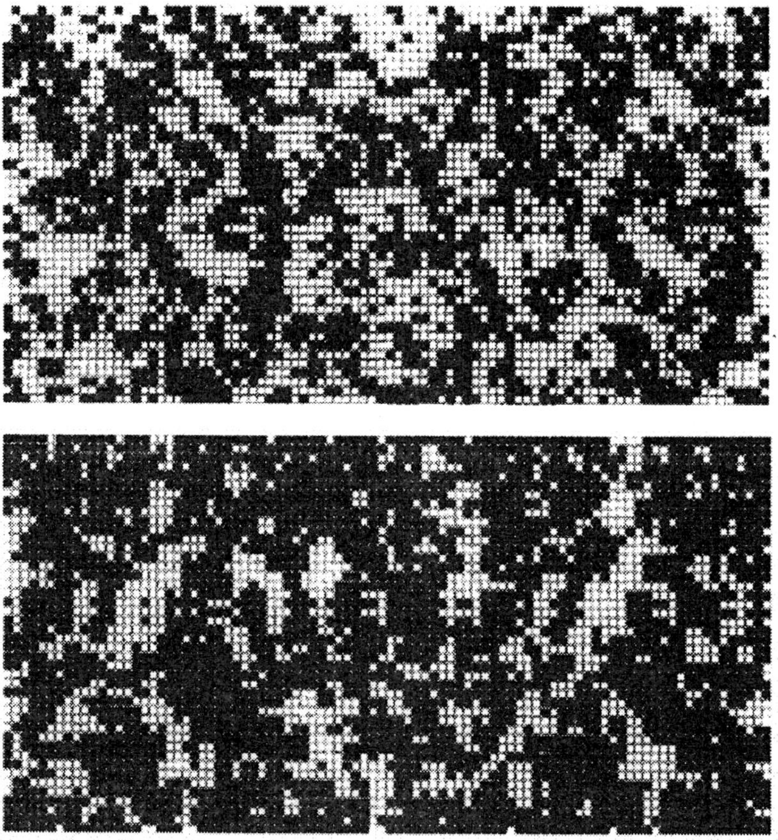

Figure 33 (reprinted from [24] and thus from [191] with the permission of Spektrum der Wissenschaft Verlagsgesellschaft mbH, Heidelberg)

The elementary magnets at the points of a lattice are indicated by small squares. A black square corresponds to the North pole (spin up), while a white square is associated with the South pole (spin down). At temperatures above the Curie point T_c (upper picture), we see nearly the same number of black and white squares, which means that the summary action of the elementary magnets is zero. Below the Curie point (lower picture), one of the two possible orientations of the elementary magnets gets the upper hand and therefore a global magnetic field emerges.

in order to prove Onsager's formula it would be desirable to have a second-order asymptotic formula for $D_n(a)$. And Szegö was able to do this and to establish his "strong" theorem (Theorem 5.2). This theorem led to significant simplifications in the proof of Onsager's formula.

The symbol a of the Toeplitz determinant in (5.9) is

$$a(t) = \left(\frac{1-At}{1-At^{-1}}\right)^{1/2}\left(\frac{1-Bt^{-1}}{1-Bt}\right)^{1/2} \qquad (t \in \mathbf{T}), \tag{5.11}$$

where A and B are positive constants depending on J_x, J_y, T:

$$A = \alpha\frac{1-\beta}{1+\beta}, \qquad B = \frac{1}{\alpha}\frac{1-\beta}{1+\beta}, \tag{5.12}$$

$$\alpha = \tanh\frac{J_x}{kT}, \qquad \beta = \tanh\frac{J_y}{kT}, \tag{5.13}$$

and k is Boltzmann's constant. Clearly, we always have $A \in (0,1)$, but it may happen that $B > 1$. Suppose for a moment that J_x, J_y, T are so that $B \in (0,1)$, i.e., suppose

$$\frac{1}{\alpha}\frac{1-\beta}{1+\beta} < 1. \tag{5.14}$$

Then each of the functions in the parantheses of (5.11) is positive and hence, so also is a. The function a belongs to C^∞ and therefore Theorem 5.2 (for positive symbols!) can be employed. We have

$$\begin{aligned}\log a(t) &= \frac{1}{2}\log(1-At) - \frac{1}{2}\log(1-At^{-1}) + \frac{1}{2}\log(1-Bt^{-1}) - \frac{1}{2}\log(1-Bt)\\ &= \sum_{k-1}^{\infty}\frac{1}{2}\left(\frac{A^k}{k} - \frac{B^k}{k}\right)t^{-k} + \sum_{k=1}^{\infty}\frac{1}{2}\left(\frac{-A^k}{k} + \frac{B^k}{k}\right)t^k\end{aligned}$$

and thus, $(\log a)_0 = 0$ and

$$\begin{aligned}\sum_{k=1}^{\infty} k(\log a)_{-k}(\log a)_k &= \frac{1}{4}\sum_{k=1}^{\infty}\frac{1}{k}\left(-A^{2k} - B^{2k} + 2A^kB^k\right)\\ &= \frac{1}{4}\log(1-A^2) + \frac{1}{4}\log(1-B^2) - \frac{1}{2}\log(1-AB).\end{aligned}$$

Consequently, $G(a) = 1$ and Theorem 5.2 gives

$$\lim_{n\to\infty} D_n(a) = \left[\frac{(1-A^2)(1-B^2)}{(1-AB)^2}\right]^{1/4}. \tag{5.15}$$

Inserting (5.12) and (5.13) in the term in the brackets of (5.15) we see that

$$\left(1-\alpha^2\left(\frac{1-\beta}{1+\beta}\right)^2\right)\left(1-\frac{1}{\alpha^2}\left(\frac{1-\beta}{1+\beta}\right)^2\right)\left(1-\left(\frac{1-\beta}{1+\beta}\right)^2\right)^{-2}$$

equals

$$\frac{\big((1+\beta)^2-\alpha^2(1-\beta)^2\big)\big(\alpha^2(1+\beta)^2-(1-\beta)^2\big)}{16\alpha^2\beta^2}$$
$$=\frac{\big((1+\beta)+\alpha(1-\beta)\big)\big(\alpha(1+\beta)-(1-\beta)\big)\big((1+\beta)-\alpha(1-\beta)\big)\big(\alpha(1+\beta)+(1-\beta)\big)}{16\alpha^2\beta^2}$$
$$=\frac{\big(4\alpha\beta-(1-\alpha^2)(1-\beta^2)\big)\big(4\alpha\beta+(1-\alpha^2)(1-\beta^2)\big)}{16\alpha^2\beta^2}$$
$$=\frac{16\alpha^2\beta^2-(1-\alpha^2)^2(1-\beta^2)^2}{16\alpha^2\beta^2}=1-\frac{(1-\alpha^2)}{4\alpha^2}\frac{(1-\beta^2)^2}{4\beta^2}$$
$$=1-\frac{1}{\sinh^2(2J_x/kT)\sinh^2(2J_y/kT)}. \tag{5.16}$$

Thus, by (5.9), (5.10), and (5.15),

$$M=M(T)=\left[1-\left(\sinh\frac{2J_x}{kT}\sinh\frac{2J_y}{kT}\right)^{-2}\right]^{1/8}. \tag{5.17}$$

This is *Onsager's formula.*

Now recall that we derived (5.17) under the assumption (5.14). From the identity (5.16) we see that (5.14) is equivalent to the condition

$$1-\left(\sinh\frac{2J_x}{kT}\sinh\frac{2J_y}{kT}\right)^{-2}>0,$$

i.e., we arrive at the conclusion that (5.17) is true whenever the term in brackets is positive. We have $M(0)=1$ and $M(T)$ decreases monotonically to zero as T increases. The spontaneous magnetization $M(T)$ becomes zero if $T=T_c$ is the solution of the equation

$$1=\sinh\frac{2J_x}{kT_c}\sin\frac{2J_y}{kT_c}. \tag{5.18}$$

In summary, from (5.18) we get the Curie point T_c, and (5.17) describes $M(T)$ for $T<T_c$. Obviously, from (5.17) it follows that $M(T)$ decays as $(T_c-T)^{1/8}$ as T approaches T_c from below.

Note. The material of this section is from McCoy and Wu's book [121] and from Fisher and Hartwig's paper [72]. The paper [25] might also be of some interest in connection with the topics discussed here.

5.3 Second-Order Trace Formulas

In accordance with (5.7), the trace of an $n\times n$ matrix $A=(a_{jk})_{j,k=1}^n$ is defined as $\operatorname{tr}A=a_{11}+\ldots+a_{nn}$. Clearly, $\operatorname{tr}T_n(a)=na_0$. We denote by

$\lambda_1^{(n)}, \ldots, \lambda_n^{(n)}$ the eigenvalues of $T_n(a)$ (counted up to algebraic multiplicity). Thus,

$$\lambda_1^{(n)} + \ldots + \lambda_n^{(n)} = na_0.$$

The strong Szegö limit theorem (Theorem 5.2) says that

$$\lim_{n\to\infty} \frac{\lambda_1^{(n)} \ldots \lambda_n^{(n)}}{G(a)^n} = E(a) \neq 0 \tag{5.19}$$

whenever $a \in W \cap B_2^{1/2}$ and $T(a)$ is invertible.

Lemma 5.5. *If $a \in PC$ and $\Omega \subset \mathbf{C}$ is any open set containing $\operatorname{sp} T(a)$, then there is an $n_0 = n_0(a, \Omega)$ such that $\operatorname{sp} T_n(a) \subset \Omega$ for all $n \geq n_0$.*

Proof. Let $\lambda \in \mathbf{C}\backslash\Omega$. Then, by Corollary 2.19,

$$\|T_n^{-1}(a-\lambda)\| \leq M(\lambda) < \infty$$

for all $n \geq m_0(\lambda)$. Hence, if $n \geq m_0(\lambda)$ and $|\mu-\lambda| < 1/M(\lambda)$, then $T_n(a-\mu)$ is invertible. A compactness argument now implies the assertion. ■

By the previous lemma, $f(T_n(a))$ is defined for all sufficiently large n if $a \in PC$ and f is analytic in some open neighborhood Ω of $\operatorname{sp} T(a)$ which has a smooth boundary $\partial\Omega$:

$$f\big(T_n(a)\big) := \frac{1}{2\pi i} \int_{\partial\Omega} f(\lambda)\big(\lambda I - T_n(a)\big)^{-1} d\lambda.$$

We then have

$$\operatorname{tr} f\big(T_n(a)\big) = f\big(\lambda_1^{(n)}\big) + \ldots + f\big(\lambda_n^{(n)}\big).$$

At least formally, (5.19) may be interpreted as a second-order asymptotic formula for $\operatorname{tr} \log T_n(a)$:

$$\operatorname{tr} \log T_n(a) = \sum_{j=1}^{n} \log \lambda_j^{(n)} = n \log G(a) + \log E(a) + o(1).$$

Theorem 5.6. *Let $a \in W \cap B_2^{1/2}$ and let Ω be an open set which contains $\operatorname{sp} T(a)$. If f is analytic in Ω, then*

$$\operatorname{tr} f\big(T_n(a)\big) = n\, G_f(a) + E_f(a) + o(1) \tag{5.20}$$

as $n \to \infty$, where $G_f(a)$ and $E_f(a)$ are certain constants. We have

$$G_f(a) = \big(f(a)\big)_0 = \frac{1}{2\pi} \int_0^{2\pi} f\big(a(e^{i\theta})\big)\, d\theta. \tag{5.21}$$

Proof. If n_0 is large enough, then $\operatorname{sp} T_n(a) \subset \Omega$ for all $n \geq n_0$ by Lemma 5.5 and $\mathcal{R}(a) \subset \operatorname{sp} T(a) \subset \Omega$ by Theorem 1.25. In particular, (5.21) is well-defined. Clearly, we may assume that Ω is bounded, that $\partial\Omega$ is a smooth curve, and that there is an open neighborhood U of $\partial\Omega$ such that $\operatorname{sp} T_n(a) \cap U = \emptyset$ for all $n \geq n_0$ and $\operatorname{sp} T(a) \cap U = \emptyset$. We can find a finite number of open disks $G_k \subset U$ covering $\partial\Omega$.

Let G be any of the disks G_k and choose branches of $\log(\lambda_j^{(n)} - \lambda)$ ($j = 1, \dots, n$) and $\log(a - \lambda)$ which depend analytically on $\lambda \in G$. The functions

$$\delta_n(\lambda) := \sum_{j=1}^{n} \log\big(\lambda_j^{(n)} - \lambda\big) - n \log G(a - \lambda) - \log E(a - \lambda)$$

are analytic in G. By (5.19), there exist integers l_n such that

$$\delta_n(\lambda) + 2\pi i l_n \to 0 \quad (\lambda \in G), \tag{5.22}$$

and a check of the proof of Theorem 5.2 shows that the convergence in (5.22) is uniform on compact subsets of G. Hence, $\delta_n'(\lambda) \to 0$ uniformly on compact subsets of G.

Divide $\partial\Omega$ into subarcs $\partial\Omega_k$ such that $\partial\Omega_k \subset G_k$ and put

$$\int_{\partial\Omega} g(\lambda)\, d\lambda := \sum_k \int_{\partial\Omega_k} g(\lambda)\, d\lambda.$$

From what was said in the preceding paragraph we obtain

$$\begin{aligned}
&\int_{\partial\Omega} f(\lambda) \frac{d}{d\lambda} \sum_{j=1}^{n} \log\big(\lambda_j^{(n)} - \lambda\big)\, d\lambda \\
&\quad = n \int_{\partial\Omega} f(\lambda) \frac{d}{d\lambda} \log G(a - \lambda)\, d\lambda + \int_{\partial\Omega} f(\lambda) \frac{d}{d\lambda} \log E(a - \lambda)\, d\lambda + o(1).
\end{aligned}$$

The left-hand side of this equality is

$$\sum_{j=1}^{n} \int_{\partial\Omega} \frac{f(\lambda)}{\lambda - \lambda_j^{(n)}}\, d\lambda = 2\pi i \sum_{j=1}^{n} f\big(\lambda_j^{(n)}\big),$$

and the first term on the right equals

$$\begin{aligned}
&n \int_{\partial\Omega} f(\lambda) \frac{d}{d\lambda} \frac{1}{2\pi} \int_0^{2\pi} \log\big(a(e^{i\theta}) - \lambda\big)\, d\theta\, d\lambda \\
&\quad = n \int_{\partial\Omega} f(\lambda) \frac{1}{2\pi} \int_0^{2\pi} \frac{d\theta}{\lambda - a(e^{i\theta})}\, d\lambda = ni \int_0^{2\pi} f\big(a(e^{i\theta})\big)\, d\theta.
\end{aligned}$$

This gives the assertion with

$$E_f(a) = \frac{1}{2\pi i} \int_{\partial\Omega} f(\lambda) \frac{d}{d\lambda} \log E(a - \lambda)\, d\lambda. \blacksquare$$

We remark that by virtue of (5.21) the formula (5.20) can also be written in the form

$$\operatorname{tr} f\big(T_n(a)\big) - \operatorname{tr} T_n\big(f(a)\big) \to E_f(a).$$

Widom [185] showed that if $a \in W \cap B_2^{1/2}$ is absolutely continuous, then

$$E_f(a) = \frac{1}{4\pi^2} \sum_{k=1}^{\infty} \int_0^{2\pi} \int_0^{2\pi} \frac{f\big(a(e^{i\theta})\big) - f\big(a(e^{i\varphi})\big)}{a(e^{i\theta}) - a(e^{i\varphi})} \times \sin\big(k(\theta-\varphi)\big)\big[a'(e^{i\varphi}) - a'(e^{i\theta})\big]\, d\theta\, d\varphi. \tag{5.23}$$

Piecewise smooth functions. For $0 < \beta < 1$, let C^β denote the set of all functions a on $\mathbf{T}$ which satisfy a Hölder condition with the exponent β:

$$|a(t_1) - a(t_2)| \le K|t_1 - t_2|^\beta \quad \text{for all } t_1, t_2 \in \mathbf{T}$$

with some constant $K < \infty$. For $m \in \mathbf{N}$, we let C^m stand for the m times continuously differentiable functions on $\mathbf{T}$. Finally, if $\alpha = m+\beta$ with $m \in \mathbf{N}$ and $0 < \beta < 1$, we let C^α refer to the m times continuously differentiable functions on $\mathbf{T}$ whose mth derivative belongs to C^β. It is well known that for every $\alpha > 0$,

$$a \in C^\alpha \implies a_n = O(1/n^\alpha) \text{ as } |n| \to \infty. \tag{5.24}$$

In particular, for every $\varepsilon > 0$, functions in $C^{1+\varepsilon}$ are absolutely continuous and belong to $W \cap B_2^{1/2}$.

We denote by PC^α $(\alpha > 0)$ the set of all functions $a \in PC$ with the following property: a has at most finitely many jumps, at $\{t_1, \ldots, t_n\}$ say, and for each connected component γ of $\mathbf{T} \setminus \{t_1, \ldots, t_n\}$ there exists a function $a_\gamma \in C^\alpha$ such that $a|\gamma = a_\gamma|\gamma$. Upon partially integrating the integral

$$a_n = \frac{1}{2\pi} \int_0^{2\pi} a(e^{i\theta}) e^{-in\theta}\, d\theta$$

we get

$$a \in C \cap PC^1 \implies a_n = \frac{1}{in}(a')_n = o\Big(\frac{1}{n}\Big). \tag{5.25}$$

Since $a' \in PC \subset L^2$ for $a \in PC^1$, we deduce from (5.25) that $C \cap PC^1$ is contained in $L^\infty \cap B_2^{1/2}$. Partial integration of the integral defining $(a')_n$ gives

$$a \in C \cap PC^2 \implies a_n = O\Big(\frac{1}{n^2}\Big). \tag{5.26}$$

This shows that $C \cap PC^2 \subset W \cap B_2^{1/2}$.

If $a \in L^\infty \cap B_2^{1/2}$, then $H(a)$ and $H(\tilde{a})$ are Hilbert-Schmidt and thus compact. From Theorem 1.16 we therefore get

$$a \in (C + H^\infty) \cap (C + \overline{H^\infty}) =: QC.$$

Functions in QC are called *quasicontinuous*. Thus, Theorems 5.2 and 5.6 do not apply to functions beyond QC.

Notes. For positive symbols a, a result like Theorem 5.6 was established by Libkind [117]. In the form cited here and with the expression (5.23) for $E_f(a)$, the theorem was proved by Widom [185]. The proof given above is Widom's. He also extended Theorem 5.6 to symbols $a \in L^\infty \cap B_2^{1/2}$.

5.4 The First Szegö Limit Theorem

As shown in the previous section, the strong Szegö limit theorem provides a second-order asymptotic formula for

$$\frac{1}{n}\operatorname{tr} f\big(T_n(a)\big) := \frac{1}{n}\sum_{j=1}^{n} f\big(\lambda_j^{(n)}\big). \tag{5.27}$$

The first Szegö limit theorem is simply concerned with the limit of (5.27). This does not imply that Szegö's first limit theorem is worse than his strong limit theorem: the latter is only applicable to "nearly continuous" symbols (to symbols in the Krein algebra $L^\infty \cap B_2^{1/2} \subset QC$), whereas the former covers much larger symbol classes.

Lemma 5.7. *Suppose $a \in L^\infty$, $T(a)$ is invertible, and $\{T_n(a)\}$ is stable. Then*

$$\lim_{n\to\infty} \frac{D_{n-1}(a)}{D_n(a)} = P_1 T^{-1}(a) P_1;$$

here, of course, $P_1 T^{-1}(a) P_1$ stands for the 1, 1 *entry of* $T^{-1}(a)$.

Proof. By Cramer's rule, $D_{n-1}(a)/D_n(a) = P_1 T_n^{-1}(a) P_1$. Propositions 2.2 and 2.4 show that this goes to $P_1 T^{-1}(a) P_1$. ■

The constant G(a). The previous lemma raises the problem of identifying $P_1 T^{-1}(a) P_1$. So assume $a \in L^\infty$ and $T(a)$ is invertible. From the proof of Theorem 2.10 we know that $P_1 T^{-1}(a) P_1 \neq 0$, and we therefore may define

$$G(a) := 1/\big(P_1 T^{-1}(a) P_1\big). \tag{5.28}$$

If $a \in W$, then Proposition 5.4 tells us that

$$1/\big(P_1 T^{-1}(a) P_1\big) = 1/\exp(\log a^{-1})_0 = \exp(\log a)_0,$$

which coincides with the right-hand side of (5.3). If $a \in L^\infty$ is real-valued and $\mathcal{R}(a) \subset (0, \infty)$, then a has a real-valued logarithm $\log a \in L^\infty$, and it is not difficult to show that then again the right-hand sides of (5.3) and (5.28) coincide. Finally, suppose $a \in L^\infty$ is locally sectorial on $\mathbf{T}$. The invertibility of $T(a)$ implies that $\text{wind}(a,0) = 0$ (Theorem 1.22). Hence $a = e^b e^{iv}$ with $b \in C$ and a real-valued $v \in L^\infty$ such that $\|v\|_\infty < \pi/2$ (Corollary 1.20). One can show that in this case (5.28) equals $\exp(\log a)_0$ with $\log a := b + iv$ (see, e.g., [39, Proposition 10.5(b)]).

Theorem 5.8 (Szegö's first limit theorem). *If $a \in L^\infty$ is locally sectorial on $\mathbf{T}$ and $T(a)$ is invertible, then*

$$\lim_{n\to\infty} \frac{D_n(a)}{D_{n-1}(a)} = G(a). \tag{5.29}$$

Proof. Immediate from Theorem 2.18, Lemma 5.7, and the definition of the constant $G(a)$. ■

Theorem 5.9 (another version of Szegö's first limit theorem). *Let $a \in L^\infty$ and let $\Omega \subset \mathbf{C}$ be an open set containing* $\operatorname{conv} \mathcal{R}(a)$. *If f is analytic in Ω, then*

$$\frac{1}{n}\operatorname{tr} f\big(T_n(a)\big) := \frac{1}{n}\sum_{j=1}^{n} f\big(\lambda_j^{(n)}\big) \to G_f(a) := \frac{1}{2\pi}\int_0^{2\pi} f\big(a(e^{i\theta})\big)\, d\theta$$

as $n \to \infty$.

Proof. From Proposition 2.17 we know that $\operatorname{sp} T_n(a) \subset \Omega$ for all $n \geq 1$. Let $G \subset \mathbf{C} \setminus \Omega$ be any open disk and choose any branch of

$$h_n(\lambda) := \sum_{j=1}^{n} \log\big(\lambda_j^{(n)} - \lambda\big)$$

which depends analytically on $\lambda \in G$. If $\lambda \in G$, then $a - \lambda$ is sectorial and hence,

$$D_n(a-\lambda)/D_{n-1}(a-\lambda) \to G(a-\lambda)$$

by Theorem 5.8. It follows that there are integers l_n such that

$$h_n(\lambda) - h_{n-1}(\lambda) - g(\lambda) + 2\pi i l_n \to 0 \quad (\lambda \in G), \tag{5.30}$$

where $g(\lambda)$ is any branch of

$$\log G(a-\lambda) = \frac{1}{2\pi}\int_0^{2\pi} \log\big(a(e^{i\theta}) - \lambda\big)\, d\theta$$

which depends analytically on $\lambda \in G$. From the proof of Lemma 5.7 and the definition of $G(a-\lambda)$ it is not difficult to see that the convergence in (5.30) is uniform on compact subsets of G.

Write $l_n = s_n - s_{n-1}$ with $s_n, s_{n-1} \in \mathbf{Z}$ and put $H_n(\lambda) := h_n(\lambda) + 2\pi i s_n$. Let K be any compact subset of G. Given any $\varepsilon > 0$, there is an n_0 such that

$$|H_j - H_{j-1} - g| < \varepsilon/2 \quad \text{on } K \text{ for all } j \geq n_0.$$

If $n \geq n_0$, then

$$\left|\frac{H_n}{n} - g\right| \leq \sum_{j=n_0}^{n} \frac{|H_j - H_{j-1} - g|}{n} + \sum_{j=1}^{n_0-1} \frac{|H_j - H_{j-1} - g|}{n} \quad \text{on } K \quad (5.31)$$

$(H_0(\lambda) := 0)$. The first term on the right-hand side of (5.31) is not greater than $(n - n_0 + 1)(\varepsilon/2)/n < \varepsilon/2$, and the second term on the right of (5.31) is smaller than $\varepsilon/2$ if only n is large enough. Thus, returning from H_n to h_n, we see that

$$(1/n)h_n(\lambda) - g(\lambda) + 2\pi i s_n/n \to 0 \qquad (5.32)$$

uniformly on compact subsets of G. We therefore can differentiate (5.32), and after multiplying the result by $f(\lambda)/(2\pi i)$ and integrating over $\partial\Omega$ (divided into pieces as in the proof of Theorem 5.6) we obtain that

$$\frac{1}{n}\sum_{j=1}^{n} f\big(\lambda_j^{(n)}\big) - \frac{1}{2\pi i}\int_{\partial\Omega} \frac{f(\lambda)}{2\pi}\int_0^{2\pi} \frac{d\theta}{\lambda - a(e^{i\theta})}\, d\lambda$$

$$= \frac{1}{n}\sum_{j=1}^{n} f\big(\lambda_j^{(n)}\big) - \frac{1}{2\pi}\int_0^{2\pi} f\big(a(e^{i\theta})\big)\, d\theta$$

must go to zero. ■

Notes. Szegö [165] proved (5.29) under the assumption that $a \in L^1$, $a \geq 0$, $\log a \in L^1$. Devinatz [54] established (5.29) for sectorial symbols. Gohberg and Feldman [80, Theorem III.2.2] used Lemma 5.7 to show that (5.29) holds if $a \in C$ and $T(a)$ is invertible. As they also had Theorem 2.18, we can say that Theorem 5.8 in the form presented here is already in [80]. In a sense, every result on the stability of $\{T_n(a)\}$ shows that the limit in (5.29) exists. Things are more difficult for unbounded symbols a, and the most general results in this direction we are aware of are due to Krein and Spitkovsky [112] and Spitkovsky [163]. Also see [39, Chapter 10].

Although the derivation of Theorem 5.9 from Theorem 5.8 looks quite complicated, it is a standard approach. The rule "differentiate-multiply-integrate" is the idea of Widom's proof of Theorem 5.6, and the other point of the proof is the well-known fact that if $D_n/D_{n-1} \to g$, then $\sqrt[n]{D_n} \to g$, which was already used by Grenander and Szegö [92, Section 5.2].

5.5 Hermitian Toeplitz Matrices

Since analytic functions cannot have "peaks", Theorem 5.9 does not tell us much about the asymptotic distribution of the eigenvalues of $T_n(a)$ as n goes to infinity. More can be said if $\operatorname{conv} \mathcal{R}(a)$ has no inner points.

Normal Toeplitz operators. A convex compact set does not have interior points if and only if it is a line segment. The Brown-Halmos result cited in Section 3.3 says that $\operatorname{conv}\mathcal{R}(a)$ is a line segment if and only if $T(a)$ is normal. Because $T^*(a) = T(\bar{a})$, it is clear that $T(a)$ is Hermitian (= Hermitean = selfadjoint) if and only if a is real-valued, i.e., if and only if $\operatorname{conv}\mathcal{R}(a) \subset \mathbf{R}$. As the spectrum of a unitary operator is always a subset of $\mathbf{T}$, we see that the only unitary Toeplitz operators are those for which $\operatorname{conv}\mathcal{R}(a)$ is a line segment contained in $\mathbf{T}$. Thus, $T(a)$ is unitary if and only if $T(a) = \mu I$ with $\mu \in \mathbf{T}$.

Theorem 5.10 (yet another version of Szegö's first limit theorem). *Let $a \in L^\infty$ and suppose $\operatorname{conv}\mathcal{R}(a)$ is a line segment. If f is any continuous function on $\operatorname{conv}\mathcal{R}(a)$, then*

$$\frac{1}{n}\operatorname{tr} f\big(T_n(a)\big) := \frac{1}{n}\sum_{j=1}^{n} f\big(\lambda_j^{(n)}\big) \to G_f(a) := \frac{1}{2\pi}\int_0^{2\pi} f\big(a(e^{i\theta})\big)\,d\theta. \quad (5.33)$$

Proof. Given $\varepsilon > 0$, there exists a polynomial p such that $|f(z) - p(z)| < \varepsilon$ for all $z \in \operatorname{conv}\mathcal{R}(a)$ (Weierstrass). Thus,

$$\frac{1}{2\pi}\int_0^{2\pi} \big|f\big(a(e^{i\theta})\big) - p\big(a(e^{i\theta})\big)\big|\,d\theta < \varepsilon,$$

$$\frac{1}{n}\sum_{j=1}^{n} \big|f\big(\lambda_j^{(n)}\big) - p\big(\lambda_j^{(n)}\big)\big|\,d\theta < \frac{1}{n}n\varepsilon = \varepsilon$$

(recall Proposition 2.17 for the last estimate), and since

$$\left|\frac{1}{n}\sum_{j=1}^{n} p\big(\lambda_j^{(n)}\big) - \frac{1}{2\pi}\int_0^{2\pi} p\big(a(e^{i\theta})\big)\,d\theta\right| < \varepsilon$$

if only n is large enough (Theorem 5.9), it follows that

$$\left|\frac{1}{n}\sum_{j=1}^{n} f\big(\lambda_j^{(n)}\big) - \frac{1}{2\pi}\int_0^{2\pi} f\big(a(e^{i\theta})\big)\right| < 3\varepsilon$$

for all sufficiently large n. ∎

If $\operatorname{conv}\mathcal{R}(a)$ is a line segment, then there is a linear map $g(z) = \alpha z + \beta$ such that $g(\operatorname{conv}\mathcal{R}(a)) = \operatorname{conv}\mathcal{R}(g(a)) \subset \mathbf{R}$. Moreover, we have the equality $g(\operatorname{sp} T_n(a)) = \operatorname{sp} T_n(g(a))$. Thus, when considering spectra of normal Toeplitz matrices we can restrict ourselves to Hermitian Toeplitz matrices.

Equal distribution in Weyl's sense. Suppose for each natural number n we are given two collections $\{\lambda_j^{(n)}\}_{j=1}^n$ and $\{\mu_j^{(n)}\}_{j=1}^n$ of real numbers. The two collections of numbers $\lambda_j^{(n)}$ and $\mu_j^{(n)}$ are said to be *equally distributed* *in the sense of Weyl* if

$$\lim_{n\to\infty} \frac{1}{n} \sum_{j=1}^{n} \left(f(\lambda_j^{(n)}) - f(\mu_j^{(n)})\right) = 0 \tag{5.34}$$

for every function $f \in C_0(\mathbf{R})$, i.e., for every continuous function f on $\mathbf{R}$ with compact support. Note that (5.34) does not imply that

$$\lim_{n\to\infty} \frac{1}{n} \sum_{j=1}^{n} f(\lambda_j^{(n)}) = \lim_{n\to\infty} \frac{1}{n} \sum_{j=1}^{n} f(\mu_j^{(n)}) \tag{5.35}$$

but that (5.34) is a consequence of (5.35).

Corollary 5.11. *Let a be a bounded, real-valued, and Riemann integrable function on the unit circle $\mathbf{T}$. Then the eigenvalues $\{\lambda_j^{(n)}\}_{j=1}^n$ of $T_n(a)$ and the values $\{a(e^{2\pi ij/n})\}_{j=1}^n$ of the symbol a at the nth unit roots are equally distributed in the sense of Weyl.*

Proof. If a is Riemann integrable, then so also is $f(a)$ $(:= f \circ a)$ for every $f \in C_0(\mathbf{R})$. Hence,

$$\frac{1}{2\pi} \int_0^{2\pi} f\big(a(e^{i\theta})\big)\, d\theta = \lim_{n\to\infty} \frac{1}{n} \sum_{j=1}^{n} f\big(a(e^{2\pi ij/n})\big).$$

Theorem 5.10 therefore shows that (5.35), and all the more (5.34), is satisfied with $\mu_j^{(n)} = a(e^{2\pi ij/n})$. ■

We remark that a bounded function is Riemann integrable if and only if it is continuous almost everywhere. In particular, functions with at most countably many discontinuities (and thus functions in PC) are Riemann integrable.

Convergence in measure. Let μ_n $(n \in \mathbf{N})$ and μ be real Borel measures on $\mathbf{R}$. One says that μ_n *converges weakly* to μ if

$$\int_{\mathbf{R}} f\, d\mu_n \to \int_{\mathbf{R}} f\, d\mu \tag{5.36}$$

for every $f \in C_0(\mathbf{R})$. Now suppose $a \in L^\infty$ is a real-valued function. As above, we denote by $\lambda_j^{(n)}$ $(j = 1, \dots, n)$ the eigenvalues of $T_n(a)$. For a Lebesgue measurable subset $\Gamma \subset \mathbf{T}$, denote by $|\Gamma|$ its Lebesgue measure (on the circle $\mathbf{T}$), and given a Borel subset $E \subset \mathbf{R}$, put

$$\mu_n(E) := \frac{1}{n} \sum_{\lambda_j^{(n)} \in E} 1, \tag{5.37}$$

$$\mu(E) := \frac{1}{2\pi} \left|\{e^{i\theta} \in \mathbf{T} : a(e^{i\theta}) \in E\}\right|. \tag{5.38}$$

In (5.37), the multiplicities of the eigenvalues are taken into account. Note that $\mu_n(E)$ is nothing but the percentage (divided by 100) of the eigenvalues of $T_n(a)$ lying in E, while $\mu(E)$ ($\in [0,1]$) is the probability of the event that $a(t)$ belongs to E for a randomly chosen equally distributed $t \in \mathbf{T}$. Also notice that

$$\mu_n(E) = \frac{1}{n}\sum_{j=1}^{n} \chi_E\big(\lambda_j^{(n)}\big), \tag{5.39}$$

$$\mu(E) = \frac{1}{2\pi}\int_0^{2\pi} \chi_E\big(a(e^{i\theta})\big)\, d\theta, \tag{5.40}$$

where χ_E is the characteristic function of E.

Corollary 5.12. *If $a \in L^\infty$ is real-valued, then the measures μ_n given by (5.37) converge weakly to the measure μ defined by (5.38).*

Proof. From (5.39) and (5.40) we easily see that if $f \in C_0(\mathbf{R})$, then

$$\int_{\mathbf{R}} f\, d\mu_n = \frac{1}{n}\sum_{j=1}^{n} f\big(\lambda_j^{(n)}\big), \tag{5.41}$$

$$\int_{\mathbf{R}} f\, d\mu = \frac{1}{2\pi}\int_0^{2\pi} f\big(a(e^{i\theta})\big)\, d\theta. \tag{5.42}$$

The assertion therefore follows from Theorem 5.10. ∎

Let $E \subset \mathbf{R}$ be an arbitrary Borel set and suppose $a \in L^\infty$ is real-valued. One can show that if $|a^{-1}(\partial E)| = 0$, where ∂E is the boundary of E, then

$$\mu_n(E) \to \mu(E), \tag{5.43}$$

that is, (5.41) converges to (5.42) also in the case where f is the characteristic function of E.

Example 5.13. Define $a : \mathbf{T} \to [-1,1]$ by $a(e^{i\theta}) = \cos\theta$. Here are four segments E_j and their measures $\mu(E_j)$ in accordance with (5.40):

$$E_1 = \left[-1, \frac{1}{2}\sqrt{3}\right], \qquad \mu(E_1) = \frac{5}{6},$$

$$E_2 = [-1, 0], \qquad \mu(E_2) = \frac{1}{2},$$

$$E_3 = \left[0, \frac{1}{2}\right], \qquad \mu(E_3) = \frac{1}{6},$$

$$E_4 = \left[\frac{1}{2}, \frac{1}{2}\sqrt{2}\right], \qquad \mu(E_4) = \frac{1}{12}.$$

Figure 34 shows that $\mu_n(E_j) \to \mu(E_j)$ as $n \to \infty$, where $\mu_n(E)$ is given by (5.39). ■

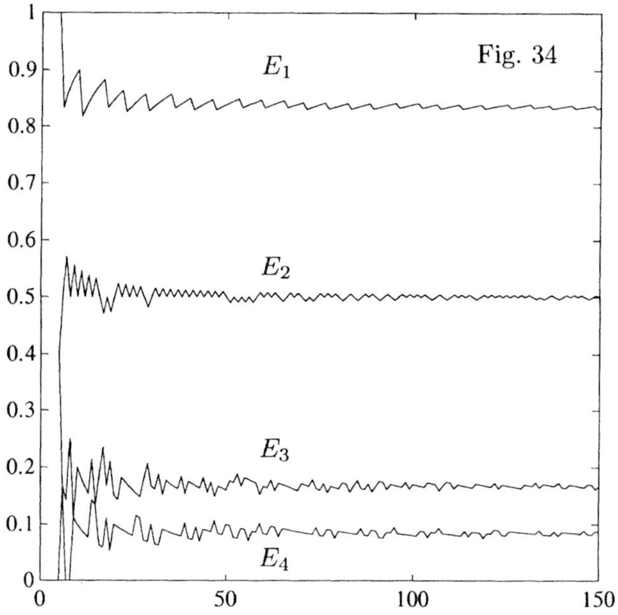

In Figure 34 we plotted $\mu_n(E_j)$ versus n for $5 \le n \le 150$. It is clearly seen that $\mu_n(E_j)$ approaches a limit as n increases.

It should be emphasized that Corollary 5.12 is stronger than Corollary 5.11: by virtue of (5.41) and (5.42), Corollary 5.12 is *equivalent* to Theorem 5.10, whereas Corollary 5.11 is an irreversible consequence of Theorem 5.10.

In connection with the above two corollaries, the following theorem on the (uniform and partial) limiting sets of $\operatorname{sp} T_n(a)$ is very instructive.

Theorem 5.14. *If $a \in PC$ is real-valued, then*

$$u\text{-}\lim_{n\to\infty} \operatorname{sp} T_n(a) = p\text{-}\lim_{n\to\infty} \operatorname{sp} T_n(a) = \operatorname{conv} \mathcal{R}(a).$$

Proof. If $M > 0$ is large enough, then $b := a + M$ is positive. Theorem 1.27 yields

$$\operatorname{sp} T(b) = \operatorname{conv} \mathcal{R}(b) = \operatorname{conv} \mathcal{R}(a) + M,$$

we clearly have $\operatorname{sp} T_n(b) = \operatorname{sp} T_n(a) + M = \{\lambda_1^{(n)} + M, \ldots, \lambda_n^{(n)} + M\}$, and since $\lambda_j^{(n)} + M > 0$ for all j (Proposition 2.17), it follows that $\Sigma(T_n(b))$ equals $\{\lambda_1^{(n)} + M, \ldots, \lambda_n^{(n)} + M\}$. Corollary 4.19 applied to b gives the assertion. ■

Example 5.15. Let

$$a(e^{i\theta}) = \begin{cases} 1 & \text{for} \quad 0 < \theta < \pi, \\ -1 & \text{for} \quad \pi < \theta < 2\pi. \end{cases}$$

In this case

$$T(a) = \frac{2}{\pi i} \begin{pmatrix} 0 & -\frac{1}{1} & 0 & -\frac{1}{3} & 0 & \dots \\ \frac{1}{1} & 0 & -\frac{1}{1} & 0 & -\frac{1}{3} & \dots \\ 0 & \frac{1}{1} & 0 & -\frac{1}{1} & 0 & \dots \\ \frac{1}{3} & 0 & \frac{1}{1} & 0 & -\frac{1}{1} & \dots \\ 0 & \frac{1}{3} & 0 & \frac{1}{1} & 0 & \dots \\ \dots & \dots & \dots & \dots & \dots & \dots \end{pmatrix}.$$

We have $\operatorname{sp} T_n(a) \subset [-1, 1]$ for every $n \geq 1$ by Proposition 2.17. Theorem 5.14 implies that for every $\lambda \in [-1, 1]$ there exist $\lambda_k^{(n)} \in \operatorname{sp} T_n(a)$ such that $\lambda_k^{(n)} \to \lambda$ as $k \to \infty$. On the other hand, Corollary 5.11 tells us that

$$\{\lambda_1^{(n)}, \dots, \lambda_n^{(n)}\} \quad \text{and} \quad \{-1, \dots, -1, 1, \dots, 1\}$$

are equally distributed in the sense of Weyl. Finally, Corollary 5.12 (or more precisely: (5.43)) says that for each $\varepsilon \in (0, 1)$,

$$\frac{1}{n} \sum_{\lambda_k^{(n)} \in [-1, -1+\varepsilon]} 1 \to \frac{1}{2}, \qquad \frac{1}{n} \sum_{\lambda_k^{(n)} \in [1-\varepsilon, 1]} 1 \to \frac{1}{2}.$$

Thus, for large n, almost half of the eigenvalues cluster near -1 and almost half of them cluster near 1. Figure 35 illustrates the situation. ∎

Note. The results of this section are essentially due to Szegö; see [92].

5.6 The Avram-Parter Theorem

The Avram-Parter formula (4.18) is a trace formula for $\operatorname{tr} f(T_n(\overline{a})T_n(a))$ and so leads to the consideration of products and thus of algebras of Toeplitz matrices.

The trace and Hilbert-Schmidt norms $\|\cdot\|_1$ and $\|\cdot\|_2$ were defined in Section 5.1. If A is a finite matrix, then $\|A\|_2$ is nothing but the Frobenius norm $\|A\|_F$ employed in Section 4.5. For arbitrary operators

$$A, B \in \mathcal{K}_2(H), \qquad C, D \in \mathcal{B}(H), \qquad K \in \mathcal{K}_1(H)$$

we have the inequalities

$$\|AB\|_1 \leq \|A\|_2 \|B\|_2, \qquad \|CKD\|_1 \leq \|C\| \, \|K\|_1 \|D\|.$$

If $K \in \mathcal{K}_1(H)$, then $|\operatorname{tr} K| \leq \|K\|_1$.

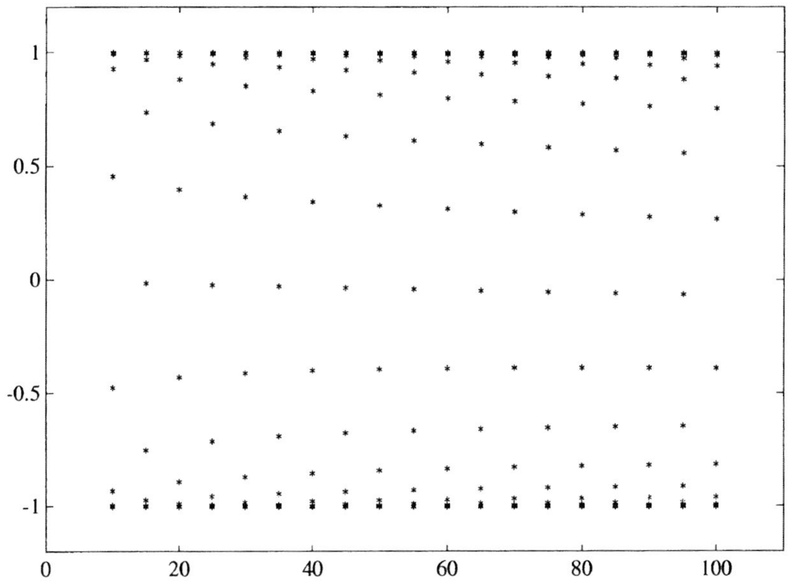

Figure 35

Let the symbol a be as in Example 5.15. Figure 35 shows the eigenvalues of $T_n(a)$ versus n for $n = 10, 15, 20, \dots, 100$. Although every point in $[-1, 1]$ is a partial limit of $\operatorname{sp} T_n(a)$ as $n \to \infty$, almost all eigenvalues are located near -1 or 1.

In what follows, we work with $H = l^2$. We denote by $\mathcal{O}$ the set of all sequences $\{K_n\}_{n=1}^{\infty}$ of operators $K_n \in \mathcal{K}_1(l^2)$ such that

$$\frac{\|K_n\|_1}{n} \to 0 \text{ as } n \to \infty.$$

Clearly, if $\{K_n\} \in \mathcal{O}$ then $(1/n) \operatorname{tr} K_n \to 0$. Here are some important examples of sequences in $\mathcal{O}$.

Lemma 5.16. *If $a, b, c \in L^\infty$, then*

$$\{P_n T(a) Q_n\}, \ \{Q_n T(a) P_n\}, \ \{P_n T(a) - T(a) P_n\}, \tag{5.44}$$

$$\{P_n H(b) H(c) P_n\}, \ \{W_n H(b) H(c) W_n\}, \tag{5.45}$$

$$\{T_n(bc) - T_n(b) T_n(c)\} \tag{5.46}$$

are sequences in $\mathcal{O}$.

Proof. We have $P_n = P_n P_n$ and thus

$$\frac{\|P_n T(a) Q_n\|_1}{n} \le \frac{\|P_n\|_2}{\sqrt{n}} \frac{\|P_n T(a) Q_n\|_2}{\sqrt{n}}.$$

Since $\|P_n\|_2 = \sqrt{n}$ and since

$$\frac{\|P_nT(a)Q_n\|_2^2}{n} = \frac{1}{n}\left(|a_{-1}|^2+2|a_{-2}|^2+\ldots+(n-1)|a_{-(n-1)}|^2+n\sum_{j=n}^{\infty}|a_{-j}|^2\right)$$

goes to zero (this follows from the sole fact that $\sum |a_k|^2 < \infty$), it results that $\{P_nT(a)Q_n\} \in \mathcal{O}$. Analogously we get that $\{Q_nT(a)P_n\} \in \mathcal{O}$, whence

$$\{P_nT(a) - T(a)P_n\} = \{P_nT(a)Q_n - Q_nT(a)P_n\} \in \mathcal{O}.$$

By (2.12),

$$P_nH(b)H(c)P_n = W_nP_nT(\tilde{b})Q_nT(c)P_nW_n,$$
$$W_nH(b)H(c)W_n = P_nT(\tilde{b})Q_nT(c)P_n,$$

and hence the membership of the sequences (5.45) in $\mathcal{O}$ follows from the fact that the first sequence in (5.44) belongs to $\mathcal{O}$. Finally, (2.13) now shows that (5.46) also lies in $\mathcal{O}$. ■

Theorem 5.17 (Avram-Parter). *If $a \in L^\infty$ and $g \in C_0(\mathbf{R})$, then*

$$\frac{1}{n}\sum_{j=1}^{n} g\big(s_j(T_n(a))\big) \to G_g(|a|) := \frac{1}{2\pi}\int_0^{2\pi} g\big(|a(e^{i\theta})|\big)\,d\theta. \tag{5.47}$$

Proof. There is a function $f \in C_0(\mathbf{R})$ such that $g(s) = f(s^2)$ for all $s \geq 0$. Thus, we have to prove that

$$\frac{1}{n}\sum_{j=1}^{n} f\big(s_j^2(T_n(a))\big) \to \frac{1}{2\pi}\int_0^{2\pi} f\big(|a(e^{i\theta})|^2\big)\,d\theta. \tag{5.48}$$

Since every $f \in C_0(\mathbf{R})$ can be uniformly approximated on $[0, \|a\|_\infty^2]$ by polynomials as closely as desired, it suffices to show (5.48) in the case where $f(\lambda) = \lambda^k$ and $k \in \mathbf{N}$.

Put $A_n = T_n(\overline{a})T_n(a)$ and $B_n = T_n(|a|^2)$. From Proposition 2.12 and Lemma 5.16 we infer that $\{A_n - B_n\} \in \mathcal{O}$. By induction on k, it can be easily verified that

$$\|A^k - B^k\|_1 \leq k\big(\|A\| + \|B\|\big)^{k-1}\|A - B\|_1$$

whenever $A, B \in \mathcal{K}_1(H)$. Thus, $\{A_n^k - B_n^k\} \in \mathcal{O}$ and we obtain

$$\frac{1}{n}\sum_{j=1}^{n} s_j^{2k}\big(T_n(a)\big) = \frac{1}{n}\operatorname{tr} A_n^k = \frac{1}{n}\operatorname{tr} B_n^k + o(1).$$

Theorem 5.10 implies that this has the limit

$$\frac{1}{2\pi}\int_0^{2\pi} |a(e^{i\theta})|^{2k}\,d\theta. \;■$$

The remark after Corollary 5.12 also applies to the situation considered here: if $a \in L^\infty$ and the set $\{t \in \mathbf{T} : |a(t)| \in \partial E\}$ has measure zero, then Theorem 5.17 gives that

$$\sigma_n(E) := \frac{1}{n} \sum_{s_k(T_n(a)) \in E} 1$$

converges to

$$\sigma(E) := \frac{1}{2\pi} \big|\{e^{i\theta} \in \mathbf{T} : |a(e^{i\theta})| \in E\}\big|.$$

Moreover, Theorem 5.17 implies that the singular values of $T_n(a)$ and the values $\{|a(e^{2\pi ik/n})|\}_{k=1}^n$ of $|a|$ at the nth unit roots are equally distributed in Weyl's sense provided a is Riemann integrable.

Theorems 5.10 and 5.17 can be refined and extended into several directions. Here is a second-order result.

Theorem 5.18 (Widom). *If a belongs to the Krein algebra $L^\infty \cap B_2^{1/2}$ and $f \in C^3[m^2, M^2]$ where $m :=$ ess inf $|a|$, $M :=$ ess sup $|a|$, then*

$$B_f(a) := f\big(T(\overline{a})T(a)\big) + f\big(T(a)T(\overline{a})\big) - 2T\big(f(|a|^2)\big)$$

is a trace class operator on l^2 and

$$\sum_{k=1}^{n} f\big(s_k^2(T_n(a))\big) = \frac{n}{2\pi} \int_0^{2\pi} f\big(|a(e^{i\theta})|^2\big)\, d\theta + \operatorname{tr} B_f(a) + o(1).$$

A proof is in [188]. It is based on the identity

$$e^{itA} - e^{itB} = i \int_0^t e^{isA}(A - B)e^{i(t-s)B}\, ds,$$

which, with A_n and B_n as in the proof of Theorem 5.17, gives

$$\begin{aligned} &e^{itA_n} - e^{itB_n} \\ &\quad = -i \int_0^t e^{itA_n}\big(P_n H(\overline{a})H(\tilde{a})P_n + W_n H(\tilde{\overline{a}})H(a)W_n\big)e^{i(t-s)B_n}\, ds. \end{aligned}$$

The latter formula is used to prove the assertion for $f(\lambda) = e^{it\lambda}$. Application of the identity

$$f(\lambda) = \int \hat{f}(t)e^{it\lambda}\, dt,$$

where $\hat{f}$ is the Fourier transform of f, then yields the assertion for general functions f.

Here is a result for unbounded symbols.

Theorem 5.19 (Tyrtyshnikov). *If $a \in L^2$ is real-valued, then* (5.33) *holds for every $f \in C_0(\mathbf{R})$, and if $a \in L^2$ is arbitrary, then* (5.47) *is valid for every $g \in C_0(\mathbf{R})$.*

This theorem was established in [176], and a full proof can be found in [177]. Tyrtyshnikov first proves that if $\{A_n\}$ and $\{B_n\}$ are two sequences of $n \times n$ matrices such that

$$\frac{1}{n}\|A_n - B_n\|_2^2 = \frac{1}{n}\|A_n - B_n\|_F^2 \to 0, \tag{5.49}$$

then their singular values and, in case both sequences are constituted by Hermitian matrices, also their eigenvalues are equally distributed in the sense of Weyl. Theorem 5.19 follows from choosing $A_n = T_n(a)$ and cleverly constructing circulants B_n such that (5.49) is satisfied.

Notes. Theorem 5.17 was established by Parter [128] for symbols $a \in L^\infty$ which are locally selfadjoint, i.e., which can be written as the product of a continuous and a real-valued function. Avram [5] proved Theorem 5.17 for general $a \in L^\infty$. SeLegue [149] introduced the set $\mathcal{O}$ and established Lemma 5.16, which does not only yield the simple proof of the Avram-Parter theorem given in the text, but will become of deciding importance in the following section. As already said, Theorems 5.18 and 5.19 are in [188] and [176], [177], respectively. *Added in Proof:* Theorem 5.19 was recently extended to L^1 symbols in [193].

5.7 The Algebraic Approach to Trace Formulas

Recently, SeLegue [149] proved a result (Corollary 5.24 below) which essentially generalizes Szegö's first limit theorem. The purpose of this section is to establish Theorem 5.23: this theorem includes both SeLegue's theorem and the Avram-Parter theorem as special cases.

Let $\mathbf{A}(L^\infty)$ stand for the smallest closed subalgebra of $\mathcal{B}(l^2)$ which contains all bounded Toeplitz operators, i.e., the set $\{T(a) : a \in L^\infty\}$. Further, we define $\mathbf{AS}(L^\infty)$ as the smallest closed subalgebra of $\mathcal{F}$ (recall Section 2.5) which contains all sequences of the form $\{P_n A P_n : A \in \mathbf{A}(L^\infty)\}$. Our aim is to compute the limit of $(1/n)\,\mathrm{tr}\, f(A_n)$ for $\{A_n\} \in \mathbf{AS}(L^\infty)$.

We first need a result on the structure of the algebra $\mathbf{A}(L^\infty)$. Put $T(L^\infty) := \{T(a) : a \in L^\infty\}$. The *quasi-commutator* of two Toeplitz operators $T(b)$ and $T(c)$ is the operator

$$T(bc) - T(b)T(c) = H(b)H(\tilde{c}). \tag{5.50}$$

We denote by $Q_T(L^\infty)$ the smallest closed two-sided ideal of $\mathbf{A}(L^\infty)$ containing the set $\{T(bc) - T(b)T(c) : b, c \in L^\infty\}$. The ideal $Q_T(L^\infty)$ is referred to as the *quasi-commutator ideal* of $\mathbf{A}(L^\infty)$.

Theorem 5.20. *The algebra* $\mathbf{A}(L^\infty)$ *decomposes into the direct sum*

$$\mathbf{A}(L^\infty) = T(L^\infty) \oplus Q_T(L^\infty).$$

The projection S_T *of* $\mathbf{A}(L^\infty)$ *onto* $T(L^\infty)$ *parallel to* $Q_T(L^\infty)$ *is given by*

$$S_T(A) := s\text{-}\lim_{n\to\infty} T(\chi_{-n})AT(\chi_n). \tag{5.51}$$

Here s-lim denotes the strong limit and $\chi_{\pm n}$ are defined by $\chi_{\pm n}(t) = t^{\pm n}$ ($t \in \mathbf{T}$). The existence of the strong limit (5.51) for every $A \in \mathbf{A}(L^\infty)$ is part of the conclusion of the theorem. For a proof see [35, Theorem 2.40 and Section 2.41] or [39, Corollary 4.3 and Proposition 4.4].

If $A \in \mathbf{A}(L^\infty)$, then $S_T(A)$ is some Toeplitz operator $T(a)$ with $a \in L^\infty$. The function a is called the *symbol* of A and is denoted by sym A. From Theorem 5.20 we infer that sym is a C^*-algebra homomorphism of $\mathbf{A}(L^\infty)$ onto L^∞.

Recall the definition of $\mathcal{O}$ given in Section 5.6. Since

$$\|K_nB_n\|_1 \le \|K_n\|_1\|B_n\|_\infty \quad \text{and} \quad \|B_nK_n\|_1 \le \|B_n\|_\infty\|K_n\|_1,$$

it is clear that $\mathcal{O} \cap \mathcal{F}$ is a two-sided ideal in the algebra $\mathcal{F}$; of course, here $\mathcal{F}$ is the algebra introduced in Section 2.5.

Lemma 5.21. *If* $A \in \mathbf{A}(L^\infty)$ *and* $K \in Q_T(L^\infty)$, *then*

$$\{P_nA - AP_n\} \in \mathcal{O} \quad \textit{and} \quad \{P_nKP_n\} \in \mathcal{O}.$$

Proof. For $A = T(a)$, the assertion follows from Lemma 5.16. Since

$$P_nBC - BCP_n = (P_nB - BP_n)C + B(P_nC - CP_n),$$

it results that $\{P_nA - AP_n\} \in \mathcal{O}$ for all A in the (non-closed) algebra generated by $\{T(a) : a \in L^\infty\}$. Finally, if $\|A - A^{(j)}\| \to 0$ as $j \to \infty$, then

$$\left\|(P_nA - AP_n) - (P_nA^{(j)} - A^{(j)}P_n)\right\|_1 \le \|P_n\|_1\,\|A - A^{(j)}\| + \|A - A^{(j)}\|\,\|P_n\|_1,$$

and as $\|P_n\|_1 = \|P_n^2\|_1 \le \|P_n\|_2^2 = n$, we see that $\{P_nA - AP_n\} \in \mathcal{O}$ for all $A \in \mathbf{A}(L^\infty)$.

If L is a quasi-commutator of the form (5.50), then $\{P_nLP_n\} \in \mathcal{O}$ due to Lemma 5.16. For $B, C \in \mathbf{A}(L^\infty)$ we have

$$\begin{aligned} P_nBLCP_n &= (P_nB - BP_n)LCP_n + BP_nLCP_n \\ &= (P_nB - BP_n)LCP_n + BP_nLP_nC - BP_nL(P_nC - CP_n), \end{aligned}$$

and from what was already proved we deduce that $\{P_nBLCP_n\} \in \mathcal{O}$. As

$$\frac{1}{n}\left\|P_nKP_n - P_nK^{(j)}P_n\right\|_1 \le \frac{\|P_n\|_2}{\sqrt{n}}\|K - K^{(j)}\|\frac{\|P_n\|_2}{\sqrt{n}}$$

and $\|P_n\|_2 = \sqrt{n}$, we obtain that $\{P_nKP_n\} \in \mathcal{O}$ for every $K \in Q_T(L^\infty)$. ■

Proposition 5.22. *If $\{A_n\} \in \mathbf{AS}(L^\infty)$, then there is a unique $a \in L^\infty$ such that*

$$A_n = T_n(a) + K_n \quad \textit{with} \quad \{K_n\} \in \mathcal{O}.$$

Proof. The algebra $\mathbf{AS}(L^\infty)$ is the closure in $\mathcal{F}$ of the set

$$\mathbf{AS}^0(L^\infty) := \Big\{ \sum_j \prod_k \{P_n A_{jk} P_n\} : A_{jk} \in \mathbf{A}(L^\infty) \Big\},$$

the sum and products being finite. Let $B, C \in \mathbf{A}(L^\infty)$. By Theorem 5.20,

$$B = T(b) + K, \quad C = T(b) + L; \quad b, c \in L^\infty; \quad K, L \in Q_T(L^\infty).$$

Thus,

$$P_n B P_n C P_n = T_n(b)T_n(c) + T_n(b)P_n L P_n + P_n K P_n T_n(c) + P_n K P_n L P_n,$$

and the sequences built up by the last three terms on the right belong to $\mathcal{O}$ due to Lemma 5.21. Taking into account Lemma 5.16 we so obtain

$$P_n B P_n C P_n = T_n(bc) - K_n \quad \text{with} \quad \{K_n\} \in \mathcal{O}.$$

Consequently, if $\{A_n\} \in \mathbf{AS}^0(L^\infty)$, then $A_n = T_n(a) + K_n$ with $a \in L^\infty$ and $\{K_n\} \in \mathcal{O}$. Moreover, it is clear that the strong limit of K_n exists and belongs to $Q_T(L^\infty)$.

We are left with showing that the set

$$\big\{ \{T_n(a) + K_n\} : a \in L^\infty, \ \{K_n\} \in \mathcal{O} \cap \mathbf{AS}(L^\infty), \ s\text{-lim}\, K_n \in Q_T(L^\infty) \big\}$$

is a closed subset of $\mathbf{AS}(L^\infty)$. So let $\{T_n(a^{(j)}) + K_n^{(j)}\}$ be in this set, suppose $\{A_n\} \in \mathbf{AS}(L^\infty)$, and

$$\sup_{n \ge 1} \|A_n - T_n(a^{(j)}) - K_n^{(j)}\| \to 0 \quad \text{as } j \to \infty.$$

We have

$$\begin{aligned} &T_n(a^{(j)}) \to T(a^{(j)}) \in T(L^\infty) \quad \text{strongly}, \\ &K_n^{(j)} \to K^{(j)} \in Q_T(L^\infty) \quad \text{strongly}, \\ &A_n \to T(a) + K \quad \text{strongly}, \quad a \in L^\infty, \quad K \in Q_T(L^\infty), \end{aligned}$$

the last fact resulting from Theorem 5.20. Clearly, $T(a^{(j)}) + K^{(j)}$ converges uniformly to $T(a) + K$ as $j \to \infty$. Since

$$\begin{aligned} &T(\chi_{-n})\big(T(a^{(j)}) + K^{(j)}\big)T(\chi_n) \to T(a^{(j)}) \quad \text{strongly}, \\ &T(\chi_{-n})\big(T(a) + K\big)T(\chi_n) \to T(a) \quad \text{strongly}, \end{aligned}$$

by Theorem 5.20, it results that $T(a^{(j)})$ converges uniformly to $T(a)$, implying that $K^{(j)}$ converges uniformly to K. Hence, as $j \to \infty$, we have

$$\{T_n(a^{(j)})\} \to \{T_n(a)\} \quad \text{and} \quad \{K_n^{(j)}\} \to \{K_n\},$$

where $K_n := A_n - T_n(a)$. Moreover, K_n converges strongly to $K \in Q_T(L^\infty)$. Since

$$\begin{aligned} \frac{\|K_n\|_1}{n} &\le \frac{\|K_n - K_n^{(j)}\|_1}{n} + \frac{\|K_n^{(j)}\|_1}{n} \\ &\le \frac{\|P_n\|_2^2}{n}\|K_n - K_n^{(j)}\| + \frac{\|K_n^{(j)}\|_1}{n} \end{aligned}$$

and $\|P_n\|_2^2 = n$, it is clear that $\{K_n\} \in \mathcal{O}$. As $A_n = T_n(a) + K_n$, the closedness of our set follows.

Finally, to show the uniqueness of a, assume $\{T_n(d)\} \in \mathcal{O}$. Then, by Lemma 5.16, $\{T_n(\chi_{-k}d)\} \in \mathcal{O}$ for every $k \in \mathbf{Z}$. It results that

$$(1/n)\operatorname{tr} T_n(\chi_{-k}d) \to 0 \quad \text{as } n \to \infty,$$

and since $(1/n)\operatorname{tr} T_n(\chi_{-k}d)$ is the kth Fourier coefficient of d, we arrive at the conclusion that $d = 0$. ■

The function $a \in L^\infty$ defined by Proposition 5.22 is called the *symbol* of the sequence $\{A_n\} \in \mathbf{AS}(L^\infty)$.

Theorem 5.23. *Let $\{A_n\} \in \mathbf{AS}(L^\infty)$ and let $a \in L^\infty$ be the symbol of $\{A_n\}$. Denote the eigenvalues of A_n by $\lambda_1^{(n)}, \dots, \lambda_n^{(n)}$. Then*

$$\frac{1}{n}\operatorname{tr} A_n^k = \frac{1}{n}\sum_{j=1}^{n}\left(\lambda_j^{(n)}\right)^k \to \frac{1}{2\pi}\int_0^{2\pi}\left(a(e^{i\theta})\right)^k d\theta$$

for every integer $k \ge 1$.

Proof. By Proposition 5.22, $A_n = T_n(a) + K_n$ with $\{K_n\} \in \mathcal{O}$. Therefore $A_n^k = T_n^k(a) + L_n$ with $\{L_n\} \in \mathcal{O}$. From Lemma 5.16 we deduce that $T_n^k(a) = T_n(a^k) + M_n$ with $\{M_n\} \in \mathcal{O}$. Consequently,

$$\frac{1}{n}\operatorname{tr} A_n^k = \frac{1}{n}\operatorname{tr}\bigl(T_n(a^k) + M_n + L_n\bigr) = \frac{1}{n}\operatorname{tr} T_n(a^k) + o(1).$$

Since $(1/n)\operatorname{tr} T_n(a^k)$ is the zeroth Fourier coefficient of a^k, we arrive at the assertion. ■

This theorem has several corollaries. First of all, we remark that Theorems 5.9, 5.10, and 5.17 were not used in the proof of Theorem 5.23. Thus, we get new proofs of these results. They can be deduced from Theorem

5.23 (along with an approximation argument) by specifying A_n as follows:

Theorem 5.9: $A_n = T_n(a)$;
Theorem 5.10: $A_n = T_n(a)$ Hermitian;
Theorem 5.17: $A_n = T_n(\overline{a})T_n(a)$.

Here are two more consequences of Theorem 5.23.

Corollary 5.24 (SeLegue). *Let $A \in \mathbf{A}(L^\infty)$ be selfadjoint and let $a = \operatorname{sym} A$ be the symbol of A. Let further $\lambda_1^{(n)}, \dots, \lambda_n^{(n)}$ be the eigenvalues of $P_n A P_n$. Then*

$$\frac{1}{n} \operatorname{tr} f(P_n A P_n) := \frac{1}{n} \sum_{j=1}^{n} f(\lambda_j^{(n)}) \to \frac{1}{2\pi} \int_0^{2\pi} f\big(a(e^{i\theta})\big)\, d\theta$$

for every function $f \in C_0(\mathbf{R})$.

Proof. This will follow from Theorem 5.23 once we have verified that the symbol of $\{P_n A P_n\}$ coincides with the symbol a of A. But since $A = T(a) + K$ with $K \in Q_T(L^\infty)$ due to Theorem 5.20 and $\{P_n K P_n\} \in \mathcal{O}$ by virtue of Lemma 5.21, we obtain from Proposition 5.22 that a is the symbol of $\{P_n A P_n\}$. ■

Corollary 5.25. *Let $A \in \mathbf{A}(L^\infty)$ and let $a = \operatorname{sym} A$ be the symbol of A. Denoting by $s_1^{(n)}, \dots, s_n^{(n)}$ the singular values of $P_n A P_n$ we have*

$$\frac{1}{n} \sum_{j=1}^{n} f(s_j^{(n)}) \to \frac{1}{2\pi} \int_0^{2\pi} f\big(|a(e^{i\theta})|\big)\, d\theta$$

for every $f \in C_0(\mathbf{R})$.

Proof. Put $A_n = (P_n A^* P_n)(P_n A P_n)$. As $A = T(a) + K$ with $K \in Q_T(L^\infty)$, we infer from Lemmas 5.21 and 5.16 that

$$A_n = T_n(\overline{a})T_n(a) + K_n = T_n(|a|^2) + L_n$$

with $\{K_n\}, \{L_n\} \in \mathcal{O}$. Consequently, the symbol of $\{A_n\}$ is $|a|^2$. Theorem 5.23 therefore gives

$$\frac{1}{n} \sum_{j=1}^{n} g\big((s_j^{(n)})^2\big) \to \frac{1}{2\pi} \int_0^{2\pi} g\big(|a(e^{i\theta})|^2\big)\, d\theta$$

for every $g \in C_0(\mathbf{R})$, which easily implies Corollary 5.25 as it is stated. ■

Notes. This section is heavily based on SeLegue's paper [149]. He established Corollary 5.24 by methods similar to those employed here. In particular, results close to Lemma 5.21 and Proposition 5.22 (though in C^*-algebraic disguise) are already in [149]. Theorem 5.23 and Corollary 5.25 seem to be new.

Douglas discovered that the algebra $\mathbf{A}(L^\infty)$ decomposes into the direct sum of $T(L^\infty)$ and the *commutator ideal* $C_T(L^\infty)$, which is the smallest closed two-sided ideal of $\mathbf{A}(L^\infty)$ containing the set

$$\{T(b)T(c) - T(c)T(b) : b, c \in L^\infty\}$$

(see [59, Theorem 7.11]). SeLegue [149] also works with $C_T(L^\infty)$ and not with $Q_T(L^\infty)$. Theorem 5.20 is essentially due to Clancey [47]. Since

$$T(b)T(c) - T(c)T(b) = \big[T(cb) - T(c)T(b)\big] - \big[T(bc) - T(b)T(c)\big],$$

it follows that $C_T(L^\infty) \subset Q_T(L^\infty)$, and as

$$\mathbf{A}(L^\infty) = T(L^\infty) \oplus C_T(L^\infty) = T(L^\infty) \oplus Q_T(L^\infty),$$

we see that actually $C_T(L^\infty) = Q_T(L^\infty)$. Nevertheless, our experience has been that using $Q_T(L^\infty)$ instead of $C_T(L^\infty)$ is often simpler and more convenient (see [141], [35], and [39]). An excellent presentation of various aspects of $\mathbf{A}(L^\infty)$, $C_T(L^\infty)$, $Q_T(L^\infty)$, and their subalgebras is in Nikolski's book [127].

That the projection S_T can explicitly be given by (5.51) was discovered by Barria and Halmos [8] and independently also by Roch and one of the authors [141]. Axler [6] writes: "The map S_T was a magical and mysterious homomorphism to me until I read the Barria-Halmos paper ... . The Barria-Halmos construction of S_T is completely different in spirit and technique from Douglas's proof that S_T exists. I knew Douglas's proof well ..., but until the Barria-Halmos paper came along, I never guessed that S_T could be explicitly constructed or that so much additional insight could be squeezed from a new approach."

We remark that the analogue of Theorem 5.20 is also true for the algebra $\mathbf{AS}(L^\infty)$ (see [39, Section 7.26]). In particular, the algebra $\mathbf{AS}(L^\infty)$ decomposes into the direct sum

$$\mathbf{AS}(L^\infty) = TS(L^\infty) \oplus Q_{TS}(L^\infty),$$

where $TS(L^\infty) := \{\{T_n(a)\} : a \in L^\infty\}$ and $Q_{TS}(L^\infty)$ is the smallest closed two-sided ideal of $\mathbf{AS}(L^\infty)$ which contains the set

$$\{\{T_n(bc)\} - \{T_n(b)\}\{T_n(c)\} : b, c \in L^\infty\}.$$

It should also be mentioned that the results of this section admit further generalization. For example, by employing the results of [141], we can extend Corollaries 5.24 and 5.25 to operators A in the smallest closed subalgebra of $\mathcal{B}(l^2)$ which contains the set $\{T(a) : a \in L^\infty\} \cup \{H(a) : a \in L^\infty\}$.

5.8 Toeplitz Band Matrices

The asymptotic distribution of the eigenvalues of Hermitian Toeplitz matrices is quite canonical (Section 5.5). We know from Section 3.5 that the pseudoeigenvalues of general Toeplitz matrices (with symbols in PC, say) also behave very well. Things change dramatically when passing to the (pure) eigenvalues of non-Hermitian Toeplitz matrices. Curiously, the behavior of the eigenvalues differs the more from "canonical distribution" the smoother the symbols are.

Before turning to Toeplitz band matrices, we mention the following easy consequence of Corollary 2.19.

Proposition 5.26. *If $a \in PC$, then*

$$u\text{-}\lim_{n\to\infty} \operatorname{sp} T_n(a) \subset p\text{-}\lim_{n\to\infty} \operatorname{sp} T_n(a) \subset \operatorname{sp} T(a).$$

Proof. Let $\lambda_0 \notin \operatorname{sp} T(a)$. Then the sequence $\{T_n(a-\lambda_0)\}$ is stable and hence, $\|T_n^{-1}(a-\lambda_0)\| \le M < \infty$ for all $n \ge n_0$. It follows that if $|\lambda-\lambda_0| < 1/(2M)$, then $\|T_n^{-1}(a-\lambda)\| \le 2M$ for all $n \ge n_0$, which shows that λ_0 has an open neighborhood $U(\lambda_0)$ such that $U(\lambda_0) \cap \operatorname{sp} T_n(a) = \emptyset$ for all $n \ge n_0$. Consequently, $\lambda_0 \notin p\text{-}\lim T_n(a)$. ■

The mystery and its resolution. Suppose $T(a)$ is a band matrix. The mystery is that the asymptotic eigenvalue distribution of $T_n(a)$ is in no obvious way related to the spectrum of $T(a)$. If $T(a)$ is lower-triangular, then

$$\operatorname{sp} T(a) = a(\overline{\mathbf{D}}), \quad \overline{\mathbf{D}} := \{\lambda \in \mathbf{C} : |\lambda| \le 1\}$$

(Theorem 1.15), while $\operatorname{sp} T_n(a)$ is obviously the singleton $\{a_0\} = \{a(0)\}$. Thus, $\operatorname{sp} T_n(a)$ does not mimic $\operatorname{sp} T(a)$ in any way unless a is a constant. The same is the case if $T(a)$ is upper-triangular.

So assume that $T(a)$ is not triangular. Then the symbol a is a trigonometric polynomial,

$$a(t) = \sum_{j=-p}^{q} a_j t^j \quad (t \in \mathbf{T}), \qquad p \ge 1,\ q \ge 1,\ a_{-p}a_q \ne 0. \tag{5.52}$$

Given $r \in (0,\infty)$, define the trigonometric polynomial a_r by

$$a_r(t) := a(rt) := \sum_{j=-p}^{q} a_j r^j t^j \quad (t \in \mathbf{T}) \tag{5.53}$$

and put $D_r := \operatorname{diag}(r, r^2, \dots, r^n)$. Since

$$T_n(a_r) = D_r T_n(a) D_r^{-1},$$

we have $\operatorname{sp} T_n(a) = \operatorname{sp} T_n(a_r)$. *Thus, if there were any reason for* $\operatorname{sp} T_n(a)$ *to mimic* $\operatorname{sp} T(a)$, *this reason would also force* $\operatorname{sp} T_n(a)$ *to mimic* $\operatorname{sp} T(a_r)$.

The equality $\operatorname{sp} T_n(a) = \operatorname{sp} T_n(a_r)$ and Proposition 5.26 give

$$u\text{-}\lim_{n\to\infty} \operatorname{sp} T_n(a) \subset p\text{-}\lim_{n\to\infty} \operatorname{sp} T_n(a) \subset \bigcap_{r\in(0,\infty)} \operatorname{sp} T(a_r). \tag{5.54}$$

Example 5.27. Let $a(t) = t + \varepsilon^2 t^{-1}$ where $\varepsilon \in (0,1)$. Then

$$a(e^{i\theta}) = (1+\varepsilon^2)\cos\theta + i(1-\varepsilon^2)\sin\theta,$$

and hence $a(\mathbf{T})$ is an ellipse. It follows that $\operatorname{sp} T(a)$ is the union of this ellipse and all its interior points. Further,

$$a_r(e^{i\theta}) = (r + \varepsilon^2/r)\cos\theta + i(r - \varepsilon^2/r)\sin\theta.$$

For $r = \varepsilon$, the image $a(\mathbf{T})$ is the line segment $[-2\varepsilon, 2\varepsilon]$, and it is easily seen that $[-2\varepsilon, 2\varepsilon]$ is in the interior of the ellipses $a_r(\mathbf{T})$ for all $r \in (0,\infty)\setminus\{\varepsilon\}$. From (5.54) we therefore see that the partial limits of $\operatorname{sp} T_n(a)$ all lie on the segment $[-2\varepsilon, 2\varepsilon]$. ■

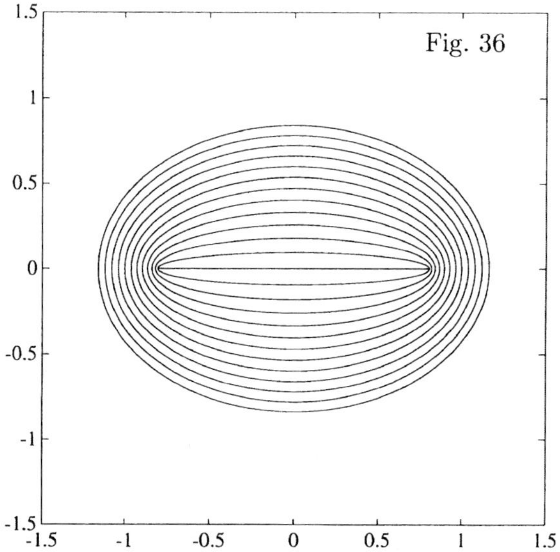

Let a be as in Example 5.27 and put $\varepsilon = 0.4$. In Figure 36 we see $a_r(\mathbf{T})$ for 13 values of r between ε and 1. In the case at hand, the set on the right of (5.54) is the line segment in the center of Figure 36.

The following theorem shows that all inclusions of (5.54) are actually equalities.

Theorem 5.28 (Schmidt-Spitzer). *If a is a trigonometric polynomial, then*

$$u\text{-}\lim_{n\to\infty} \operatorname{sp} T_n(a) = p\text{-}\lim_{n\to\infty} \operatorname{sp} T_n(a) = \bigcap_{r\in(0,\infty)} \operatorname{sp} T(a_r). \tag{5.55}$$

The set (5.55) *is always a finite union of analytic arcs each pair of which have at most endpoints in common, and* (5.55) *is always connected.*

For a proof we refer to [148]. The connectedness of the set (5.55) was established by Ullman [178]. In Figures 37 and 38 we plotted the 200 eigenvalues of $T_{200}(a)$ for 12 trigonometric polynomials a. Figure 39 provides one more beautiful example. The finite unions of analytic arcs, i.e., the sets (5.55) are clearly visible in these pictures.

The proof of Theorem 5.28 is based on a formula for the determinants of truncated Toeplitz band matrices and on still another description of the set $\bigcap_{r\in(0,\infty)} \operatorname{sp} T(a_r)$. These two results are of independent interest and will be cited now.

The next theorem first appeared in Schmidt and Spitzer's paper [148] and is a modification of a formula by Widom [180].

Theorem 5.29. *Let a be given by* (5.52) *and write*

$$a(t) = a_q t^{-p} \prod_{j=1}^{p+q} (t - \varrho_j) \quad (t \in \mathbf{T}).$$

If the zeros $\varrho_1, \ldots, \varrho_{p+q}$ are pairwise distinct then for every $n \geq 1$,

$$D_n(a) = \sum_M C_M w_M^n, \tag{5.56}$$

where the sum is taken over all $\binom{p+q}{p}$ subsets $M \subset \{1, 2, \ldots, p+q\}$ of cardinality $|M| = p$ and, with $\overline{M} := \{1, 2, \ldots, p+q\} \setminus M$,

$$w_M := (-1)^q a_q \prod_{j\in\overline{M}} \varrho_j, \qquad C_M := \prod_{j\in\overline{M}} \varrho_j^p \prod_{\substack{j\in\overline{M}\\ k\in M}} (\varrho_j - \varrho_k)^{-1}.$$

A proof is also in [51], [22], and [35, Theorem 6.29].

Example 5.30. Let

$$a(t) = a_{-1}t^{-1} + a_0 + t = t^{-1}(t-\varrho)(t-\sigma) \quad (t \in \mathbf{T}).$$

If $\varrho \neq \sigma$, Theorem 5.29 gives

$$D_n(a) = \frac{\sigma}{\sigma - \varrho}(-\sigma)^n + \frac{\varrho}{\varrho - \sigma}(-\varrho)^n = (-1)^n \frac{\sigma^{n+1} - \varrho^{n+1}}{\sigma - \varrho}$$

for all $n \geq 1$. ■

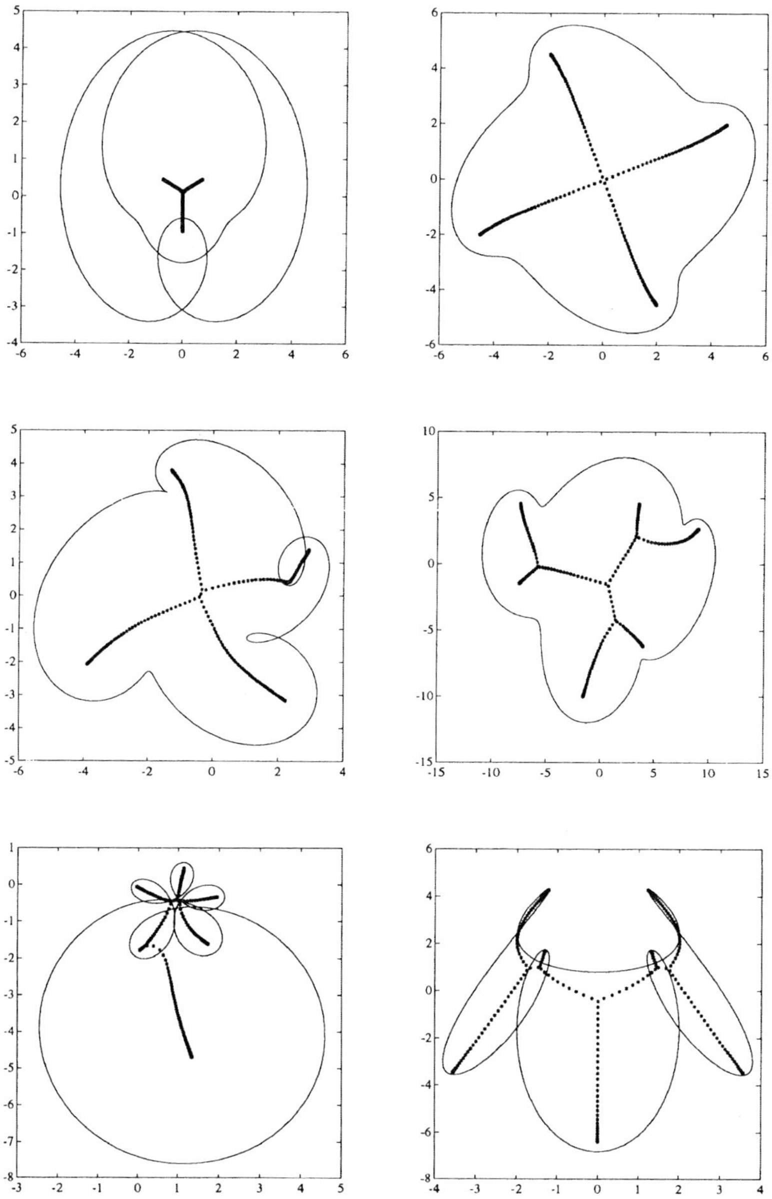

Figure 37

The essential ranges $\mathcal{R}(a)$ of several trigonometric polynomials a and the 200 eigenvalues of the matrix $T_{200}(a)$.

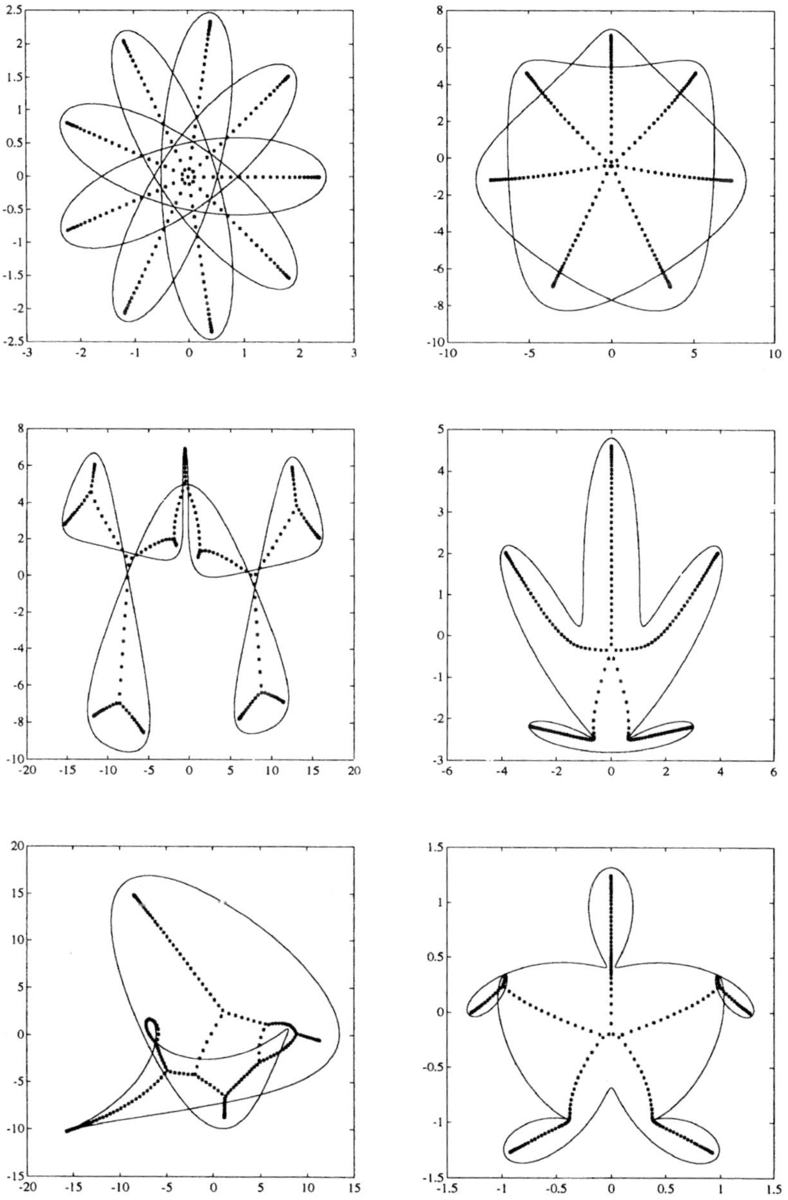

Figure 38

The essential ranges $\mathcal{R}(a)$ of several trigonometric polynomials a and the 200 eigenvalues of the matrix $T_{200}(a)$.

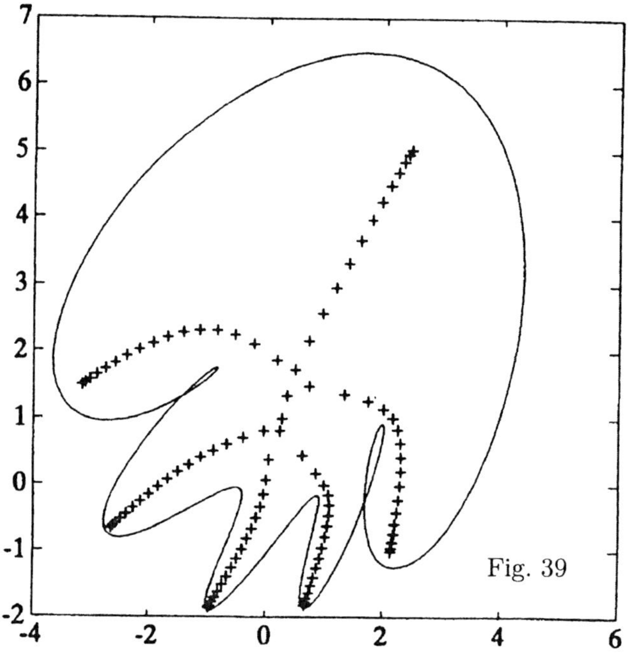

Fig. 39

This nice picture was produced by Michael Eiermann in 1992 when he taught us how to compute the eigenvalues of Toeplitz matrices with the help of matlab. He then picked the coefficients of the symbol by a random procedure, and we have not been able to recover the trigonometric polynomial whose range is seen in Figure 39 and thus to repeat the computation. We nevertheless hope the reader enjoys Eiermann's jellyfish.

Note that if $z^p a(z)$ has multiple zeros, then $D_n(a)$ can be found by first perturbing a and by subsequently passing to an appropriate limit in the formula delivered by Theorem 5.29 for the perturbed a. For instance, if

$$a(t) = t^{-1}(t-\varrho)^2,$$

we obtain from Example 5.30 that

$$D_n(a) = (-1)^n \lim_{\sigma\to\varrho} \frac{\sigma^{n+1}-\varrho^{n+1}}{\sigma-\varrho} = (-1)^n (n+1)\varrho^n.$$

Example 5.31: Szegö's strong limit theorem for band matrices. Clearly, a symbol a as in Theorem 5.29 generates an invertible Toeplitz operator if and only if the zeros $\varrho_1, \ldots, \varrho_{p+q}$ can be indexed so that

$$|\varrho_1| \le \ldots \le |\varrho_p| < 1 < |\varrho_{p+1}| \le \ldots \le |\varrho_{p+q}|$$

(Theorem 1.15). The leading term in the asymptotics of (5.56) is then given for $M = \{1, \ldots, p\}$. Thus, (5.56) becomes

$$D_n(a) = (-1)^{qn} a_q^n \Big(\prod_{j \in \overline{M}} \varrho_j \Big)^{n+p} \prod_{\substack{j \in \overline{M} \\ k \in M}} (\varrho_j - \varrho_k)^{-1} \Big(1 + O\Big(\Big| \frac{\varrho_p}{\varrho_{p+1}} \Big|^n \Big)\Big). \tag{5.57}$$

Let us compare (5.57) with Theorem 5.2. We have

$$a(t) = (-1)^q a_q \prod_{k \in M} \Big(1 - \frac{\varrho_k}{t}\Big) \prod_{j \in \overline{M}} \varrho_j \prod_{j \in \overline{M}} \Big(1 - \frac{t}{\varrho_j}\Big),$$

and hence $\log a(t)$ equals

$$\log \Big((-1)^q a_q \prod_{j \in \overline{M}} \varrho_j \Big) - \sum_{k \in M} \sum_{s \geq 1} \frac{1}{s} \frac{\varrho_k^s}{t^s} - \sum_{j \in \overline{M}} \sum_{s \geq 1} \frac{1}{s} \frac{t^s}{\varrho_j^s}.$$

This gives

$$(\log a)_s = -\frac{1}{s} \sum_{j \in \overline{M}} \frac{1}{\varrho_j^s}, \qquad (\log a)_{-s} = -\frac{1}{s} \sum_{k \in M} \varrho_k^s,$$

and thus,

$$G(a) := \exp(\log a)_0 = (-1)^q a_q \prod_{j \in \overline{M}} \varrho_j,$$

while for the constant $E(a)$ we obtain

$$\begin{aligned} E(a) &:= \exp \sum_{s \geq 1} s (\log a)_s (\log a)_{-s} \\ &= \exp \sum_{s \geq 1} \frac{1}{s} \sum_{\substack{j \in \overline{M} \\ k \in M}} \frac{\varrho_k^s}{\varrho_j^s} \\ &= \exp \sum_{\substack{j \in \overline{M} \\ k \in M}} \log \Big(1 - \frac{\varrho_k}{\varrho_j} \Big)^{-1} \\ &= \prod_{\substack{j \in \overline{M} \\ k \in M}} \Big(1 - \frac{\varrho_k}{\varrho_j} \Big)^{-1} \\ &= \Big(\prod_{j \in \overline{M}} \varrho_j^p \Big) \prod_{\substack{j \in \overline{M} \\ k \in M}} (\varrho_j - \varrho_k)^{-1}. \end{aligned}$$

Consequently, Theorem 5.2 gives (5.57) with $o(1)$ in place of the remainder $O(|\varrho_p/\varrho_{p+1}|^n)$. ■

Here is the other main ingredient of the proof of Theorem 5.28.

Theorem 5.32. *Let a be given by (5.52) and index the roots $z_1(\lambda), \ldots, z_{p+q}(\lambda)$ of $z^p(a(z) - \lambda)$ so that $0 < |z_1(\lambda)| \le |z_2(\lambda)| \le \ldots \le |z_{p+q}(\lambda)|$. Then*

$$\bigcap_{r \in (0,\infty)} \operatorname{sp} T(a_r) = \{\lambda \in \mathbf{C} : |z_p(\lambda)| = |z_{p+1}(\lambda)|\}.$$

A proof is in [148]. We remark that despite its beauty, this theorem is effectively applicable to a few rare cases only (e.g., try doing Example 5.27 with the help of this theorem!).

Finally, Hirschman [103] was able to associate with every trigonometric polynomial a a measure μ_a supported on the set (5.55) such that

$$\frac{1}{n}\sum_{j=1}^{n} f(\lambda_j^{(n)}) \to \int_{\mathbf{C}} f(\lambda)\, d\mu_a(\lambda) \tag{5.58}$$

for every $f \in C_0(\mathbf{C})$. Of course, here $\lambda_1^{(n)}, \ldots, \lambda_n^{(n)}$ are the eigenvalues of $T_n(a)$. This measure is rather complicated in the general case. In the case where a is as in Example 5.27, Hirschman showed that (see Figure 40)

$$d\mu_a(x) = (1/\pi)(4\varepsilon^2 - x^2)^{-1/2}\, dx, \quad x \in [-2\varepsilon, 2\varepsilon]. \tag{5.59}$$

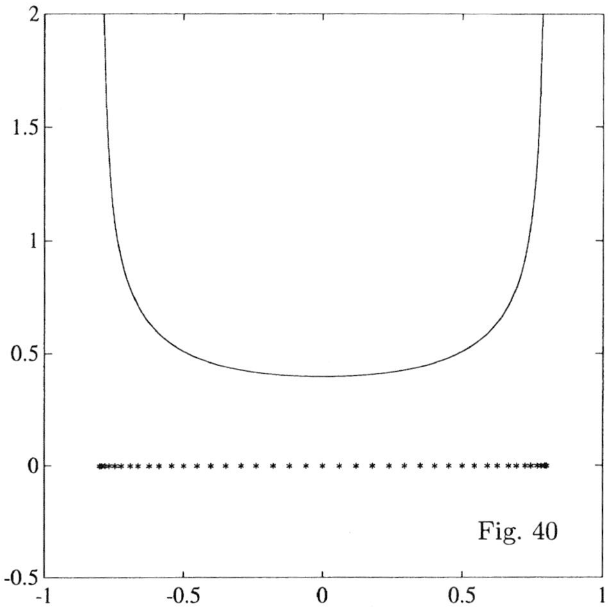

Fig. 40

Let $a(t) = t + \varepsilon^2 t^{-1}$ and $\varepsilon = 0.4$. Figure 40 shows the density function (5.59) and the eigenvalues of $T_{61}(a)$.

Preconditioning. Let a be the trigonometric polynomial (5.52) and define a_r by (5.53). Then

$$T_n(a_r) = D_r T_n(a) D_r^{-1} \quad \text{where} \quad D_r := \operatorname{diag}(r, r^2, \ldots, r^n).$$

Hence, $T_n(a)$ and $T_n(a_r)$ have the same eigenvalues. Clearly, we may think of $T_n(a_r)$ as resulting from $T_n(a)$ by some kind of preconditioning: an appropriate choice of r gives a matrix $T_n(a_r)$ whose numerically computed eigenvalues agree much better with the actual eigenvalues of $T_n(a)$ than those obtained by feeding $T_n(a)$ itself into the computer.

The strategy for choosing r is to produce a symbol a_r for which the spectrum of $T(a_r)$ is as "small" (or "thin") as possible. The spectrum of $T(a_r)$ can in turn be easily read off from the range $\mathcal{R}(a_r)$ (Theorems 1.15 or 1.17) and plotting $\mathcal{R}(a_r)$ for a few values of r is no problem.

Example 5.33. Let a be the symbol considered in Example 5.27,

$$a(t) = t + \varepsilon^2 t^{-1} \quad (t \in \mathbf{T}).$$

Suppose $\varepsilon = 0.4$. Using matlab to compute the eigenvalues of $T_n(a)$ we obtain Figure 41a for $n = 40$ and Figure 41b for $n = 60$. We know that the actual eigenvalues of a are distributed on the segment $[-2\varepsilon, 2\varepsilon] = [-0.8, 0.8]$ with the density (5.59); also see Figure 40. Thus, Figure 41a can be accepted, but Figure 41b is absolutely wrong.

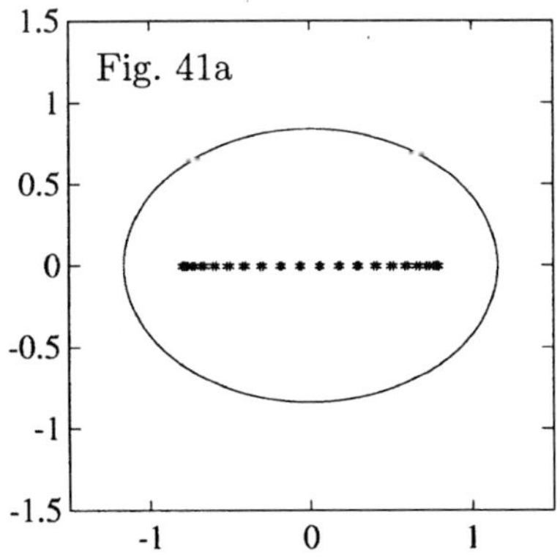

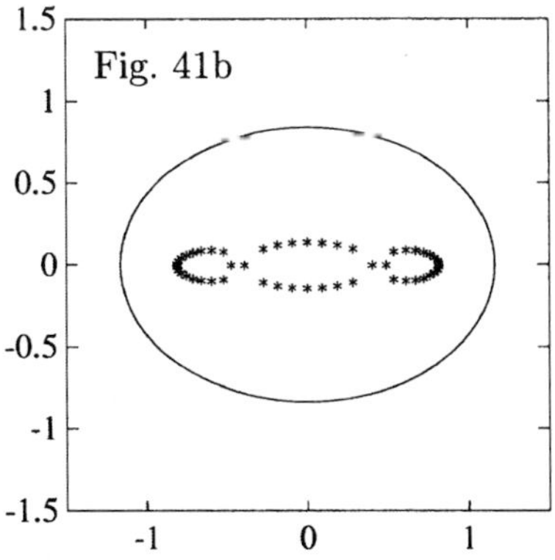

If the symbol a is bad in the sense that the interior of the set sp$T(a)$ is a heavy set, matlab may give absolutely the wrong eigenvalues. In the case at hand, the set sp $T(a)$ is an ellipse.

Choosing $r = \varepsilon = 0.4$, we get $a_\varepsilon(e^{i\theta}) = \varepsilon(e^{i\theta} + e^{-i\theta}) = 2\varepsilon \cos\theta$. This is a real-valued function and hence $T_n(a_\varepsilon)$ is Hermitian. While matlab yields wrong eigenvalues of $T_n(a)$ for $n = 60$, it gives the correct eigenvalues of $T_n(a_\varepsilon)$ even in the case where n is very large. ■

Example 5.34. Let a be the symbol of Example 3.16:

$$a(t) = -t^4 - (3+2i)t^{-3} + it^2 + t^{-1} + 10t + (3+i)t^2 + 4t^3 + it^4.$$

The range $\mathcal{R}(a)$ is the fish of Figure 16, and the set (5.55) consists of five analytic arcs; the 100 eigenvalues plotted in Figure 16 all lie on (or at least very close to) these five arcs. The wrong eigenvalue distributions shown in Figures 18, 19, 20, 21 stem from the fact that $\operatorname{sp} T(a)$, i.e., the boundary and the interior of the fish, is a rather heavy set.

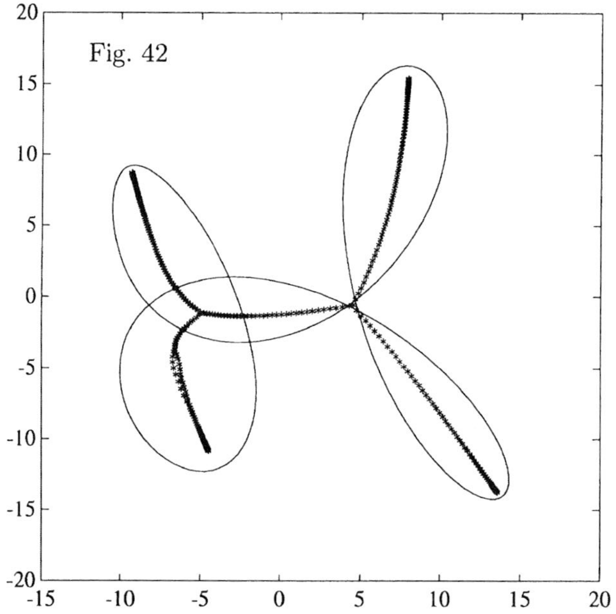

The range $\mathcal{R}(a_r)$ and the 300 eigenvalues of $T_{300}(a_r)$ as they appear on the computer's screen.

In Figure 42 we plotted $\mathcal{R}(a_r)$ for $r = 0.8$. Obviously, $\operatorname{sp} T(a_r)$ is much "thinner" than $\operatorname{sp} T(a)$. Computing the 200 eigenvalues of $T_{200}(a)$ with matlab shows a distribution as in Figure 16 and thus the correct result (recall that without preconditioning we would get Figure 18). Using matlab to compute the 300 eigenvalues of $T_{300}(a)$ gives the distribution plotted in Figure 42. Although the distribution of Figure 42 is slightly garbled in the lower tail fin, it is still very close to the truth (Figure 16) and much better than Figure 22, which exhibits the erroneous eigenvalues obtained without preconditioning. ■

We learned the trick of replacing $T_n(a)$ by $T_n(a_r)$ from Beam and Warming's paper [17]. Of course, this is merely a baby version of preconditioning, but it is extremely easy to use and works astonishingly well.

The preceding two examples are concerned with preconditioning in order to compute eigenvalues. There is a vast literature devoted to preconditioning in connection with the problem of solving large systems with Toeplitz matrices. Here are a few sample articles: the pioneering work is Chan and Strang [44], several circulant preconditioners are compared in Ku and Jay [113] as well as in Strela and Tyrtyshnikov [164], one of the most recent survey articles is Jin [105], and block Toeplitz systems are, for instance, studied in Chan and Jin [45].

A good and fairly elementary introduction to several asymptotic properties of the solutions of large systems with Toeplitz band matrices is Berg's book [19].

5.9 Rational Symbols

For Toeplitz matrices with rational symbols $a \in L^\infty$ the situation is similar to the one of the preceding section, although the technical details cause much more trouble.

Let $a \in L^\infty$ be a rational function. Suppose $T(a)$ is not triangular. Then a is a constant multiple of a function of the form

$$\prod_{j=1}^{r}(t-\varrho_j)\prod_{j=1}^{s}\left(1-\frac{t}{\mu_j}\right)^{-1}\prod_{j=1}^{p}(t-\nu_j)^{-1} \quad (t \in \mathbf{T}),$$

where $\{\mu_1, \ldots, \mu_s\} \subset \mathbf{C}\setminus\overline{\mathbf{D}}$ and $\{\nu_1, \ldots, \nu_p\} \subset \mathbf{D}$. We can without loss of generality assume that $r \geq p$, since otherwise we can pass to adjoint matrices.

Theorem 5.35 (Day). *Let a be as in the preceding paragraph. If the numbers $\varrho_1, \ldots, \varrho_r$ are pairwise distinct, then for every $n \geq 1$,*

$$D_n(a) = \sum_M C_M w_M^n,$$

where the sum is over all $\binom{r}{p}$ subsets $M \subset \{1, \ldots, r\}$ of cardinality $|M| = p$ and, with $\overline{M} := \{1, \ldots, r\}\setminus M$, $P := \{1, \ldots, p\}$, $S := \{1, \ldots, s\}$, the constants are given by

$$w_M := (-1)^{r-p}\prod_{j\in\overline{M}}\varrho_j,$$

$$C_M := \prod_{\substack{j\in\overline{M}\\ l\in P}}(\varrho_j-\nu_l)\prod_{\substack{i\in S\\ k\in M}}(\mu_i-\varrho_k)\prod_{\substack{i\in S\\ l\in P}}(\mu_i-\nu_l)^{-1}\prod_{\substack{j\in\overline{M}\\ k\in M}}(\varrho_j-\varrho_k)^{-1}.$$

This theorem was established by Day [51], another (simpler) proof is in [35, Theorem 6.29]. Determinants of rationally generated Toeplitz matrices have been studied by many people since Day's paper. Nowadays one knows various (seemingly and really) different formulas for such determinants. Their authors include Bart, Gohberg, Kaashoek, van Schagen [9], [81], [83], Berg [18], Gorodetsky [90], [91], Høholdt, Justesen [104], Tismenetsky [168], and Trench [173]. The report [22] is a survey on this topic.

In order to determine the limiting sets of $\operatorname{sp} T_n(a)$, we have to consider

$$a(z) - \lambda = G_m(z,\lambda) \prod_{j=1}^{s} \left(1 - \frac{z}{\mu_j}\right)^{-1} \prod_{j=1}^{p} (z - \nu_j)^{-1}$$

with

$$G_m(z,\lambda) := \prod_{j=1}^{r} (z - \varrho_j) - \lambda \prod_{j=1}^{s} \left(1 - \frac{z}{\mu_j}\right) \prod_{j=1}^{p} (z - \nu_j).$$

If $\lambda \neq 0$ and $\lambda \neq \prod_{j=1}^{s}(-\mu_j)$, then the polynomial $G_m(\cdot,\lambda)$ has the degree $m := \max\{r, s+p\}$.

Theorem 5.36 (Day). *Let a be as in Theorem 5.35. Suppose $\lambda \neq 0$ and $\lambda \neq \prod_{j=1}^{s}(-\mu_j)$. Label the zeros $z_1(\lambda), \ldots, z_m(\lambda)$ of $G_m(z,\lambda)$ so that*

$$|z_1(\lambda)| \le |z_2(\lambda)| \le \ldots \le |z_m(\lambda)|.$$

Then both u-$\lim \operatorname{sp} T_n(a)$ and p-$\lim \operatorname{sp} T_n(a)$ equal

$$\operatorname{clos} \big\{ \lambda \in \mathbf{C} \setminus \{0\} : |z_p(\lambda)| = |z_{p+1}(\lambda)| \big\}. \tag{5.60}$$

For a proof see [51] and [52]. We remark that the closure in the last equality is only caused by the fact that we artificially excluded two values of λ. Day also pointed out that the limiting sets are finite unions of analytic arcs each pair of which have at most endpoints in common and that the limiting sets are always connected. Furthermore, Day [52] showed that there is a limiting measure μ_a supported on (5.60) such that (5.58) holds for every $f \in C_0(\mathbf{C})$. In contrast to the case of trigonometric polynomials, this limiting measure may possess up to two atoms. The symbols a which give rise to atoms and the weights of the atoms are identified in [52].

For a long time it had been an open question whether the limiting set (5.60) can separate the plane. Now examples are known which show that this can happen, i.e., which show that the complement of the limiting set (5.60) may indeed be disconnected. One such example is in Figure 5.3(c) of Beam and Warming's paper [17]. Another example is the symbol in the lower left picture of Figure 38 (the "smiling shark"); the explicit form of the latter symbol is

$$a(t) = -2.8t^{-4} + it^{-3} - 4it^{-1} - 2 - i - 2t + 8it^2 - 5it^3 + it^4.$$

One might conjecture that if $\operatorname{sp} T(a)$ has a hole, then the limiting set $\lim \operatorname{sp} T_n(a)$ surrounds this hole and therefore separates the plane. However, as shown by Figure 43, this is not true in general.

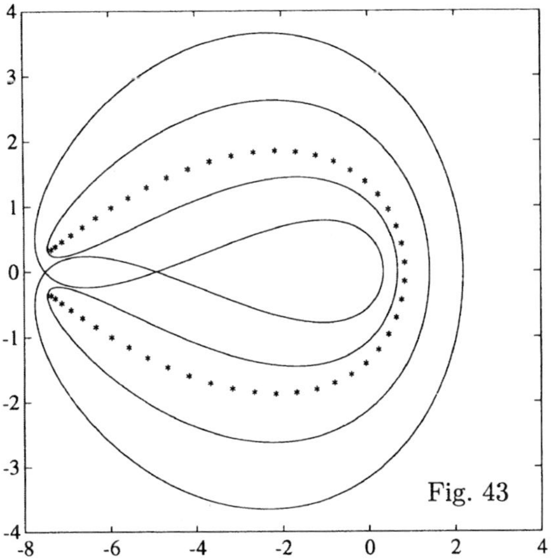

Fig. 43

This figure shows the range of the trigonometric polynomial

$$a(t) = (-0.7t^{-1} + 1 + t)^3$$

(outer curve), the range of a_r for $r = 0.9$ (inner curve), and the eigenvalues of $T_{100}(a)$. We see that $\operatorname{sp} T(a)$ has a hole but that $\lim \operatorname{sp} T_n(a)$ does not separate the plane. This should not come as a surprise, since $\operatorname{sp} T(a_r)$ is seen to have no hole.

Finally, the continuous analogues of Toeplitz operators are Wiener-Hopf integral operators. In [2] and [40], Wiener-Hopf integral operators with rational symbols were discovered whose spectrum is an ellipse (including its interior) and whose limiting set of the spectra of the truncations contains a circle and thus separate the plane.

5.10 Continuous Symbols

Unlike the situation in Chapters 1 to 4, continuous symbols are the most refractory symbols in connection with eigenvalues.

Canonical eigenvalue distribution. Following Widom [189], [190], we say that the eigenvalues of the finite truncations $T_n(a)$ of the infinite

Toeplitz matrix $T(a)$ are *canonically distributed* if

$$\frac{1}{n}\operatorname{tr} f\big(T_n(a)\big) := \frac{1}{n}\sum_{j=1}^{n} f\big(\lambda_j^{(n)}\big) \to \frac{1}{2\pi}\int_0^{2\pi} f\big(a(e^{i\theta})\big)\,d\theta \tag{5.61}$$

for every function $f \in C_0(\mathbf{C})$, i.e., for every continuous function f on $\mathbf{C}$ with compact support. In measure-theoretic language, this is equivalent to saying that μ_n *converges weakly* to μ where μ_n and μ are defined by (5.37) and (5.38) for Borel subsets $E \subset \mathbf{C}$. With these measures μ_n and μ, the validity of (5.61) for all $f \in C_0(\mathbf{C})$ also implies that $\mu_n(E) \to \mu(E)$ for every open or closed disk $E \subset \mathbf{C}$ whenever $|a^{-1}(\partial E)| = 0$.

We know from Theorem 5.10 that (5.61) holds for every $f \in C_0(\mathbf{C})$ if $a \in L^\infty$ and $\operatorname{conv}\mathcal{R}(a)$ is a line segment. Theorem 5.9 tells us that if a is an arbitrary function in L^∞, then (5.61) is valid for every f analytic in some open neighborhood of $\operatorname{conv}\mathcal{R}(a)$, which is, of course, far from the truth of (5.61) for every $f \in C_0(\mathbf{C})$. Widom [189], [190] nevertheless raised the intrepid conjecture that except in rare cases (5.61) is really true for every $f \in C_0(\mathbf{C})$. For example, such a rare case takes place if a extends analytically a little into either the interior or the exterior of $\mathbf{T}$. This happens in particular if a is a rational function, and the preceding two sections show that in this case (5.61) may indeed be false (although the μ_n's have a weak limit μ_a, which, however, is in general different from the measure μ defined by (5.38)).

In this section we record some results which support Widom's conjecture and which are *all* taken from Widom's papers [189], [190].

A key result for understanding the behavior of the eigenvalues of large Toeplitz matrices is the following theorem.

Theorem 5.37 (Widom). *Let $a \in L^\infty$ and suppose there exists a measurable and almost everywhere finite function $g : \mathbf{C} \to \mathbf{R}$ such that*

$$(1/n)\log|D_n(a-\lambda)| \to g(\lambda)$$

for almost all $\lambda \in \mathbf{C}$. Then $|g|$ is locally integrable, $\nu := \Delta g$ (where Δ is the distributional Laplacian) is a measure, and the measure μ_n given by (5.37) *converges weakly to the measure $\nu/(2\pi)$, that is,*

$$\frac{1}{n}\operatorname{tr} f\big(T_n(a)\big) := \frac{1}{n}\sum_{j=1}^{n} f\big(\lambda_j^{(n)}\big) \to \frac{1}{2\pi}\int_{\mathbf{C}} f(\lambda)\,d\nu(\lambda)$$

for every $f \in C_0(\mathbf{C})$.

If $a \in L^\infty$ and $\lambda \notin \mathcal{R}(a)$, then $|a-\lambda|$ is obviously bounded away from zero and hence $G(|a-\lambda|)$ is well defined (see Section 5.4).

Corollary 5.38. *Let $a \in L^\infty$ and assume the (two-dimensional Lebesgue) measure of $\mathcal{R}(a)$ is zero. If*

$$(1/n)\log|D_n(a-\lambda)| \to \log G(|a-\lambda|) \tag{5.62}$$

for almost all $\lambda \in \mathbf{C}\backslash\mathcal{R}(a)$, then the measure μ_n given by (5.37) converges weakly to the measure μ given by (5.38), i.e., (5.61) holds for every f in $C_0(\mathbf{C})$.

These two results are explicitly stated in [190]. They can be proved by the argument used in the proof of [189, Lemma 5.1]. We remark that Corollary 5.38 follows from Theorem 5.37 and the fact that if $\mathcal{R}(a)$ has measure zero, then

$$\frac{1}{2\pi}\Delta_\lambda \log G(|a-\lambda|) = \frac{1}{2\pi}\Delta_\lambda \frac{1}{2\pi}\int_0^{2\pi} \log|a(e^{i\theta}) - \lambda|\, d\theta = \mu(\lambda).$$

Example 5.39: Toeplitz operators with thin spectrum. If $a \in L^\infty$ and $\operatorname{conv}\mathcal{R}(a)$ is a line segment, then $\mathcal{R}(a)$ has measure zero and

$$\frac{|D_n(a-\lambda)|}{|D_{n-1}(a-\lambda)|} \to G(|a-\lambda|) \tag{5.63}$$

for all $\lambda \in \mathbf{C}\setminus\operatorname{conv}\mathcal{R}(a)$ (Theorem 5.8) and thus for almost all λ in the set $\mathbf{C}\setminus\mathcal{R}(a)$. From (5.63) it follows that (5.62) holds. Thus, Corollary 5.38 implies a canonical eigenvalue distribution of $\{T_n(a)\}$, which is in accordance with Theorem 5.10.

Now suppose $a \in C$, $\mathcal{R}(a) = a(\mathbf{T})$ has measure zero, and $\operatorname{sp} T(a) = \mathcal{R}(a)$. This happens, for example, if $\mathcal{R}(a)$ does not separate the plane or is, for instance, the case if $a = b$ is given by (3.24). We deduce again from Theorem 5.8 that (5.63) is true for all $\lambda \in \mathbf{C}\backslash\mathcal{R}(a)$, which implies that (5.62) with $h = 0$ is valid for all $\lambda \in \mathbf{C}\backslash\mathcal{R}(a)$. Consequently, the eigenvalues of $\{T_n(a)\}$ are canonically distributed by virtue of Corollary 5.38. See Figure 44.

We remark that in the case where $\mathcal{R}(a)$ does not separate the plane the same conclusion can be drawn by the approximation argument of the proof of Theorem 5.10 with the Weierstrass theorem replaced by Runge's theorem (see, e.g., [74, p. 76]). ■

To tackle general functions $a \in C$, we have to deal with (5.62) in the situation where $\lambda \notin \mathcal{R}(a)$ and $\operatorname{wind}(a, \lambda) \neq 0$ and are thus led to Toeplitz determinants generated by symbols with nonvanishing index. Asymptotic formulas for such determinants are in [73], [35, Theorem 6.24], and [39, Theorem 10.43], but these formulas do not tell us whether (5.62) holds or not. The appropriate formulas were established by Widom [189].

Assumptions and notations. Recall the definition of PC^α given in Section 5.3 and put $PC^\infty = \bigcap_{\alpha>0} PC^\alpha$. Thus, $a \in PC^\infty$ if and only if there are at most finitely many points $t_1, \ldots, t_N \in \mathbf{T}$ such that a is C^∞ on $\mathbf{T}\backslash\{t_1, \ldots, t_N\}$ and the one-sided limits $a^{(m)}(t_j \pm 0)$ of the mth derivative exist for every $m \geq 0$ and every $j \in \{1, \ldots, N\}$. Also let C^∞ denote the functions $a : \mathbf{T} \to \mathbf{C}$ which have derivatives of all orders.

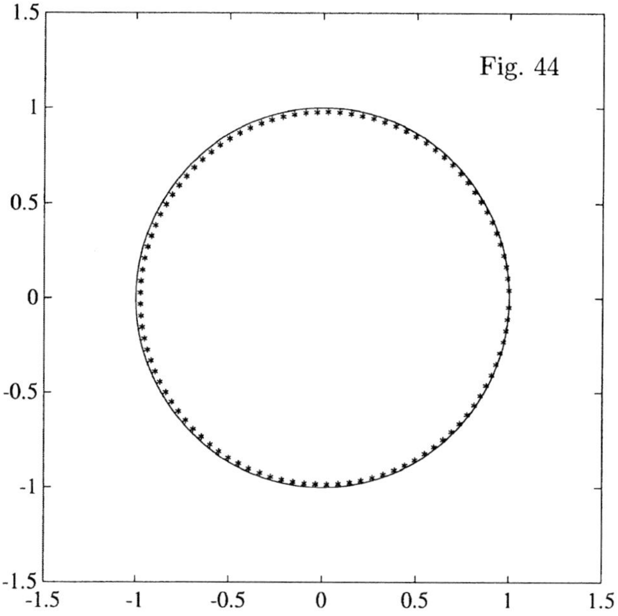

The symbol a given by the right of (3.24) traverses the unit circle once counterclockwise and once clockwise. Hence, $\operatorname{sp} T(a)$ is the unit circle and thus as thin set. In Figure 44 we see that the 100 eigenvalues of $T_{100}(a)$ indeed all lie very close to the unit circle.

Suppose $a \in C \cap (PC^\infty \setminus C^\infty)$, i.e., let a be continuous and piecewise C^∞ but assume a is not C^∞. Then there are finitely many $t_j \in \mathbf{T}$ $(j = 1, \ldots, N)$ at each of which there is a smallest integer $m_j \geq 1$ such that

$$a^{(m_j)}(t_j - 0) \neq a^{(m_j)}(t_j + 0)$$

and a is C^∞ on $\mathbf{T} \setminus \{t_1, \ldots, t_N\}$.

Assume further that a has no zeros on $\mathbf{T}$ and let $\varkappa := \operatorname{wind}(a, 0)$. Then $a(t) = t^\varkappa a_0(t)$ with $\operatorname{wind}(a_0, 0) = 0$. Put

$$H_j := \exp\bigl(i \operatorname{sign} \varkappa \log a_0(t_j)\bigr) \frac{a_0^{(m_j)}(t_j + 0) - a_0^{(m_j)}(t_j - 0)}{a_0(t_j)}.$$

Also let

$$\mathcal{K} := \bigl\{(k_1, \ldots, k_N) \in \mathbf{Z}^N : k_j \geq 0,\ k_1 + \ldots + k_N = |\varkappa|\bigr\},$$

$$Q := \min\Bigl\{ \sum_{j=1}^{N} k_j(k_j + m_j) : (k_1, \ldots, k_N) \in \mathcal{K} \Bigr\},$$

$$\mathcal{K}^* := \left\{ (k_1, \ldots, k_N) \in \mathcal{K} : \sum_{j=1}^{N} k_j(k_j + m_j) = Q \right\}.$$

The terms $G(a_0)$ and $E(a_0)$ are defined as in Section 5.1. Note that, by (5.26), $a \in W \cap B_2^{1/2}$ and hence the series defining $E(a_0)$ converges.

Theorem 5.40 (Widom). *Under the above assumptions,*

$$D_n(a) = (-1)^{n\varkappa} G(a_0)^n E(a_0) n^{-Q}$$
$$\times \left\{ \sum_{(k_1,\ldots,k_N)\in\mathcal{K}^*} C_{k_1,\ldots,k_N} \prod_{j=1}^{N} H_j^{k_j} \left(t_j^{k_j}\right)^n + O\left(\frac{\log n}{n}\right) \right\},$$

where $C_{k_1,\ldots,k_N}$ *are nonzero constants depending only on the* k_j, m_j, t_j.

A proof is in [189].

Corollary 5.41. *Let* $a \in C \cap (PC^\infty \setminus C^\infty)$ *and assume that for each* $\lambda \notin \mathcal{R}(a)$ *with* $\varkappa := \text{wind}(a, \lambda) \neq 0$ *the set* $\mathcal{K}^*$ *is a singleton. Then* (5.61) *holds for every* $f \in C_0(\mathbf{C})$.

Proof. Take $\lambda \in \mathbf{C} \setminus \mathcal{R}(a)$ and apply Theorem 5.40 to the function $a - \lambda$. The sum in the braces consists of a single term and $|D_n(a - \lambda)|$ equals

$$G(|a - \lambda|)^n E(|a - \lambda|) n^{-Q} \left\{ \left| C_{k_1,\ldots,k_N} \prod_{j=1}^{N} H_j^{k_j} \right| + O\left(\frac{\log n}{n}\right) \right\}.$$

Since all constants in this expression are nonzero, it follows that (5.62) is valid with $h = 0$. Corollary 5.38 therefore implies the assertion. ■

Note that the hypothesis of Corollary 5.41 is satisfied if a has exactly one singularity ($N = 1$). Such a case is illustrated in Figure 45.

Other situations in which Corollary 5.41 implies canonical eigenvalue distribution are discussed in [189]. Notice, however, that Corollary 5.41 does not provide any insight into the asymptotic eigenvalue distribution of $\{T_n(a)\}$ if $a \in C^\infty$.

Here is another instructive result. It concerns symbols whose range is a smooth Jordan curve.

Theorem 5.42 (Widom). *Suppose* $a \in C^1$, *a is injective, and* $a'(t) \neq 0$ *for all* $t \in \mathbf{T}$. *Denote by D the bounded component of* $\mathbf{C} \setminus \mathcal{R}(a)$. *If there exists a compact subset* $K \subset D$ *such that* $u\text{-lim sp}\, T_n(a) \subset K$, *then a extends analytically to an annulus* $r < |z| < 1$ (*resp.*, $1 < |z| < r$) *if* $\text{wind}(a, K)$ *equals* 1 (*resp.*, -1). *Conversely, if a does extend analytically as described then there is a compact subset* $K \subset D$ *such that* $p\text{-lim sp}\, T_n(a) \subset K$.

For a proof see [189, Theorem 6.1].

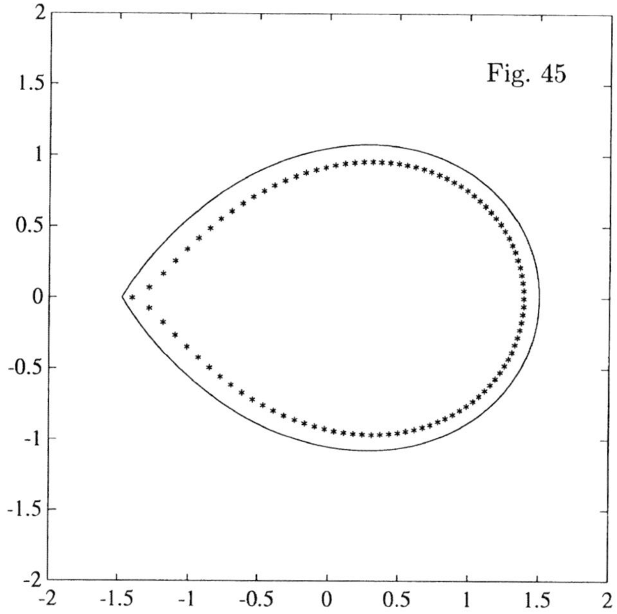

Figure 45 shows the range $\mathcal{R}(a)$ of $a(e^{i\theta}) = (\theta^4/100 + 1)e^{i\theta}$ (where $\theta \in (-\pi, \pi]$) and the 100 eigenvalues of $T_{100}(a)$. The symbol a is continuous but not C^1, so that, in accordance with Corollary 5.41, the eigenvalues are canonically distributed.

5.11 Fisher-Hartwig Determinants

The strong Szegö limit theorem gives an asymptotic expression for $D_n(a)$ provided

(i) the symbol a is bounded and nearly continuous, e.g., $a \in L^\infty \cap B^2_{1/2} \subset (C + H^\infty) \cap (C + \overline{H^\infty}) =: QC$;

(ii) a has no "zeros" on $\mathbf{T}$, i.e., $a \in GL^\infty$;

(iii) a satisfies an index zero condition, i.e., $\operatorname{Ind} T(a) = 0$.

In 1968, Fisher and Hartwig [72] raised a conjecture about the asymptotic behavior of $D_n(a)$ for a large class of symbols a which meet none of the three requirements (i), (ii), or (iii).

The pure jump. Let $\beta \in \mathbf{C} \setminus \mathbf{Z}$. The function φ_β given by

$$\varphi_\beta(e^{i\theta}) := e^{i\beta(\theta - \pi)}, \quad \theta \in [0, 2\pi) \tag{5.64}$$

belongs to PC and has a single jump at $e^{i\theta} = 1$. With ψ_β as in Example 1.7, we have

$$\varphi_\beta(e^{i\theta}) = -\frac{\sin \pi\beta}{\pi}\psi_{-\beta}(e^{i\theta}).$$

Whether one prefers ψ or φ is mainly a question of habit. The Fourier coefficients of ψ look at little better than those of φ, but in connection with Fisher-Hartwig determinants the φ is more convenient. Therefore we henceforth also use the φ.

For $t_0 = e^{i\theta_0} \in \mathbf{T}$, put

$$\varphi_{\beta,t_0}(e^{i\theta}) := \varphi_\beta\big(e^{i(\theta-\theta_0)}\big) = e^{i\beta(\theta-\theta_0-\pi)}, \quad \theta \in [\theta_0, \theta_0 + 2\pi). \tag{5.65}$$

Thus, $\varphi_{\beta,t_0} \in PC$ has a jump at $t_0 = e^{i\theta_0}$ and

$$\varphi_{\beta,t_0}(t_0 + 0) = e^{-\pi i\beta}, \quad \varphi_{\beta,t_0}(t_0 - 0) = e^{\pi i\beta}.$$

A direct computation shows that

$$(\varphi_{\beta,t_0})_n = \frac{\sin \pi\beta}{\pi} e^{-in\theta_0} \frac{1}{\beta - n} \quad (n \in \mathbf{Z}).$$

From Theorem 1.23 (or Example 1.24) we infer that $T(\varphi_{\beta,t_0})$ is Fredholm if and only if $\operatorname{Re}\beta - 1/2 \notin \mathbf{Z}$; in that case

$$\operatorname{Ind} T(\varphi_{\beta,t_0}) = -k \iff k - 1/2 < \operatorname{Re}\beta < k + 1/2.$$

In particular, $T(\varphi_{\beta,t_0})$ is invertible if and only if $|\operatorname{Re}\beta| < 1/2$.

The pure modulus singularity. For $\alpha \in \mathbf{C}$ and $t_0 = e^{i\theta_0} \in \mathbf{T}$, define ω_{α,t_0} by

$$\omega_{\alpha,t_0}(e^{i\theta}) := |e^{i\theta} - e^{i\theta_0}|^{2\alpha} = 2^{2\alpha}\left|\sin\frac{\theta-\theta_0}{2}\right|^{2\alpha}, \quad \theta \in [0, 2\pi).$$

The function ω_{α,t_0} has a zero at t_0 if $\operatorname{Re}\alpha > 0$, a pole at t_0 if $\operatorname{Re}\alpha < 0$, and a discontinuity of oscillating type at t_0 if $\operatorname{Re}\alpha = 0$ but $\alpha \neq 0$. An example of a zero is plotted in Figure 46. Obviously,

$$\omega_{\alpha,t_0} \in L^1 \iff \operatorname{Re}\alpha > -\frac{1}{2}, \qquad \omega_{\alpha,t_0} \in L^\infty \iff \operatorname{Re}\alpha \geq 0.$$

One can show that $T(\omega_{\alpha,t_0})$ is Fredholm on l^2 if and only if $\operatorname{Re}\alpha = 0$, in which case $T(\omega_{\alpha,t_0})$ is invertible. If $\operatorname{Re}\alpha > -1/2$, the Fourier coefficients of ω_{α,t_0} are given by

$$(\omega_{\alpha,t_0})_n = e^{-in\theta_0}\frac{\Gamma(1+2\alpha)}{\Gamma(\alpha - n + 1)\Gamma(\alpha + n + 1)} = O\left(\frac{1}{n^{1+2\operatorname{Re}\alpha}}\right);$$

here Γ is the Gamma function and $(\omega_{\alpha,t_0})_n$ is understood to be zero if $\alpha - n + 1$ or $\alpha + n + 1$ is in $\{0, -1, -2, \ldots\}$.

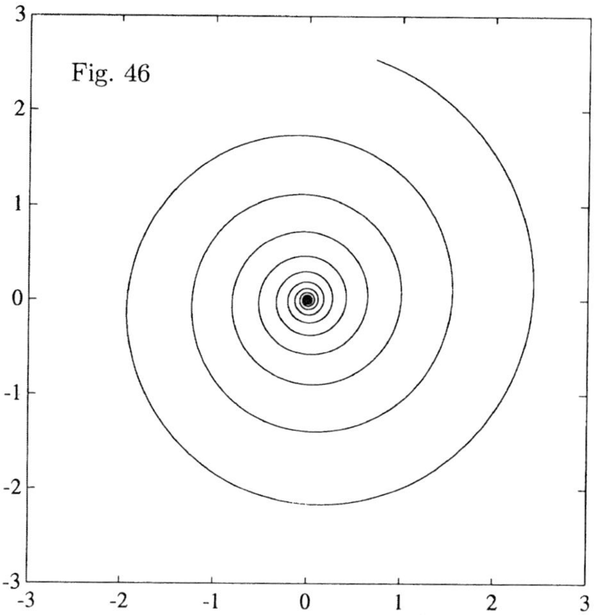

Figure 46 shows $\mathcal{R}(\omega_{\alpha,1})$ for $\alpha = 0.7 + 10i$. We see a logarithmic spiral. As θ moves from $-\pi$ to π, the point $\omega_{\alpha,1}(e^{i\theta})$ traverses this spiral first from the outer endpoint to the origin and then back from the origin to the outer endpoint.

The pure Fisher-Hartwig singularity. This is a function of the form $\omega_{\alpha,t_0}\,\varphi_{\beta,t_0}$ where $t_0 \in \mathbf{T}$, $\alpha, \beta \in \mathbf{C}$, and $\operatorname{Re}\alpha > -1/2$. Note that for $\beta \in \mathbf{Z}$ we define φ_{β,t_0} also by (5.65); in this case

$$\varphi_{\beta,t_0}(e^{i\theta}) = e^{-i\beta(\theta_0+\pi)}e^{i\beta\theta}, \qquad \theta \in [0, 2\pi)$$

is a continuous function and $\operatorname{wind}(\varphi_{\beta,t_0}, 0) = \beta$. We can write

$$\omega_{\alpha,t_0}(t)\,\varphi_{\beta,t_0}(t) = \left(1 - \frac{t_0}{t}\right)^{\alpha-\beta}\left(1 - \frac{t}{t_0}\right)^{\alpha+\beta}, \qquad t \in \mathbf{T},$$

with appropriate branches of the functions on the right, and after some computations we get

$$(\omega_{\alpha,t_0}\,\varphi_{\beta,t_0})_n = \left(-\frac{1}{t_0}\right)^n \frac{\Gamma(1+2\alpha)}{\Gamma(\alpha+\beta-n+1)\Gamma(\alpha-\beta+n+1)} = O\left(\frac{1}{n^{1+2\operatorname{Re}\alpha}}\right),$$

where $(\omega_{\alpha,t_0}\,\varphi_{\beta,t_0})_n = 0$ if $\alpha + \beta - n + 1$ or $\alpha - \beta + n + 1$ belongs to $\{0, -1, -2, \ldots\}$ (see, e.g., [39, Lemma 6.18]).

The Barnes G-function. The entire function G defined by

$$G(z+1) = (2\pi)^{z/2} e^{-z(z+1)/2 - Cz^2/2} \prod_{n=1}^{\infty} \left\{ \left(1+\frac{z}{n}\right)^n e^{-z+z^2/(2n)} \right\},$$

where $C = 0.577\ldots$ is Euler's constant, is referred to as the *Barnes G-function.* It appears frequently in formulas for Toeplitz determinants. This has its reason on the identity $G(z+1) = \Gamma(z)G(z)$, where $\Gamma(z)$ is the Gamma function. We remark that $G(0) = 0$, $G(1) = 1$, and

$$G(n) = (n-2)!\,(n-1)!\ldots 1!\,0! \quad \text{if} \quad n \in \{2, 3, \ldots\}.$$

One can show that for arbitrary $\gamma, \delta \in \mathbf{C}$,

$$\frac{G(n)G(n+\gamma+\delta)}{G(n+\gamma)G(n+\delta)} \sim n^{\gamma\delta} \quad \text{as} \quad n \to \infty, \tag{5.66}$$

where here and in the following $x_n \sim y_n$ means that $y_n \neq 0$ for all sufficiently large n and $x_n/y_n \to 1$ as $n \to \infty$.

The determinants $D_n(a) = \det T_n(a)$ are well-defined for every pure Fisher-Hartwig singularity $a = \omega_{\alpha,t_0}\varphi_{\beta,t_0}$ (note that $a \in L^1$ because $\operatorname{Re}\alpha > -1/2$). As the following theorem shows, they can be evaluated "exactly" and asymptotically estimated precisely.

Theorem 5.43. *If* $\operatorname{Re}\alpha > -1/2$ *and neither* $\alpha+\beta$ *nor* $\alpha-\beta$ *is a negative integer, then* $D_n(\omega_{\alpha,t_0}\varphi_{\beta,t_0})$ *equals*

$$\frac{G(1+\alpha+\beta)G(1+\alpha-\beta)}{G(1+2\alpha)} \frac{G(n+1)G(n+1+2\alpha)}{G(n+1+\alpha+\beta)G(n+1+\alpha-\beta)}$$

for all $n \geq 1$. *Furthermore, we have*

$$D_n(\omega_{\alpha,t_0}\varphi_{\beta,t_0}) \sim \frac{G(1+\alpha+\beta)G(1+\alpha-\beta)}{G(1+2\alpha)} n^{\alpha^2-\beta^2} \quad \text{as } n \to \infty. \tag{5.67}$$

On the other hand, if $\operatorname{Re}\alpha > -1/2$ *and* $\alpha+\beta$ *or* $\alpha-\beta$ *is a negative integer, then*

$$D_n(\omega_{\alpha,t_0}\varphi_{\beta,t_0}) = 0 \quad \text{for all } n \geq 1.$$

For $\alpha = 0$, this result is due to Fisher and Hartwig [72]; note that $D_n(\varphi_{\beta,t_0})$ is a so-called Cauchy determinant. In the form stated here, the theorem was first proved in our paper [38] (also see [39, Theorem 10.54]). The proof is based on an ingenious factorization of $T_n(\omega_{\alpha,t_0}\varphi_{\beta,t_0})$, which was discovered by R. Duduchava and S. Roch (see [39, Theorem 6.20 and formula 7.77(2)]). Of course, (5.67) is immediate from (5.66).

Again the Ising model. What we encounter in applications is not the pure Fisher-Hartwig singularity, but symbols which decompose into the product of a finite number of Fisher-Hartwig singularities and a "nice" function (i.e., a function subject to the above conditions (i), (ii), and (iii)).

Let us look again at the two-dimensional Ising model (Section 5.2). The spontaneous magnetization M is given by

$$M^2 = M^2(T) = \lim_{n\to\infty} D_n(a), \tag{5.68}$$

and in order to show that the model also works for $T \geq T_c$ (i.e., even beyond the phase transition), we have to prove that (5.68) is zero for $T \geq T_c$. The symbol a is given by (5.11), (5.12), and (5.13).

From (5.16) we see that if $T > T_c$, then $0 < A < 1 < B$. In this case we can write $a(t) = t^{-1}b(t)$ with

$$b(t) := \left(\frac{1-At}{1-At^{-1}}\right)^{1/2}\left(\frac{1-B^{-1}t}{1-B^{-1}t^{-1}}\right)^{1/2}, \quad t \in \mathbf{T}.$$

Clearly, b is a positive C^∞ function. The function a is also C^∞, it has no zeros on $\mathbf{T}$, but $\text{wind}(a,0) = -1$, Thus, we arrive at the problem of computing Toeplitz determinants for symbols with nonvanishing index. In [73] and [39, Theorem 10.43] it is shown that if $b \in C^\alpha$ (recall Section 5.3), $b(t) \neq 0$ for $t \in \mathbf{T}$, and $\text{wind}(b,0) = 0$, then

$$D_n(a) = C(a)(-1)^n G(b)^n \big(c_n + O(n^{-3\alpha})\big)\big(1 + O(n^{1-2\alpha})\big),$$

where $C(a)$ is some nonzero constant and $c = b_- b_+^{-1}$, $b = b_- G(b) b_+$ being the normalized Wiener-Hopf factorization (see below). In the case at hand, we have $G(b) = 1$,

$$\begin{aligned} b_-(t) &= (1-At^{-1})^{-1/2}(1-B^{-1}t^{-1})^{-1/2}, \\ b_+(t) &= (1-At)^{1/2}(1-B^{-1}t)^{1/2}. \end{aligned}$$

Obviously, $c_n \to 0$, and hence (5.68) vanishes indeed for $T > T_c$.

If $T = T_c$, then (5.16) implies that $0 < A < B = 1$. Since

$$\left(\frac{1-e^{-i\theta}}{1-e^{i\theta}}\right)^{1/2} = \big(-e^{-i\theta}\big)^{1/2} = \big(e^{i(\pi-\theta)}\big)^{1/2} = e^{i(-1/2)(\theta-\pi)}$$

is just $\varphi_{-1/2}(e^{i\theta}) = \varphi_{-1/2,1}(e^{i\theta})$, we get

$$a(t) = \varphi_{-1/2,1}(t)b(t), \qquad b(t) := \left(\frac{1-At}{1-At^{-1}}\right)^{1/2} \tag{5.69}$$

and thus arrive at the product of a pure jump and a nice function. Note that the jump is especially vicious: $T(\varphi_{-1/2,1})$ is not even Fredholm.

Normalized Wiener-Hopf factorization. Let $b \in GW \cap B_2^{1/2}$ and suppose $\text{wind}(b,0) = 0$. Put

$$b_-(t) := \exp\left(\sum_{n=1}^{\infty}(\log b)_{-n}t^{-n}\right), \qquad b_+(t) := \exp\left(\sum_{n=1}^{\infty}(\log b)_n t^n\right),$$

$$G(b) := \exp(\log b)_0, \qquad E(b) := \exp \sum_{n=1}^{\infty} n(\log b)_n (\log b)_{-n}. \tag{5.70}$$

We know from Section 5.1 that the series defining $E(b)$ converges absolutely. We have

$$b = b_- G(b) b_+ ,$$

and this representation is referred to as the *normalized Wiener-Hopf factorization* of b (recall Section 1.5).

Theorem 5.44. *Suppose* $a = \omega_{\alpha,t_0} \varphi_{\beta,t_0} b$ *where* $t_0 \in \mathbf{T}$, $\alpha, \beta \in \mathbf{C}$, $\operatorname{Re}\alpha > -1/2$, $b \in GW \cap B_2^{1/2}$, $\operatorname{wind}(b,0) = 0$. *Let* $b = b_- G(b) b_+$ *be the normalized Wiener-Hopf factorization of* b. *In addition, assume*

$$b \in C^\infty. \tag{5.71}$$

Then

$$D_n(a) \sim G(b)^n n^{\alpha^2-\beta^2} E(a) \quad \textit{as} \quad n \to \infty \tag{5.72}$$

where

$$E(a) := E(b) \frac{G(1+\alpha+\beta)G(1+\alpha-\beta)}{G(1+2\alpha)} b_+(t_0)^{-\alpha+\beta} b_-(t_0)^{-\alpha-\beta}. \tag{5.73}$$

Such a result was conjectured by Fisher and Hartwig [72]. Note that if a is given by (5.69), then $\alpha = 0$, $\beta = -1/2$, *and* $G(b) = 1$, and hence

$$D_n(a) \sim E(a) n^{-1/4} \quad \text{as} \quad n \to \infty,$$

which shows that $M(T_c) = 0$.

The proof of Theorem 5.44 has been a challenge for almost three decades now. Special cases of the theorem were proved by several authors, including:

Widom 1973 [184]: α, β real, $|\beta| < 1/2$ (without determination of $E(a)$);

Widom 1973 [184]: $\beta = 0$;

Basor 1978 [10]: $\operatorname{Re}\beta = 0$;

Basor '79 [11], Blekher '82 [20], Böttcher '82 [21]: $\alpha = 0$, $|\operatorname{Re}\beta| < 1/2$;

Böttcher, Silbermann 1986 [38]: $\operatorname{Re}\alpha \ge 0$, $|\operatorname{Re}\beta| < \operatorname{Re}\alpha + 1$;

Libby 1990 [118]: $\alpha = 0$, $|\operatorname{Re}\beta| < 5/2$;

Ehrhardt, Silbermann 1996 [70]: in the form stated here.

In [70], the above theorem is even proved under the sole assumption that $2\alpha \notin \{-1,-2,-3,\ldots\}$; for $2\,\mathrm{Re}\,\alpha \le -1$ and $2\alpha \notin \{-1,-2,-3,\ldots\}$ we have to interpret $\omega_{\alpha,t_0}\varphi_{\beta,t_0}b$ in a distributional sense in order to make the matrices $T_n(\omega_{\alpha,t_0}\varphi_{\beta,t_0}b)$ well-defined. A full proof of Theorem 5.44 is in [70].

We wish to point out two refinements of the theorem.

Refinement 5.45. *Let K be a compact subset of $\{(\alpha,\beta) \in \mathbf{C}\times\mathbf{C} : \mathrm{Re}\,\alpha > -1/2\}$, let $\varepsilon \in (0,1)$, and let b be as in Theorem 5.44. Then there exists a constant $C = C(K,b,\varepsilon) < \infty$ such that*

$$\left|\frac{D_n(\omega_{\alpha,t_0}\varphi_{\beta,t_0}b)}{G(b)^n n^{\alpha^2-\beta^2}} - E(a)\right| \le \frac{C}{n^{1-\varepsilon}}$$

for all $(\alpha,\beta) \in K$ and all $n \ge 1$; here $G(b)$ and $E(a)$ are as in Theorem 5.44.

For a proof see [70].

Condition (5.71) can be essentially relaxed, and we refer the reader to the works cited above and to [35] and [39] for precise results. When studying asymptotic eigenvalue distribution we have to deal with the case where

$$b \in C \cap PC^\infty. \tag{5.74}$$

We know from (5.26) that (5.74) implies that

$$b_n = O(1/n^2) \quad \text{as} \quad n \to \infty \tag{5.75}$$

(note that (5.75) cannot be improved even if the PC^2 in (5.26) is replaced by PC^∞). The following result concerns the case where $\alpha = 0$ and $|\mathrm{Re}\,\beta| < 1$ and is applicable to symbols b satisfying (5.75).

Refinement 5.46. *Suppose $a = \varphi_{\beta,t_0}b$ where $t_0 \in \mathbf{T}$, $|\mathrm{Re}\,\beta| < 1$, $b \in GW$, and $\mathrm{wind}(b,0) = 0$. In addition, assume there is an $\varepsilon > 0$ such that*

$$\sum_{|n|\ge 1} |n|^{1/2}|b_n| + \sum_{|n|\ge 1} |n|^{2+\varepsilon}|b_n|^2 < \infty \ \textit{if}\ 0 \le |\mathrm{Re}\,\beta| < \frac{1}{2}, \tag{5.76}$$

$$\sum_{|n|\ge 1} |n|^{|\mathrm{Re}\,\beta|+\varepsilon}|b_n| + \sum_{|n|\ge 1} |n|^{2|\mathrm{Re}\,\beta|+1+\varepsilon}|b_n|^2 < \infty \ \textit{if}\ \frac{1}{2} \le |\mathrm{Re}\,\beta| < 1. \tag{5.77}$$

Then (5.72) and (5.73) hold with $\alpha = 0$.

With (5.77) replaced by the stronger condition

$$\sum_{|n|\ge 1} |n|\,|b_n| + \sum_{|n|\ge 1} |n|^{3+\varepsilon}|b_n|^2 < \infty \ \text{ if }\ \frac{1}{2} \le |\mathrm{Re}\,\beta| < 1, \tag{5.78}$$

this was proved in [38, Theorem 5.4]. The proof given there also works with the spaces $l_2^{\pm(1/2+\varepsilon/4)}$ substituted by $l_2^{\pm(q/2+\varepsilon/4)}$ where q is any number such that $|\mathrm{Re}\,\beta| < q \le 1$, which gives the desired result with (5.77) in place of (5.78).

We now turn to symbols with a finite number of Fisher-Hartwig singularities.

Trace class Hankel operators. We know from Section 5.1 that the two Hankel operators $H(b)$ and $H(\tilde{b})$ are Hilbert-Schmidt if and only if b belongs to the Besov space $B_2^{1/2}$. The characterization of trace class Hankel operators is much more difficult. In 1980, Peller [130] proved the following criterion: if $b \in L^\infty$, then $H(b)$ and $H(\tilde{b})$ are trace class operators on l^2 if and only if $b \in B_1^1$ where B_1^1 is the Besov space of all functions $b \in L^1$ for which

$$\int_{-\pi}^{\pi} \frac{1}{|t|^2} \int_{-\pi}^{\pi} \left| b\big(e^{i(\theta+t)}\big) - 2b(e^{i\theta}) + b\big(e^{i(\theta-t)}\big) \right| d\theta\, dt < \infty.$$

We remark that $C^2 \subset L^\infty \cap B_1^1 \subset W$ (see, e.g., [35, Lemmas 6.47 and 6.49] for elementary "operator" proofs of these two inclusions).

Theorem 5.47. *Let $t_1, \ldots, t_N \in \mathbf{T}$ be pairwise distinct points and let $\alpha_1, \ldots, \alpha_N, \beta_1, \ldots, \beta_N$ be complex numbers satisfying*

$$|\mathrm{Re}\,\alpha_j| < 1/2, \qquad |\mathrm{Re}\,\beta_j| < 1/2 \quad (j = 1, \ldots, N). \tag{5.79}$$

Suppose

$$a = b \prod_{j=1}^{N} \omega_{\alpha_j, t_j}\, \varphi_{\beta_j, t_j},$$

where $b \in L^\infty \cap B_1^1$ does not vanish on $\mathbf{T}$ and $\mathrm{wind}(b, 0) = 0$. *Let $b = b_- G(b) b_+$ be the normalized Wiener-Hopf factorization of b. Then*

$$D_n(a) \sim G(b) n^Q E(a) \quad \text{as} \quad n \to \infty, \tag{5.80}$$

where $Q = (\alpha_1^2 - \beta_1^2) + \ldots + (\alpha_N^2 - \beta_N^2)$,

$$\begin{aligned} E(a) := E(b) &\prod_{j=1}^{N} b_+(t_j)^{-\alpha_j+\beta_j} b_-(t_j)^{-\alpha_j-\beta_j} \\ &\times \prod_{j=1}^{N} \frac{G(1+\alpha_j+\beta_j) G(1+\alpha_j-\beta_j)}{G(1+2\alpha_j)} \\ &\times \prod_{j \ne k} \left(1 - \frac{t_j}{t_k}\right)^{-(\alpha_j-\beta_j)(\alpha_k+\beta_k)}, \end{aligned}$$

and $G(b)$ and $E(b)$ are given by (5.70).

A result of this type was again conjectured by Fisher and Hartwig [72] and subsequently confirmed in a series of special cases. Under the sole assumption that $\operatorname{Re}\alpha_j > -1/2$ for all j and with some more smoothness of b, the theorem was proved by:

Widom 1973 [184]: $\beta_j = 0$ for all j;

Basor 1978 [10]: $\operatorname{Re}\beta_j = 0$ for all j;

Silbermann 1981 [153]: $\alpha_j = \beta_j$ for all j or $\alpha_j = -\beta_j$ for all j.

For $|\operatorname{Re}\alpha_j| < 1/2$, the theorem was established by:

Basor 1979 [11], Böttcher 1982 [21]: $\alpha_j = 0$ for all j;

Böttcher, Silbermann 1983 [35]: $\alpha_j\beta_j = 0$ for all j;

Böttcher, Silbermann 1985 [37]: in the form stated here.

More details and a full proof are in [37] and [39].

We remark that the restriction (5.79) is an essential one. In [33] we considered the case where $\alpha_j \pm \beta_j \in \mathbf{Z}$ for all j and determined the asymptotics of $D_n(a)$ in this case: it turns out that $D_n(a)$ behaves as in (5.80) in rare cases only (see also Theorems 5.29 and 5.35).

Ambiguous symbols. To convince the reader of the relevancy of conditions (5.79), let us consider the symbol

$$a = \varphi_{\beta_1,t_1}\varphi_{\beta_2,t_2},$$

where $t_1, t_2 \in \mathbf{T}$ are distinct points. Let us assume for a moment that for almost all $\beta_1, \beta_2 \in \mathbf{C}$ it were true that

$$D_n(a) \sim C_1(a)^n n^{-\beta_1^2-\beta_2^2} C_2(a) \quad \text{as} \quad n \to \infty$$

with certain nonzero constants $C_1(a)$, $C_2(a)$. This is just what the Fisher-Hartwig conjecture predicts. As first pointed out by Widom and made explicit by Basor and Tracy [15], for every $\varkappa \in \mathbf{Z}$ the function a can also be written in the form

$$a = \varphi_{\beta_1+\varkappa,t_1}\varphi_{\beta_2-\varkappa,t_2}(t_1/t_2)^{\varkappa},$$

and hence, our assumption would imply that

$$D_n(a) \sim (t_1/t_2)^{n\varkappa}\tilde{C}_1(a)^n n^{-(\beta_1+\varkappa)^2-(\beta_2-\varkappa)^2}\tilde{C}_2(a) \quad \text{as} \quad n \to \infty$$

with $\tilde{C}_j := C_j\big(a(t)(t_1/t_2)^{\varkappa}\big)$. As there are certainly $\varkappa \in \mathbf{Z}$ such that

$$\beta_1^2 + \beta_2^2 \neq (\beta_1 + \varkappa)^2 + (\beta_2 - \varkappa)^2,$$

this is impossible.

The Basor-Tracy conjecture. This is a reformulation and generalization of the Fisher-Hartwig conjecture, which was proposed by Basor and Tracy in [15] (see also [13] and [14]). It reads as follows.

Let $t_1, \ldots, t_N \in \mathbf{T}$ be distinct points. Suppose

$$a = b_k \prod_{j=1}^{N} \omega_{\alpha_j^{(k)}, t_j} \varphi_{\beta_j^{(k)}, t_j}$$

for values $\alpha_1^{(k)}, \ldots, \alpha_N^{(k)}, \beta_1^{(k)}, \ldots, \beta_N^{(k)} \in \mathbf{C}$ and smooth functions b_k such that $\operatorname{wind}(b_k, 0) = 0$; note that if $N \geq 2$ and a admits one such representation, then a has countably many such representations. Define

$$Q(k) := \sum_{j=1}^{N} \left((\alpha_j^{(k)})^2 - (\beta_j^{(k)})^2 \right),$$

$$Q := \max_k \operatorname{Re} Q(k), \quad \mathcal{K} := \{k : \operatorname{Re} Q(k) = Q\}.$$

One can easily see that $|G(b_k)| =: |G(b)|$ is independent of k. Basor and Tracy conjectured that

$$D_n(a) = \sum_{k \in \mathcal{K}} G(b_k)^n n^{Q(k)} E_k + o\big(|G(b)|^n n^Q\big) \tag{5.81}$$

as $n \to \infty$. Here E_k are certain constants depending on $b_k, \alpha_j^{(k)}, \beta_j^{(k)}, t_j$. Since $|n^{Q(k)}| = n^{\operatorname{Re} Q(k)}$, we can also write (5.81) in the form

$$D_n(a) \doteq |G(b)|^n n^Q \sum_{k \in \mathcal{K}} \left(E_k \omega_k^n \sigma_k^{\log n} \right) + o\big(|G(b)|^n n^Q\big)$$

with constants $\omega_k \in \mathbf{T}$ and $\sigma_k \in \mathbf{T}$. This conjecture fits all the known cases. Some significant progress in proving the Basor-Tracy conjecture has recently been made by Ehrhardt [69].

5.12 Piecewise Continuous Symbols

We finally consider the asymptotic eigenvalue distribution of $T_n(a)$ in case $a \in PC$. In the case where a has more than one jump, this problem still involves lots of mysteries (see, e.g., [13] and [14]), but the results of Section 5.10 indicate that $\operatorname{sp} T_n(a)$ should behave canonically if $a \in PC$ has a single jump.

Example 5.48. Put $\omega = e^{i2\pi/3}$. Hence, $1, \omega, \omega^2$ are three third-unit roots. Define

$$a(e^{i\theta}) = \begin{cases} e^{i\theta} + i & \text{for} \quad 0 \leq \theta < 2\pi/3, \\ e^{i\theta} + i\omega & \text{for} \quad 2\pi/3 \leq \theta < 4\pi/3 \\ e^{i\theta} + i\omega^2 & \text{for} \quad 4\pi/3 < \theta < 2\pi. \end{cases}$$

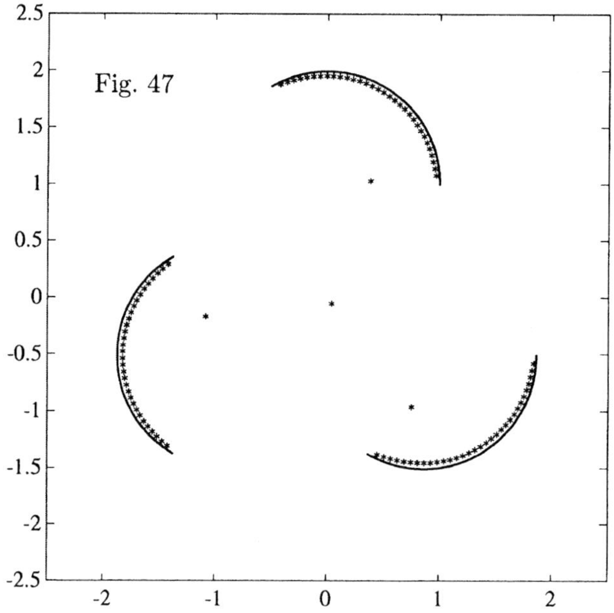

The 100 eigenvalues of $T_{100}(a)$ for the symbol a as in Example 5.48.

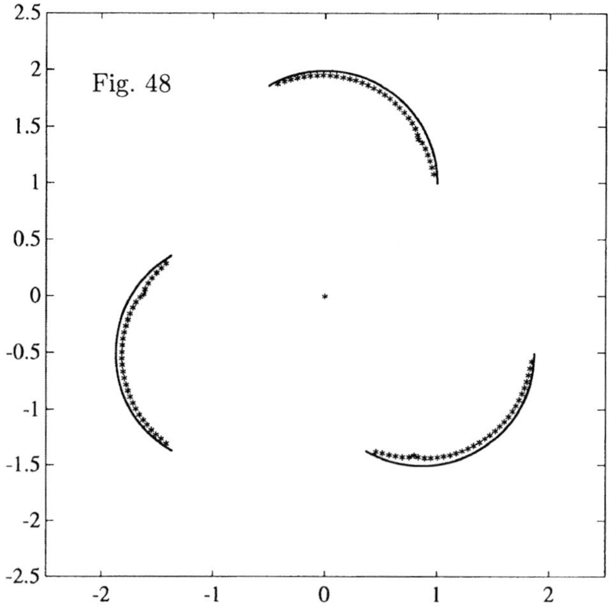

The 101 eigenvalues of $T_{101}(a)$ for the symbol a as in Example 5.48.

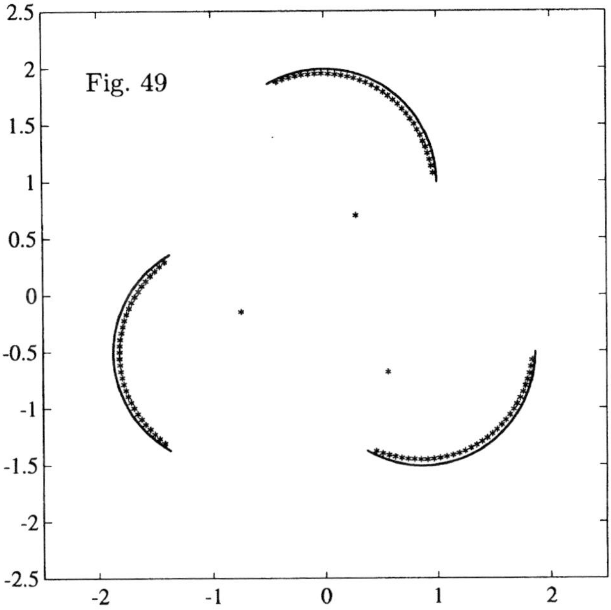

The 102 eigenvalues of $T_{102}(a)$ for the symbol a as in Example 5.48.

The function a has three jumps on the unit circle. The essential range is seen in Figures 47 to 49. These figures also show the eigenvalues of $T_n(a)$ as given by matlab for n about 100 as given by matlab. It results that apart from a few stray eigenvalues almost all eigenvalues cluster along the essential range of a. Such an observation was probably first made by Basor and Morrison [13], [14]. ■

We now turn to symbols with a single jump. Thus, suppose a belongs to $PC \setminus C$ and has exactly one jump, at $t_0 \in \mathbf{T}$ say. For $\lambda \in \mathbf{C}\setminus\mathcal{R}(a)$, choose any continuous argument of $a(t) - \lambda$ on $\mathbf{T}\setminus\{t_0\}$ and put

$$\{\arg(a-\lambda)\} := \arg\big(a(t_0-0)-\lambda\big) - \arg\big(a(t_0+0)-\lambda\big).$$

As usual, let $\lambda_1^{(n)}, \ldots, \lambda_n^{(n)}$ be the eigenvalues of $T_n(a)$.

Theorem 5.49. *Suppose $a \in PC^2\setminus C$ has exactly one jump and*

$$\frac{1}{2\pi}\big|\{\arg(a-\lambda)\}\big| < 1 \quad \text{for each} \quad \lambda \in \mathbf{C}\setminus\mathcal{R}(a). \tag{5.82}$$

Then for every $f \in C_0(\mathbf{C})$,

$$\frac{1}{n}\operatorname{tr} f\big(T_n(a)\big) := \frac{1}{n}\sum_{j=1}^{n} f\big(\lambda_j^{(n)}\big) \;\to\; \frac{1}{2\pi}\int_0^{2\pi} f\big(a(e^{i\theta})\big)\,d\theta \tag{5.83}$$

Proof. Without loss of generality assume that a has its jump at the point 1. Fix $\lambda \in \mathbf{C} \setminus \mathcal{R}(a)$. A little thought reveals that there is a $\beta(\lambda) \in \mathbf{C}$ such that

$$e^{2\pi i \beta(\lambda)} = \frac{a(1-0)-\lambda}{a(1+0)-\lambda}$$

and $a - \lambda = \varphi_{\beta(\lambda)} b_\lambda$ with some $b_\lambda \in C \cap PC^2$ such that $\text{wind}(b_\lambda, 0) = 0$. Condition (5.82) guarantees that $|\text{Re}\, \beta(\lambda)| < 1$. Since $(b_\lambda)_n = O(1/n^2)$ by (5.26), we deduce from Refinement 5.46 that

$$D_n(a-\lambda) \sim G(b_\lambda)^n n^{-\beta(\lambda)^2} E(a-\lambda) \quad \text{as} \quad n \to \infty \tag{5.84}$$

with $G(b_\lambda) = \exp(\log b_\lambda)_0$ and some nonzero constant $E(a-\lambda)$. We have

$$\begin{aligned} \log |G(b_\lambda)| &= \log G(|b_\lambda|) = \big(\log |b_\lambda|\big)_0, \\ \log G(|a-\lambda|) &= \log G(|\varphi_{\beta(\lambda)} b_\lambda|) = \big(\log |\varphi_{\beta(\lambda)} b_\lambda|\big)_0 \\ &= \big(\log |\varphi_{\beta(\lambda)}|\big)_0 + \big(\log |b_\lambda|\big)_0 = \big(\log |b_\lambda|\big)_0, \end{aligned}$$

and hence, by (5.84), $(1/n) \log |D_n(a-\lambda)| \to \log |G(b_\lambda)| = \log G(|a-\lambda|)$. Corollary 5.38 now implies (5.83). ■

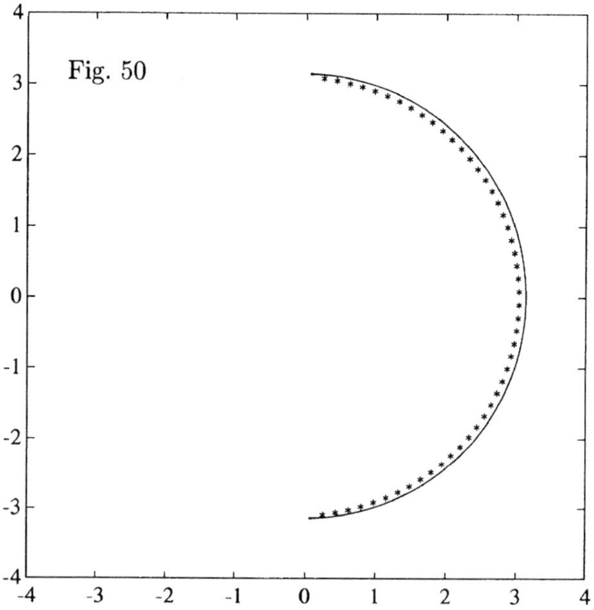

In Figure 50 we plotted the essential range of $\psi_{1/2}$ and the 50 eigenvalues of $T_{50}(\psi_{1/2})$. As predicted by Theorem 5.49, the eigenvalues cluster along the essential range.

Example 5.50: Cauchy-Toeplitz matrices. Let $a = \psi_\beta$ with ψ_β as in Example 1.7 or, which is essentially the same, $a = \varphi_\beta$ with φ_β given by (5.64). The hypothesis of Theorem 5.49 is satisfied whenever $|\mathrm{Re}\,\beta| < 1$. Thus, in this case the eigenvalues of $\{T_n(a)\}$ are canonically distributed. The case $\beta = 1/2$ is covered by Theorem 5.49 and illustrated by Figure 50. In Figure 51 we consider two values of β for which Theorem 5.49 cannot be used. ■

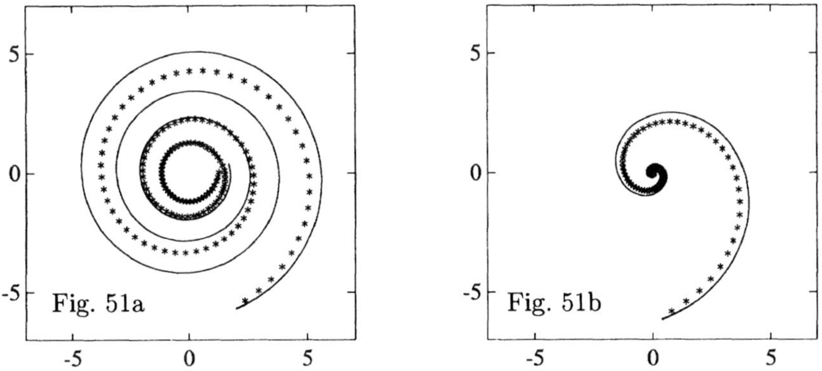

These pictures show the essential range $\mathcal{R}(\psi_\beta)$ and the 150 eigenvalues of $T_{150}(\psi_\beta)$ for $\beta = 3.25 + 0.2i$ (Figure 51a) and $\beta = 3.25 + i$ (Figure 51b). Although Theorem 5.49 is not applicable to the two symbols considered here, the pictures indicate canonical eigenvalue distribution.

Notes. In principle, Theorem 5.49 was established by Widom [189]. More results on the eigenvalue distribution of Toeplitz matrices with piecewise continuous symbols are in Basor and Morrison's papers [13] and [14]. In this connection we also recommend reading Morrison's beautiful article [124].

The proof of Theorem 5.49 is heavily based on Refinement 5.46. The nasty additional hypothesis (5.82) could be removed once we would know that Theorem 5.44 (for $\alpha = 0$) holds with (5.71) replaced (5.74), which would give (5.83) for all $a \in PC^\infty \setminus C$ with a single jump. It seems that Libby's results [118] give Theorem 5.44 for $\alpha = 0$ and $|\mathrm{Re}\,\beta| < 5/2$ with $b \in C \cap PC^\infty$.

6
Block Toeplitz Matrices

6.1 Infinite Matrices

A Toeplitz matrix is constant along the parallels to the main diagonal. Matrices whose entries in the parallels to the main diagonal form periodic sequences (with the same period N) are referred to as *block Toeplitz matrices*. Equivalently, A is a block Toeplitz matrix if and only if

$$A = \begin{pmatrix} a_0 & a_{-1} & a_{-2} & \dots \\ a_1 & a_0 & a_{-1} & \dots \\ a_2 & a_1 & a_0 & \dots \\ \dots & \dots & \dots & \dots \end{pmatrix}, \tag{6.1}$$

where $\{a_k\}_{k\in\mathbf{Z}}$ is a sequence of $N \times N$ matrices, $a_k \in \mathcal{B}(\mathbf{C}^N)$ for all $k \in \mathbf{Z}$.

We may consider the infinite matrix (6.1) as an operator on l^2, but it will be more convenient to study block matrices with $N \times N$ blocks on the $\mathbf{C}^N$-valued l^2 space l^2_N of all sequences $x : \mathbf{N} \to \mathbf{C}^N$ such that

$$\|x\| := \left(\sum_{n=1}^{\infty} \|x_n\|^2 \right)^{1/2} < \infty,$$

where $\|x_n\|^2 = \|(x_n^{(1)}, \dots, x_n^{(n)})\|^2 = |x_n^{(1)}|^2 + \dots + |x_n^{(n)}|^2$.

Boundedness and norm. We denote by $L^\infty_{N\times N}$ the algebra of all matrix functions $a : \mathbf{T} \to \mathbf{C}^{N\times N}$ whose entries belong to L^∞. Multiplication by a matrix function $a \in L^\infty_{N\times N}$ is a bounded operator on the direct sum of N

copies of L^2 endowed with the norm

$$\|f\| = \|(f_1, \dots, f_N)\| = \left(\sum_{j=1}^{N} \|f_j\|_2^2 \right)^{1/2};$$

the norm of this multiplication operator is denoted by $\|a\|_\infty$. The algebra $L^\infty_{N\times N}$ with the norm $\|\cdot\|_\infty$ is a C^*-algebra. Given a subset $A \subset L^\infty$, we let $A_{N\times N}$ stand for the $N \times N$ matrix functions with entries in A. Clearly, $C_{N\times N}$ and $PC_{N\times N}$ are C^*-subalgebras of $L^\infty_{N\times N}$.

Using Theorem 1.9 it is easy to see that the matrix (6.1) induces a bounded operator on l^2_N if and only if there is a matrix function $a \in L^\infty_{N\times N}$ such that

$$a_k = \frac{1}{2\pi} \int_0^{2\pi} a(e^{i\theta}) e^{-ik\theta}\, d\theta \quad (k \in \mathbf{Z}).$$

In that case the matrix (6.1) is denoted by $T(a)$ and the matrix function a is called the *symbol* of $T(a)$. Obvious modifications of the arguments of Sections 1.2 and 1.3 yield the equality

$$\|T(a)\| = \|a\|_\infty \quad \text{for every} \quad a \in L^\infty_{N\times N}.$$

Invertibility. Given a block Toeplitz matrix $T(a)$, it is in general *very difficult* to decide whether it is invertible. The reason is that Coburn's lemma (Theorem 1.10) has no analogue in the block case.

Example 6.1. Let $a(t) = \operatorname{diag}(t, t^{-1})$ $(t \in \mathbf{T})$. Then

$$T(a) = \begin{pmatrix} 0 & 0 & 0 & 0 & 0 & 0 & \\ 0 & 0 & 0 & 1 & 0 & 0 & \\ 1 & 0 & 0 & 0 & 0 & 0 & \\ 0 & 0 & 0 & 0 & 0 & 1 & \\ 0 & 0 & 1 & 0 & 0 & 0 & \\ 0 & 0 & 0 & 0 & 0 & 0 & \\ & & & & & & \ddots \end{pmatrix}.$$

It follows that $\operatorname{Ker} T(a)$ is constituted by the sequences $\{0, x_2, 0, 0, \dots\}$ $(x_2 \in \mathbf{C})$ and that $\operatorname{Im} T(a)$ consists of the sequences in l^2 whose first term is zero. Thus, $T(a)$ is Fredholm of index

$$\dim \operatorname{Ker} T(a) - \dim\big(l^2/\operatorname{Im} T(a)\big) = 1 - 1 = 0,$$

but $T(a)$ is obviously not invertible. ■

Remark 6.2: block Toeplitz matrices as matrices with Toeplitz blocks. Let $a = (a_{jk})_{j,k=1}^N \in L^\infty_{N\times N}$. The Hilbert space l^2_N of all $\mathbf{C}^N$-valued l^2 sequences is canonically isometrically isomorphic to the direct sum of N copies of l^2. Accordingly, the block Toeplitz matrix $T(a)$ on l^2_N is unitarily equivalent to the operator

$$\begin{pmatrix} T(a_{11}) & \dots & T(a_{1N}) \\ \vdots & & \vdots \\ T(a_{N1}) & \dots & T(a_{NN}) \end{pmatrix} : l^2 \oplus \dots \oplus l^2 \to l^2 \oplus \dots \oplus l^2.$$

The latter operator is given by an operator matrix of dimension N whose entries are Toeplitz operators. For example, if $a(t) = \operatorname{diag}(t, t^{-1})$ is as in Example 6.1, we may also think of $T(a)$ as given by

$$\begin{pmatrix} T(\chi_1) & 0 \\ 0 & T(\chi_{-1}) \end{pmatrix} : l^2 \oplus l^2 \to l^2 \oplus l^2$$

(recall that $\chi_\varkappa(t) := t^\varkappa$ for $t \in \mathbf{T}$). From this representation it is immediate that $\dim \operatorname{Ker} T(a) = 1$ and $\dim(l^2/\operatorname{Im} T(a)) = 1$. ∎

The associated operator. For $a \in L^\infty_{N\times N}$, we define the *associated symbol* $\tilde{a} \in L^\infty_{N\times N}$ by $\tilde{a}(t) := a(t^{-1})$ $(t \in \mathbf{T})$. The operator $T(\tilde{a})$ is called the *associated operator* of $T(a)$. In the scalar case $(N = 1)$, $T(\tilde{a})$ is the transposed operator of $T(a)$ and hence $T(\tilde{a})$ is invertible if and only if $T(a)$ is. This is no longer true for $N > 1$.

Example 6.3. Let

$$a(t) = \begin{pmatrix} t & 1 \\ 0 & t^{-1} \end{pmatrix} \qquad (t \in \mathbf{T}).$$

The equation $T(a)x = y$ can be written in the form

$$\begin{pmatrix} 0 & 1 & 0 & 0 & 0 & 0 & \\ 0 & 0 & 0 & 1 & 0 & 0 & \\ 1 & 0 & 0 & 1 & 0 & 0 & \\ 0 & 0 & 0 & 0 & 0 & 1 & \\ 0 & 0 & 1 & 0 & 0 & 1 & \\ 0 & 0 & 0 & 0 & 0 & 0 & \\ & & & & & & \ddots \end{pmatrix} \begin{pmatrix} \xi_1 \\ \eta_1 \\ \xi_2 \\ \eta_2 \\ \xi_3 \\ \eta_3 \\ \vdots \end{pmatrix} = \begin{pmatrix} \varphi_1 \\ \psi_1 \\ \varphi_2 \\ \psi_2 \\ \varphi_3 \\ \psi_3 \\ \vdots \end{pmatrix}$$

and the unique solution is

$$\eta_1 = \varphi_1, \quad \eta_2 = \psi_1, \quad \eta_3 = \psi_2, \quad \dots$$
$$\xi_1 = \varphi_2 - \psi_1, \quad \xi_2 = \varphi_3 - \psi_2, \quad \xi_3 = \varphi_4 - \psi_3, \quad \dots.$$

Thus, $T(a)$ is invertible. We have

$$\tilde{a}(t) = \begin{pmatrix} t^{-1} & 1 \\ 0 & t \end{pmatrix} \qquad (t \in \mathbf{T}),$$

so

$$T(\tilde{a}) = \begin{pmatrix} 0 & 1 & 1 & 0 & 0 & 0 & \\ 0 & 0 & 0 & 0 & 0 & 0 & \\ 0 & 0 & 0 & 1 & 1 & 0 & \\ 0 & 1 & 0 & 0 & 0 & 0 & \\ 0 & 0 & 0 & 0 & 0 & 1 & \\ 0 & 0 & 0 & 1 & 0 & 0 & \\ & & & & & & \ddots \end{pmatrix},$$

and since the second row contains only zeros, the operator $T(\tilde{a})$ is not surjective. ■

The Hartman-Wintner theorem was extended by Simonenko [159] to the matrix case.

Theorem 6.4. *If $a \in L^\infty_{N\times N}$ and $T(a)$ is Fredholm on the space l^2_N, then a is in $GL^\infty_{N\times N}$.*

A proof is in [39, Theorem 2.93], for example.

Continuous symbols. There is an effectively verifiable Fredholm criterion for block Toeplitz operators with continuous symbols.

Given $a \in L^\infty_{N\times N}$, the Hankel operators $H(a)$ and $H(\tilde{a})$ are defined on l^2_N by the matrices A and $\tilde{A}$ of (1.17), respectively. Theorem 1.16 implies that $H(a)$ and $H(\tilde{a})$ are compact whenever $a \in C_{N\times N}$. The formula of Proposition 1.12 remains true in the matrix case:

$$T(ab) = T(a)T(b) + H(a)H(\tilde{b}) \quad \text{for} \quad a, b \in L^\infty_{N\times N}. \tag{6.2}$$

A matrix function $a \in C_{N\times N}$ is invertible in $L^\infty_{N\times N}$ (and thus in $C_{N\times N}$) if and only if its determinant $\det a$ has no zeros on $\mathbf{T}$. Thus, if $\det a(t) \neq 0$ for $t \in \mathbf{T}$, then (6.2) shows that $T(a^{-1})$ is a regularizer of $T(a)$, which yields the Fredholmness of $T(a)$. Conversely, Theorem 6.4 tells us that $\det a$ cannot have zeros on $\mathbf{T}$ if $T(a)$ is Fredholm. Taking into account the well-known fact that a matrix function $a \in GC_{N\times N}$ is homotopic to the diagonal matrix function $\operatorname{diag}(\det a, 1, \ldots, 1)$ within $GC_{N\times N}$, we therefore arrive at the following result.

Theorem 6.5 (Gohberg). *Let $a \in C_{N\times N}$. Then $T(a)$ is Fredholm on l^2_N if and only if $\det a$ has no zeros on $\mathbf{T}$. In that case*

$$\operatorname{Ind} T(a) = -\operatorname{wind}(\det a, 0).$$

This is a classical theorem, and we refer to the monographs [48], [80], [81], and [120] for historical notes.

Wiener-Hopf factorization. We define $W_{N\times N}$ and $W^{\pm}_{N\times N} := (W_{\pm})_{N\times N}$ in the natural manner. The block analogue of Theorems 1.14 and 1.15 is as follows.

Theorem 6.6 (Gohberg and M.G. Krein). *Let $a \in W_{N\times N}$ and suppose* $\det a$ *does not vanish on* $\mathbf{T}$. *Then there exist integers*

$$\varkappa_1 \le \varkappa_2 \le \ldots \le \varkappa_N \tag{6.3}$$

and matrix functions $a_- \in GW^-_{N\times N}$, $a_+ \in GW^+_{N\times N}$ such that

$$a(t) = a_-(t)\operatorname{diag}(t^{\varkappa_1}, t^{\varkappa_2}, \ldots, t^{\varkappa_N})\, a_+(t) \quad \textit{for all} \quad t \in \mathbf{T}.$$

The integers (6.3) *are uniquely determined. We have*

$$\dim \operatorname{Ker} T(a) = \sum_{\varkappa_j<0} |\varkappa_j|, \qquad \dim\big(l^2_N/\operatorname{Im} T(a)\big) = \sum_{\varkappa_j>0} \varkappa_j,$$

and in particular,

$$\operatorname{Ind} T(a) = -(\varkappa_1 + \varkappa_2 + \ldots + \varkappa_N).$$

In this form, the theorem was established by Gohberg and Krein [85]. Full proofs are also in [48], [80], [81], and [120], for example.

The numbers (6.3) are called the *right partial indices* of the matrix function a, while their sum $\varkappa_1 + \ldots + \varkappa_N$ is referred to as the *total index* of a. Comparing Theorems 6.5 and 6.6 we see that

$$\varkappa_1 + \ldots + \varkappa_N = \operatorname{wind}(\det a, 0). \tag{6.4}$$

Clearly, $T(a)$ is invertible if and only if all its right partial indices are zero. Notice, however, that determining the partial indices of a matrix function is in general a highly nontrivial problem.

The right partial indices of the associated matrix function $\tilde{a}$ are called the *left partial indices* of a. If $\nu_1 \le \nu_2 \le \ldots \le \nu_N$ are the left partial indices of a, then, obviously,

$$a(t) = b_+(t)\operatorname{diag}(t^{\nu_1}, t^{\nu_2}, \ldots, t^{\nu_N})\, b_-(t) \quad (t \in \mathbf{T})$$

with $b_+ \in GW^+_{N\times N}$ and $b_- \in GW^-_{N\times N}$. The left and right partial indices of a matrix function need not coincide. However, (6.4) implies that their sums are the same. Obviously, $T(a)$ and $T(\tilde{a})$ are both invertible if and only if all left and all right partial indices equal zero.

Local sectoriality. Given $a \in L^\infty_{N\times N}$ and $\tau \in \mathbf{T}$, we define the *local essential range* $\mathcal{R}_\tau(a)$ as in Section 1.7:

$$\mathcal{R}_\tau(a) = \bigcap_{U\in\mathcal{U}_\tau} \mathcal{R}_U(a)$$

where $\mathcal{R}_U(a) \subset \mathcal{B}(\mathbf{C}^N)$ is the set of essential values taken by $a|U$.

A matrix function $a \in L^\infty_{N\times N}$ is said to be *analytically locally sectorial* on $\mathbf{T}$ if for each $\tau \in \mathbf{T}$ there are matrices $b_\tau, c_\tau \in G\mathcal{B}(\mathbf{C}^N)$ and a number $\varepsilon_\tau > 0$ such that

$$\operatorname{Re}(b_\tau \lambda c_\tau x, x) \geq \varepsilon_\tau \|x\|^2 \text{ for all } \lambda \in \mathcal{R}_\tau(a) \text{ and all } x \in \mathbf{C}^N. \tag{6.5}$$

One calls a matrix function $a \in L^\infty_{N\times N}$ *geometrically locally sectorial* on $\mathbf{T}$ if

$$\operatorname{conv} \mathcal{R}_\tau(a) := \big\{(1-\mu)\lambda_1 + \mu\lambda_2 : \mu \in [0,1];\ \lambda_1, \lambda_2 \in \mathcal{R}_\tau(a)\big\}$$

consists only of invertible matrices. In the scalar case, analytic and geometric local sectorialities are equivalent to the local sectoriality as was defined in Section 1.7. In the matrix case, there are differences. It is easily seen that analytic local sectoriality implies geometric local sectoriality (see, e.g., [39, Proposition 3.2]). Azoff and Clancey [7] showed that the converse is not true for $N > 1$. However, Clancey [46] proved that if $\operatorname{conv} \mathcal{R}_\tau(a)$ is a line segment for every $\tau \in \mathbf{T}$, then analytic and geometric local sectorialities are equivalent. Note that this happens in particular if $a \in PC_{N\times N}$, in which case $\mathcal{R}_\tau(a) = \{a(t-0),\ a(t+0)\}$.

Theorem 6.7 (Simonenko). *If $a \in L^\infty_{N\times N}$ is analytically locally sectorial on $\mathbf{T}$, then $T(a)$ is Fredholm on l^2_N.*

This theorem was established in [159], proofs are also in [39], [48], [120]. We remark that for $N > 1$ the theorem is no longer true with analytic local sectoriality replaced by geometric local sectoriality (see [7]).

Theorem 6.8. *Let $a \in PC_{N\times N}$. Then the following are equivalent:*

(i) *$T(a)$ is Fredholm on l^2_N;*

(ii) *a is analytically locally sectorial on $\mathbf{T}$;*

(iii) *a is geometrically locally sectorial on $\mathbf{T}$;*

(iv) $\det\big((1-\mu)a(t-0) + \mu a(t+0)\big) \neq 0 \ \forall\, \mu \in [0,1] \ \forall\, t \in \mathbf{T}$;

(v) *$a(t \pm 0)$ is invertible for all $t \in \mathbf{T}$ and there is no $t \in \mathbf{T}$ such that $a^{-1}(t-0)a(t+0)$ has eigenvalues on $(-\infty, 0]$.*

If $T(a)$ is Fredholm, then

$$\operatorname{Ind} T(a) = -\operatorname{wind}\big(\det(a^\#), 0\big),$$

where $a^\#$ is the continuous closed and naturally oriented curve in $\mathcal{B}(\mathbf{C}^N)$ which results from the essential range of a by filling in the line segments

$$\big\{(1-\mu)a(t-0) + \mu a(t+0) : \mu \in [0,1]\big\}$$

between the endpoints of each jump.

This theorem is essentially due to Simonenko, Gohberg, and Krupnik. Proofs are in [35], [39], [48], [81], and [120]. The latter book also contains detailed historical notes.

We emphasize that in general $\det(a^\#)$ differs from $(\det a)^\#$.

6.2 Finite Section Method and Stability

Finite section method. For $n \geq 1$, we define the projection P_n on l^2_N by

$$P_n : \{x_1, x_2, x_3, \dots\} \mapsto \{x_1, \dots, x_n, 0, 0, \dots\}$$

(here $x_k \in \mathbf{C}^N$). Thus, if $A \in \mathcal{B}(l^2_N)$ is given by an infinite matrix, then $P_n A P_n | \operatorname{Im} P_n$ is an $nN \times nN$ matrix, which may also be viewed as an $n \times n$ matrix whose entries are $N \times N$ matrices.

In what follows we identify $nN \times nN$ matrices (or, equivalently, $n \times n$ matrices constituted by $N \times N$ blocks) with operators on both

$$\mathbf{C}^{nN} \cong \mathbf{C}^n_N := \mathbf{C}^n \oplus \dots \oplus \mathbf{C}^n$$

and on $\operatorname{Im} P_n$. A sequence $\{A_n\}_{n=1}^\infty$ of matrices $A_n \in \mathcal{B}(\operatorname{Im} P_n)$ is called an *approximating sequence* for $A \in \mathcal{B}(l^2_N)$ if $A_n \to A$ strongly as $n \to \infty$. As in Section 2.1, we write $A \in \Pi\{A_n\}$ and say that the *approximating method* $\{A_n\}$ *is applicable* to A if the matrices A_n are invertible for all sufficiently large n and the (unique) solutions $x^{(n)} \in \operatorname{Im} P_n$ $(n \geq n_0)$ of the equations $A_n x^{(n)} = P_n y$ converge in l^2_N to a solution of the equation $Ax = y$. In the case where $A_n = P_n A P_n | \operatorname{Im} P_n$, we speak of the *finite section method.*

Propositions 2.2 and 2.4 remain valid in the block case: if $\{A_n\}$ is an approximating sequence for $A \in \mathcal{B}(l^2_N)$, then

$A \in \Pi\{A_n\}$

$\Longleftrightarrow$ A is invertible, A_n is invertible for all sufficiently large n, and

$$A_n^{-1} \; (:= A_n^{-1} P_n) \to A^{-1} \quad \text{strongly}$$

$\Longleftrightarrow$ A is invertible and $\{A_n\}$ is stable.

Of course, a sequence $\{A_n\}$ is said to be *stable* if

$$\limsup_{n \to \infty} \|A_n^{-1}\| < \infty.$$

Proposition 2.3 also remains literally true in the matrix case.

Truncated block Toeplitz matrices. For $a \in L^\infty_{N \times N}$, we put

$$T_n(a) = P_n T(a) P_n | \operatorname{Im} P_n = \begin{pmatrix} a_0 & \dots & a_{-(n-1)} \\ \vdots & \ddots & \vdots \\ a_{n-1} & \dots & a_0 \end{pmatrix}.$$

Note that $T_n(a)$ is actually an $nN \times nN$ matrix. The operator W_n is defined on l_N^2 by

$$W_n : \{x_1, x_2, x_3, \ldots\} \mapsto \{x_n, \ldots, x_1, 0, 0, \ldots\}$$

(where, again, $x_k \in \mathbf{C}^N$). It is easily seen that

$$W_n T_n(a) W_n = T_n(\tilde{a}). \tag{6.6}$$

Proposition 2.12 holds in the matrix case without any changes. In particular,

$$T_n(ab) = T_n(a)T_n(b) + P_n H(a)H(\tilde{b})P_n + W_n H(\tilde{a})H(b)W_n \tag{6.7}$$

for every $a, b \in L^\infty_{N\times N}$.

The role of the associated Toeplitz operator. Let $a \in L^\infty_{N\times N}$. On defining $a^* \in L^\infty_{N\times N}$ by

$$a^*(t) = \big(a(t)\big)^* \quad (t \in \mathbf{T}), \tag{6.8}$$

the asterisk on the right-hand side noting passage to the adjoint matrix (= complex conjugate of the transposed matrix), we have

$$T^*(a) := \big(T(a)\big)^* = T(a^*), \qquad T_n^*(a) := \big(T_n(a)\big)^* = T_n(a^*).$$

In the scalar case, $a^* = \overline{a}$.

Now suppose $\{T_n(a)\}$ is stable. As $T_n(a) \to T(a)$ and $T_n(a^*) \to T(a^*)$ strongly, we deduce from the matrix analogue of Proposition 2.3 that $T(a)$ is invertible. By (6.6),

$$\|T_n^{-1}(\tilde{a})\| = \|W_n T_n^{-1}(a) W_n\| = \|T_n^{-1}(a)\|,$$

and hence $\{T_n(\tilde{a})\}$ is also stable. Consequently, $T(\tilde{a})$ must be invertible, too. Example 6.3 tells us that $T(\tilde{a})$ is not automatically invertible if $T(a)$ is invertible ($N > 1$).

Moral: in the block case the invertibility of only $T(a)$ can never suffice to guarantee the stability of $\{T_n(a)\}$; we have to require more, at least the invertibility of also $T(\tilde{a})$.

Now take any $a \in GL^\infty_{N\times N}$. We claim that

$$T(\tilde{a}) \text{ is invertible} \iff T(a^{-1}) \text{ is invertible}. \tag{6.9}$$

This can be checked by using (6.2) and (the block analogue of) (5.2): if $T(\tilde{a})$ is invertible, then

$$T(a) - H(a)T^{-1}(\tilde{a})H(\tilde{a}) \text{ is the inverse of } T(a^{-1}),$$

while if $T(a^{-1})$ is invertible, then

$$T(\tilde{a}^{-1}) - H(\tilde{a}^{-1})T^{-1}(a^{-1})H(a^{-1}) \text{ is the inverse of } T(\tilde{a}).$$

Another way of convincing oneself of the truth of (6.9) is as follows. Define the projections P and Q on $l^2_N(\mathbf{Z})$ by

$$P : \{x_k\}_{k=-\infty}^{\infty} \mapsto \{\ldots, 0, 0, x_0, x_1, \ldots\},$$
$$Q : \{x_k\}_{k=-\infty}^{\infty} \mapsto \{\ldots, x_{-2}, x_{-1}, 0, 0, \ldots\}$$

($x_k \in \mathbf{C}^N$). We then have

$$T(a) \cong PaP|\mathrm{Im}\, P, \qquad T(\tilde{a}^{-1}) \cong Qa^{-1}Q|\mathrm{Im}\, Q,$$
$$T(\tilde{a}) \cong QaQ|\mathrm{Im}\, Q, \qquad T(a^{-1}) \cong Pa^{-1}P|\mathrm{Im}\, P,$$

where a and a^{-1} stand for the block Laurent operators on $l^2_N(\mathbf{Z})$ with the symbols a and a^{-1}, respectively, and $A \cong B$ means that A and B have the same matrix in the standard bases of the spaces. Lemma 2.9 therefore implies that the invertibility of $T(a)$ (resp., $T(\tilde{a})$) is equivalent to the invertibility of $T(\tilde{a}^{-1})$ (resp., $T(a^{-1})$).

Continuous symbols. Here is the block case analogue of Theorems 2.11 and 2.14.

Theorem 6.9. *Let $a \in C_{N\times N}$. Then $\{T_n(a)\}$ is stable if and only if $T(a)$ and $T(\tilde{a})$ are invertible. In that case the operators*

$$K(a) := T^{-1}(a) - T(a^{-1}) \quad \text{and} \quad K(\tilde{a}) := T^{-1}(\tilde{a}) - T(\tilde{a}^{-1})$$

are compact and

$$\begin{aligned} T_n^{-1}(a) &= P_n T^{-1}(a) P_n + W_n K(\tilde{a}) W_n + C_n \\ &= T_n(a^{-1}) + P_n K(a) P_n + W_n K(\tilde{a}) W_n + C_n, \end{aligned}$$

where $\|C_n\| \to 0$ as $n \to \infty$.

Proof. First of all we remark that Theorem 2.10 remains true for $a \in L^\infty_{N\times N}$; its scalar case proof works in the matrix case without changes. If $T(a)$ and $T(\tilde{a})$ are invertible, then $a \in GC_{N\times N}$ by Theorem 6.4 and the operator $T(a^{-1})$ is invertible by virtue of (6.9). The rest of the proof in an almost literal repetition of the proofs of Theorems 2.11 and 2.14. ∎

The algebraic approach. The method of Section 2.4 is not applicable to symbols $a \in PC_{N\times N}$, because for matrix functions an equality like $d_+d_-r = d_+rd_-$ is in general not true. Fortunately, the approach of Sections 2.5 and 2.7 works in the block case as nicely as in the scalar case.

We define the matrix case analogues of $\mathcal{F}, \mathcal{N}, \mathcal{S}, \mathcal{J}$ as in Section 2.5. The only difference is that we now consider sequences $\{A_n\}_{n=1}^{\infty}$ of $nN \times nN$ matrices or, equivalently, of $n \times n$ matrices whose entries are $N \times N$ matrices. Proposition 2.20, Lemmas 2.21 and 2.22, and Theorem 2.23, as well as their proofs given in Section 2.5, remain true in the matrix case.

At the present moment, we confine ourselves to stating the following criterion.

Theorem 6.10. *Let $a \in C_{N\times N}$ and let $K \in \mathcal{K}(l_N^2)$. Then $\{T_n(a)+P_nKP_n\}$ is stable if and only if $T(a)+K$ and $T(\tilde{a})$ are invertible.*

Proof. If $T(a)$ is Fredholm, then $a^{-1} \in C_{N\times N}$ by virtue of Theorem 6.4, and the identity (6.7) shows that $\{T_n(a^{-1})\} + \mathcal{J}$ is the inverse of

$$\{T_n(a)\} + \mathcal{J} = \{T_n(a) + P_nKP_n\} + \mathcal{J}$$

in $\mathcal{S}/\mathcal{J}$. As $W_nKW_n \to 0$ strongly, the assertion is a straightforward consequence of (the block case version of) Theorem 2.23. ■

Localization. From (6.2) and Theorem 1.16 we know that

$$T(a)T(c) - T(c)T(a) \ \text{ is compact} \tag{6.10}$$

whenever $a \in L^\infty$ and $c \in C$. This is not true for arbitrary $a \in L^\infty_{N\times N}$ and $c \in C_{N\times N}$. However, if c is a diagonal matrix with equal entries on the diagonal, then $ac = ca$ for every matrix a. Consequently, (6.10) holds for every $a \in L^\infty_{N\times N}$ and every $c = \mathrm{diag}(\varphi,\ldots,\varphi)$ with $\varphi \in C$.

In analogy to Section 2.7, we define Σ as the set of all $\{A_n\} \in \mathcal{S}$ satisfying

$$\big\{A_nT_n(c) - T_n(c)A_n\big\} \in \mathcal{J}$$

for all $c = \mathrm{diag}(\varphi,\ldots,\varphi)$ with $\varphi \in C$. As in the scalar case, we obtain that $\{T_n(a)\} \in \Sigma$ for every $a \in L^\infty_{N\times N}$. Put $\Sigma^\pi := \Sigma/\mathcal{J}$ and $\{A_n\}^\pi := \{A_n\} + \mathcal{J}$. For $\tau \in \mathbf{T}$, let J_τ stand for the smallest closed two-sided ideal of Σ^π containing the set

$$\big\{\{T_n(c)\}^\pi : c = \mathrm{diag}\,(\varphi,\ldots,\varphi),\ \varphi \in C,\ \varphi(\tau) = 0\big\} \tag{6.11}$$

and denote by $\{A_n\}^\pi_\tau$ the coset $\{A_n\}^\pi + J_\tau$ of the quotient algebra Σ^π/J_τ. Since (6.11) is a central subalgebra of Σ^π whose maximal ideal space can be identified with $\mathbf{T}$ (Proposition 2.34), we can proceed as in Section 2.7 to see that Theorem 2.35 is literally true in the block case.

The left inclusion in Theorem 2.36 also holds in the block case (see [39, Corollary 3.64]). The right inclusion can be replaced by the following.

Theorem 6.11. *Let $a \in L^\infty_{N\times N}$ and let $\tau \in \mathbf{T}$. Suppose there exist matrices $b_\tau, c_\tau \in G\mathcal{B}(\mathbf{C}^N)$ and a number $\varepsilon_\tau > 0$ such that (6.5) is in force. Then $\{T_n(a)\}^\pi_\tau$ is invertible.*

Proof. Taking into account the definition of $\mathcal{R}_\tau(a)$, we can assume that

$$\mathrm{Re}\big(b_\tau a(t)c_\tau x, x\big) \geq (\varepsilon_\tau/2)\|x\|^2$$

for all $x \in \mathbf{C}^N$ and almost all t in some neighborhood $U \in \mathcal{U}_\tau$ of τ. Pick any $\lambda_0 \in \mathcal{R}_\tau(a)$ and define $a_U \in L^\infty_{N\times N}$ by

$$a_U(t) := \begin{cases} a(t) & \text{for} \quad t \in U, \\ \lambda_0 & \text{for} \quad t \in \mathbf{T}\backslash U. \end{cases}$$

Clearly, for all $x \in \mathbf{C}^N$ and almost all $t \in \mathbf{T}$,

$$\mathrm{Re}\big(b_\tau a_U(t) c_\tau x, x\big) \geq (\varepsilon_\tau/2)\|x\|^2.$$

Consequently, for all $z \in \mathbf{C}^N$ and almost all $t \in \mathbf{T}$,

$$\begin{aligned}\mathrm{Re}\big(a_U(t)c_\tau(b_\tau^*)^{-1}b_\tau^* z, b_\tau^* z\big) &= \mathrm{Re}\big(a_U(t)c_\tau z, b_\tau^* z\big) \\ &= \mathrm{Re}\big(b_\tau a_U(t)c_\tau z, z\big) \geq (\varepsilon_\tau/2)\|z\|^2 \geq (\varepsilon_\tau/2)\|b_\tau^*\|^{-2}\|b_\tau^* z\|^2,\end{aligned}$$

whence $\mathrm{Re}\,(a_U(t)d_\tau x, x) \geq \delta_\tau \|x\|^2$ with $d_\tau := c_\tau(b_\tau^*)^{-1}$ and $\delta_\tau := (\varepsilon_\tau/2)\,\|b_\tau^*\|^{-2}$. Put

$$\alpha_\tau := \operatorname*{ess\,sup}_{t\in\mathbf{T}} \|a_U(t)d_\tau\| \; (>0), \qquad \gamma_\tau := \delta_\tau \alpha_\tau^{-2}.$$

If $x \in \mathbf{C}^N$ and $\|x\| = 1$, then

$$\begin{aligned}\big\|\big(I - a_U(t)\gamma_\tau d_\tau\big)x\big\|^2 &= 1 - 2\gamma_\tau \mathrm{Re}\big(a_U(t)d_\tau x, x\big) + \gamma_\tau^2\big\|a_U(t)d_\tau x\big\|^2 \\ &\leq 1 - 2\gamma_\tau\delta_\tau + \gamma_\tau^2\alpha_\tau^2 = 1 - \delta_\tau^2\alpha_\tau^{-2} < 1.\end{aligned}$$

It follows that $\|I - a_U\gamma_\tau d_\tau\|_\infty < 1$ and hence,

$$\big\|\{P_n - T_n(a_U)\gamma_\tau d_\tau P_n\}\big\| \leq \|I - a_U\gamma_\tau d_\tau\|_\infty < 1.$$

Since $\{P_n\}$ is the identity of Σ, we see that $\{T_n(a_U)\}$ is invertible in Σ. Thus, $\{T_n(a_U)\}_\tau^\pi$ is all the more invertible in Σ^π/J_τ. As $\{T_n(a_U)\}_\tau^\pi = \{T_n(a)\}_\tau^\pi$, we arrive at the invertibility of $\{T_n(a)\}_\tau^\pi$. ■

Corollary 6.12. *Let $a \in PC_{N\times N}$ and let $K \in \mathcal{K}(l^2_N)$. Then $\{T_n(a) + P_nKP_n\}$ is stable if and only if $T(a)+K$ and $T(\tilde{a})$ are invertible.*

Proof. If $T(a)$ is Fredholm, we deduce from Theorem 6.8 that a is analytically locally sectorial on $\mathbf{T}$. This in conjunction with Theorem 6.11 implies that $\{T_n(a)\}_\tau^\pi$ is invertible in Σ^π/J_τ for every $\tau \in \mathbf{T}$. From Theorem 2.29 we obtain the invertibility of $\{T_n(a)\}^\pi = \{T_n(a) + P_nKP_n\}^\pi$ in $\Sigma/\mathcal{J}$ and thus in the algebra $\mathcal{S}/J$. The block case version of Theorem 2.23 completes the proof. ■

Corollary 6.13. *If $a \in PC_{N\times N}$ and $K \in \mathcal{K}(l^2_N)$, then*

$$\operatorname*{p\text{-}lim}_{n\to\infty} \mathrm{sp}\big(T_n(a) + P_nKP_n\big) \subset \mathrm{sp}\big(T(a)+K\big) \cup \mathrm{sp}\,T(\tilde{a}).$$

Proof. This follows from Corollary 6.12 by the argument of the proof of Proposition 5.26. ■

Notes. The first half of Theorem 6.9 is Gohberg and Feldman's [80], and the second half is Widom's [185]. The algebraic approach was developed and Theorem 6.10 as well as Corollary 6.12 were established in [154].

6.3 Norms of Inverses and Pseudospectra

Let $\mathbf{A}(PC_{N\times N})$ be the smallest closed subalgebra of $\mathcal{B}(l^2_N)$ containing the set $\{T(a) : a \in PC_{N\times N}\}$ and let $\mathbf{S}(PC_{N\times N})$ stand for the smallest closed subalgebra of $\mathcal{F}$ which contains $\{\{T_n(a)\} : a \in PC_{N\times N}\}$. As in Section 3.3, put

$$\mathbf{A}^\pi(PC_{N\times N}) := \mathbf{A}(PC_{N\times N})/\mathcal{K}(l^2_N), \qquad \mathbf{S}^\pi(PC_{N\times N}) := \mathbf{S}(PC_{N\times N})/\mathcal{J}.$$

Theorems 3.10 and 3.3 imply that the map

$$\mathbf{S}^\pi(PC_{N\times N}) \to \mathbf{A}^\pi(PC_{N\times N}), \quad \{A\} + \mathcal{J} \mapsto \big(s\text{-}\lim_{n\to\infty} A_n\big) + \mathcal{K}(l^2_N)$$

is a well-defined isometric C^*-algebra isomorphism of $\mathbf{S}^\pi(PC_{N\times N})$ onto $\mathbf{A}^\pi(PC_{N\times N})$. Proceeding as in Section 3.3, we therefore get the following generalizations of Theorems 3.11 and 3.12 and of Corollary 6.13.

Theorem 6.14. *A sequence $\{A_n\} \in \mathbf{S}(PC_{N\times N})$ is stable if and only if the strong limits*

$$A := s\text{-}\lim_{n\to\infty} A_n \quad \text{and} \quad \tilde{A} := s\text{-}\lim_{n\to\infty} W_n A_n W_n \tag{6.12}$$

are invertible.

Corollary 6.15. *If $\{A_n\} \in \mathbf{S}(PC_{N\times N})$ and A and $\tilde{A}$ are given by* (6.12), *then*

$$\lim_{n\to\infty} \|A_n\| = \max\{\|A\|, \|\tilde{A}\|\}.$$

In particular, if $a \in PC_{N\times N}$ then

$$\lim_{n\to\infty} \|T_n^{-1}(a)\| = \max\{\|T^{-1}(a)\|, \|T^{-1}(\tilde{a})\|\}. \tag{6.13}$$

We remark that unlike the scalar case, the right-hand side of (6.13) cannot be replaced by $\|T^{-1}(a)\|$. Also notice that (6.13) has to be interpreted as $\|T_n^{-1}(a)\| \to \infty$ in case $T(a)$ or $T(\tilde{a})$ is not invertible.

Corollary 6.16. *If $\{A_n\} \in \mathbf{S}(PC_{N\times N})$, then for each $\varepsilon > 0$,*

$$u\text{-}\lim_{n\to\infty} \operatorname{sp}_\varepsilon A_n = p\text{-}\lim_{n\to\infty} \operatorname{sp}_\varepsilon A_n = \operatorname{sp}_\varepsilon A \cup \operatorname{sp}_\varepsilon \tilde{A},$$

where $\operatorname{sp}_\varepsilon$ *refers to the* ε*-pseudospectrum and* A *and* $\tilde{A}$ *are defined by* (6.12). *In particular, for* $a \in PC_{N\times N}$ *we have*

$$u\text{-}\lim_{n\to\infty} \operatorname{sp}_\varepsilon T_n(a) = p\text{-}\lim_{n\to\infty} \operatorname{sp}_\varepsilon T_n(a) = \operatorname{sp}_\varepsilon T(a) \cup \operatorname{sp}_\varepsilon T(\tilde{a}) \tag{6.14}$$

for each $\varepsilon > 0$.

The proof is the same as the proof of Theorem 3.17 and Corollary 3.18. Notice that the right-hand side of (6.14) cannot be simplified to $\operatorname{sp}_\varepsilon T(a)$ in the case $N > 1$.

Of course, the conclusion of Corollary 3.13(a), (b) and Corollary 3.18(a), (b) also remain literally true in the block case.

Note. Theorem 6.14 is in principle already in [36], Corollaries 6.15 and 6.16 emerged in [23], [145], and [155].

6.4 Distribution of Singular Values

An $nN \times nN$ matrix A_n has nN singular values

$$0 \le s_1(A_n) \le s_2(A_n) \le \ldots \le s_{nN}(A_n) = \|A_n\|.$$

The lowest singular value. Suppose we are given a sequence $\{A_n\} \in \mathcal{F}$ of $nN \times nN$ matrices. As in Section 4.2, we see that

$$\liminf_{n\to\infty} s_1(A_n) > 0 \iff \{A_n\} \text{ is stable.}$$

In particular, from Corollary 6.12 we infer that if a is in $PC_{N\times N}$ and K is an operator in $\mathcal{K}(l^2_N)$, then

$$\liminf_{n\to\infty} s_1\big(T_n(a) + P_n K P_n\big) > 0 \iff T(a) + K \text{ and } T(\tilde{a}) \text{ are invertible.}$$

Upper singular values. The proof of Theorem 4.13 can be easily extended to the block case. It results that if $a \in L^\infty_{N\times N}$, then for each $k \ge 0$,

$$\lim_{n\to\infty} s_{nN-k}\big(T_n(a)\big) = \|T(a)\| = \|a\|_\infty.$$

Limiting sets. The results of Section 4.6 can be carried over to the block case without difficulty. Proceeding exactly as in the proof of Theorem 4.16 we obtain that if $\{B_n\} \in \mathbf{S}(PC_{N\times N})$ and $B_n^* = B_n$ for all n, then

$$u\text{-}\lim_{n\to\infty} \operatorname{sp} B_n = p\text{-}\lim_{n\to\infty} \operatorname{sp} B_n = \operatorname{sp} B \cup \operatorname{sp} \tilde{B}.$$

Letting $B_n = A_n^* A_n$ with $\{A_n\} \in \mathbf{S}(PC_{N\times N})$ we get

$$u\text{-}\lim_{n\to\infty} \Sigma(A_n) = p\text{-}\lim_{n\to\infty} \Sigma(A_n) = \Sigma(A) \cup \Sigma(\tilde{A}),$$

where $\Sigma(A_n) := \{s_1(A_n), \ldots, s_{nN}(A_n)\}$ and $\Sigma(D) := \text{sp}((D^*D)^{1/2})$. In particular, if $a \in PC_{N\times N}$ then

$$u\text{-}\lim_{n\to\infty} \Sigma\big(T_n(a)\big) = p\text{-}\lim_{n\to\infty} \Sigma\big(T_n(a)\big) = \Sigma\big(T(a)\big) \cup \Sigma\big(T(\tilde{a})\big).$$

In the block case, $\Sigma(T_n(a)) \cup \Sigma(T(\tilde{a}))$ need not coincide with $\Sigma(T(a))$.

The splitting property. The methods of Section 4.3 are *not* applicable to the block case. The extension of Theorem 4.5 to matrix symbols is as follows.

Theorem 6.17. *Let $a \in PC_{N\times N}$ and suppose $T(a)$ is Fredholm on l^2_N. This implies that $T(\tilde{a})$ is also Fredholm. Put*

$$k := \dim \operatorname{Ker} T(a) + \dim \operatorname{Ker} T(\tilde{a}). \tag{6.15}$$

Then the singular values of $\{T_n(a)\}$ have the k-splitting property, that is,

$$\lim_{n\to\infty} s_k\big(T_n(a)\big) = 0 \quad \text{and} \quad \liminf_{n\to\infty} s_{k+1}\big(T_n(a)\big) > 0.$$

If $a \in PC_{N\times N}$ and $T(a)$ is not Fredholm, then

$$\lim_{n\to\infty} s_k\big(T_n(a)\big) = 0 \quad \text{for each} \quad k \geq 1.$$

This result was established by Roch and one of the authors in [143]. The assertion concerning non-Fredholmness can be shown by the arguments of the proof of Proposition 4.10. However, the proof of the first part of Theorem 6.17 is highly nontrivial and goes beyond the scope of this book. A full proof can be found in [143].

We remark that, by Theorem 1.10, in the case $N = 1$ we have

$$\dim \operatorname{Ker} T(a) + \dim \operatorname{Ker} T(\tilde{a}) = |\operatorname{Ind} T(a)|,$$

so that in the scalar case Theorem 6.17 indeed yields Theorem 4.5.

6.5 Asymptotic Moore-Penrose Inversion

Moore-Penrose sequences. A sequence $\{A_n\} \in \mathcal{F}$ of $nN \times nN$ matrices A_n is called a *Moore-Penrose sequence* if there exists a sequence $\{B_n\} \in \mathcal{F}$ of $nN \times nN$ matrices such that (4.36) and (4.37) hold.

Theorem 6.18. *A sequence $\{A_n\} \in \mathcal{F}$ is a Moore-Penrose sequence if and only if*

$$\Sigma(A_n) = \big\{s_1(A_n), s_2(A_n), \ldots, s_{nN}(A_n)\big\}$$

has the splitting property, i.e., if and only if there are $c_n \to 0$ and $d > 0$ such that $\Sigma(A_n) \subset [0, c_n] \cup [d, \infty)$ for all $n \geq 1$.

This can be proved in the same manner as Theorem 4.30. Of course, Corollary 4.31 also extends to the block case. This observation together with Theorems 6.17 and 6.18 give the implication (ii) $\Rightarrow$ (iii) of Theorem 4.32 and the strong convergences $B_n \to T^+(a)$ and $B_n^* \to (T^+(a))^*$ for $a \in PC_{N\times N}$. However, there is still another way to this conclusion. Here it is.

Given a closed subspace M of l^2_N, we denote by P_M the orthogonal projection of l^2_N onto M. From Theorem 4.24 we know that if $A \in \mathcal{B}(l^2_N)$ is a normally solvable operator, then $A^*A + P_{\mathrm{Ker}\,A}$ is invertible and A^+ equals $(A^*A + P_{\mathrm{Ker}\,A})^{-1}A^*$.

Theorem 6.19. *Let* $\{A_n\} \in \mathbf{S}(PC_{N\times N})$ *and put*

$$A := s\text{-}\lim_{n\to\infty} A_n, \qquad \tilde{A} := s\text{-}\lim_{n\to\infty} W_n A_n W_n.$$

If A *is Fredholm, then* $\tilde{A}$ *is also Fredholm, the sequence* $\{R_n\}$ *given by*

$$R_n := A_n^* A_n + P_n P_{\mathrm{Ker}\,A} P_n + W_n P_{\mathrm{Ker}\,\tilde{A}} W_n$$

is stable, and (4.36) *and* (4.37) *are satisfied with* $B_n := R_n^{-1} A_n^*$.

Proof. Suppose A is Fredholm. The Fredholmness of $\tilde{A}$ follows without difficulty from the isometric isomorphy of $\mathbf{S}^\pi(PC_{N\times N})$ and $\mathbf{A}^\pi(PC_{N\times N})$ (see Section 6.3). Since $P_{\mathrm{Ker}\,A}$ and $P_{\mathrm{Ker}\,\tilde{A}}$ are finite-rank operators, the sequence $\{R_n\}$ belongs to $\mathbf{S}^\pi(PC_{N\times N})$. Clearly,

$$R_n \to A^*A + P_{\mathrm{Ker}\,A}, \qquad W_n R_n W_n \to \tilde{A}^*\tilde{A} + P_{\mathrm{Ker}\,\tilde{A}},$$

and as A and $\tilde{A}$ are Fredholm and thus normally solvable, these two strong limits are invertible due to Theorem 4.24. From Theorem 6.14 we therefore obtain that $\{R_n\}$ is stable. Thus, $\{B_n\} \in \mathbf{S}(PC_{N\times N})$.

By Propositions 2.2 and 2.4 and by Corollary 4.22, $B_n \to A^+$ and $W_n B_n W_n \to (\tilde{A})^+$ strongly. Hence,

$$\begin{aligned} &A_n B_n A_n - A_n \to AA^+A - A = 0, \\ &W_n A_n B_n A_n W_n - W_n A_n W_n \to \tilde{A}(\tilde{A})^+\tilde{A} - \tilde{A} = 0, \end{aligned}$$

and Corollary 6.15 implies that

$$\lim_{n\to\infty} \|A_n B_n A_n - A_n\| = \max\big\{\|0\|, \|0\|\big\} = 0.$$

This proves the first relation of (4.36). The remaining three relations of (4.36) and (4.37) can be verified analogously. ∎

Recall the definition of a^* by (6.8).

Corollary 6.20. *If* $a \in PC_{N\times N}$ *and* $T(a)$ *is Fredholm, then* $T(\tilde{a})$ *is also Fredholm and*

$$B_n := \big(T_n(a^*)T_n(a) + P_n P_{\mathrm{Ker}\,T(a)} P_n + W_n P_{\mathrm{Ker}\,T(\tilde{a})} W_n\big)^{-1} T_n(a^*)$$

satisfies (4.40), (4.41), *i.e.*, $\{B_n\} + \mathcal{N}$ *is the Moore-Penrose inverse of* $\{T_n(a)\} + \mathcal{N}$.

Proof. Immediate from Theorem 6.19. ■

Since we did not use Theorem 6.17 in the proof of Corollary 6.20, this corollary and Theorem 6.18 show that $\Sigma(T_n(a))$ has the splitting property (but we do not obtain the k-splitting property with k as in (6.15) in this way).

Exact Moore-Penrose sequences. As in Section 4.10, we call a sequence $\{A_n\} \in \mathcal{F}$ of $nN \times nN$ matrices an *exact Moore-Penrose sequence* if (4.34), (4.35) hold with some sequence $\{B_n\} \in \mathcal{F}$ of $nN \times nN$ matrices B_n.

Theorem 6.21. *A sequence* $\{A_n\} \in \mathcal{F}$ *is an exact Moore-Penrose sequence if and only if there is a* $d > 0$ *such that*

$$\Sigma(A_n) \subset \{0\} \cup [d, \infty) \quad \textit{for all} \quad n \geq 1. \tag{6.16}$$

If $\{A_n\} \in \mathcal{F}_{cc}$, *then* (6.16) *is satisfied if and only if* $A = \text{s-lim}\, A_n$ *is normally solvable and* $A_n^+ \to A^+$ *strongly.*

This theorem can be proved as its scalar version (Theorem 4.33 and Corollary 4.34).

The question whether $\{T_n(a)\}$ with $a \in PC$ is an exact Moore-Penrose sequence is answered by Theorem 4.36. Things are much more complicated in the matrix case. Here is a sufficient condition.

Theorem 6.22. *Let* $a \in PC_{N\times N}$, *suppose* $T(a)$ *is Fredholm, and assume there is an* $n_0 \geq 1$ *such that*

$$\operatorname{Ker} T(a) \subset \operatorname{Im} P_{n_0} \quad \textit{and} \quad \operatorname{Ker} T(\tilde{a}) \subset \operatorname{Im} P_{n_0}. \tag{6.17}$$

Then

$$P_{\operatorname{Ker} T_n(a)} = P_n P_{\operatorname{Ker} T(a)} P_n + W_n P_{\operatorname{Ker} T(\tilde{a})} W_n \tag{6.18}$$

for all sufficiently large n *and*

$$T_n^+(a) = \left(T_n(a^*)T_n(a) + P_{\operatorname{Ker} T_n(a)}\right)^{-1} T_n(a^*) \tag{6.19}$$

converges strongly to $T^+(a)$ *as* $n \to \infty$. *In particular,* $\{T_n(a)\}$ *is an exact Moore-Penrose sequence.*

Proof outline. A full proof is in [156, Proposition 3.3]. The equality (6.19) is nothing but a special case of Theorem 4.24. From (6.18), (6.19), and Corollary 6.20 we obtain

$$\limsup_{n\to\infty} \|T_n^+(a)\| < \infty.$$

Hence, $\{T_n^+(a)\} \in \mathcal{F}$ is the Moore-Penrose inverse of $\{T_n(a)\} \in \mathcal{F}$. Since $\{T_n(a)\} \in \mathcal{F}_{cc}$, we deduce from Corollary 4.22 that $\{T_n^+(a)\}$ also belongs to $\mathcal{F}_{cc}$. This implies that $T_n^+(a) \to T^+(a)$ strongly. ■

Conjecture 6.23. *If $a \in PC_{N\times N}$, $T(a)$ is Fredholm, and $\{T_n(a)\}$ is an exact Moore-Penrose sequence, then* (6.17) *holds.*

In the scalar case, this conjecture is confirmed by Theorem 4.36. Notice that in the case $N > 1$ condition (6.17) is *not* equivalent to the existence of an $m_0 \geq 1$ such that $(a^{-1})_0 = 0$ for all $m \geq m_0$ or for all $m \leq -m_0$ (to see this, consider $a = \operatorname{diag}(b, c)$).

Note. Theorems 6.18 and 6.21 were established by Roch and Silbermann in [143], all other results of this section (and also Conjecture 6.23) are from Silbermann [156].

6.6 Trace Formulas

The approach to trace formulas presented in Section 5.7 extends to the block case.

Let $\mathcal{O}$ stand for the collection of all sequences $\{K_n\}$ of operators K_n in $\mathcal{K}(l^2_N)$ such that $\|K_n\|_1/n \to 0$ as $n \to \infty$, where $\|\cdot\|_1$ denotes the trace norm. If a, b, c are in $L^\infty_{N\times N}$, then the sequences (5.44), (5.45), (5.46) belong to $\mathcal{O}$. The quasi-commutator ideal $Q_T(L^\infty_{N\times N})$ of $\mathbf{A}(L^\infty_{N\times N})$ is defined as the smallest closed two-sided ideal of $\mathbf{A}(L^\infty_{N\times N})$ which contains the set

$$\{T(bc) - T(b)T(c) : b, c \in L^\infty_{N\times N}\}.$$

Theorem 5.20 remains true in the matrix case: the algebra $\mathbf{A}(L^\infty_{N\times N})$ decomposes into the direct sum

$$\mathbf{A}(L^\infty_{N\times N}) = T(L^\infty_{N\times N}) \oplus Q_T(L^\infty_{N\times N})$$

and the projection S_T of $\mathbf{A}(L^\infty_{N\times N})$ onto $T(L^\infty_{N\times N})$ parallel to $Q_T(L^\infty_{N\times N})$ is given by

$$S_T(A) = \operatorname*{s\text{-}lim}_{n\to\infty} T(\chi_{-n}I)AT(\chi_n I),$$

where I is the $N \times N$ identity matrix (see, e.g., [39, Corollary 4.3 and Proposition 4.4]). Lemma 5.21 is also valid in the block case.

Let $\mathbf{AS}(L^\infty_{N\times N})$ denote the smallest closed subalgebra of the algebra $\mathcal{F}$ which contains all sequences of the form

$$\{P_n A P_n : A \in \mathbf{A}(L^\infty_{N\times N})\}.$$

The block case analogue of Proposition 5.22 says that if $\{A_n\}$ is a sequence in $\mathbf{AS}(L^\infty_{N\times N})$, then there is a unique $a \in L^\infty_{N\times N}$ such that $A_n = T_n(a) + K_n$

with $\{K_n\} \in \mathcal{O}$; the function a is referred to as the *symbol* of the sequence $\{A_n\}$.

Here is Theorem 5.23 in the block case. In the following results, we denote by $\lambda_1^{(n)}, \ldots, \lambda_{nN}^{(n)}$ the eigenvalues of the matrices under consideration.

Theorem 6.24. *If* $\{A_n\} \in \mathbf{AS}(L^\infty_{N\times N})$ *and* $a \in L^\infty_{N\times N}$ *is the symbol of* $\{A_n\}$, *then*

$$\frac{1}{nN}\operatorname{tr} A_n^k = \frac{1}{nN}\sum_{j=1}^{nN}\left(\lambda_j^{(n)}\right)^k \to \frac{1}{N}\frac{1}{2\pi}\int_0^{2\pi}\operatorname{tr}\left(a(e^{i\theta})\right)^k d\theta$$

for every integer $k \geq 1$.

The proof is the same as the proof of Theorem 5.23: we have

$$\frac{1}{nN}\operatorname{tr} A_n^k = \frac{1}{nN}\operatorname{tr} T_n(a^k) + o(1),$$

and $\operatorname{tr} T_n(a^k) = n\operatorname{tr}(a^k)_0$, where $(a^k)_0 \in \mathcal{B}(\mathbf{C}^N)$ is the zeroth Fourier coefficient of a^k.

Corollary 6.25 (SeLegue's theorem in the block case). *Let* A *in* $\mathbf{A}(L^\infty_{N\times N})$ *be selfadjoint and let* $S_T(A) = T(a)$ $(a \in L^\infty_{N\times N})$. *Then*

$$\frac{1}{nN}\operatorname{tr} f(P_nAP_n) := \frac{1}{nN}\sum_{j=1}^{nN} f\left(\lambda_j^{(n)}\right) \to \frac{1}{N}\frac{1}{2\pi}\int_0^{2\pi}\operatorname{tr} f\left(a(e^{i\theta})\right) d\theta$$

for every function $f \in C_0(\mathbf{R})$.

This can be proved as Corollary 5.24. Note that $p(a(e^{i\theta}))$ is well-defined for almost all $e^{i\theta} \in \mathbf{T}$ in case p is a polynomial. Given $f \in C_0(\mathbf{R})$, there is a sequence $\{p_k\}$ of polynomials p_k which converge uniformly to f on the spectrum of $a = a^*$ in $L^\infty_{N\times N}$. Therefore $\{p_k(a)\}$ is a Cauchy sequence in $L^\infty_{N\times N}$ by the spectral theorem for selfadjoint operators (see, e.g., [139, Theorem VII.1(g)]). It follows that $e^{i\theta} \mapsto f(a(e^{i\theta}))$ is a well-defined function in $L^\infty_{N\times N}$ for every $f \in C_0(\mathbf{R})$.

Corollary 6.26 (Szegö's first limit theorem in the block case). *Let* $a \in L^\infty_{N\times N}$ *and suppose* $a = a^*$ *(i.e.,* $a_k^* = a_{-k}$ *for all* $k \in \mathbf{Z}$*). Then*

$$\frac{1}{nN}\operatorname{tr} f(T_n(a)) := \frac{1}{nN}\sum_{j=1}^{nN} f\left(\lambda_j^{(n)}\right) \to \frac{1}{N}\frac{1}{2\pi}\int_0^{2\pi}\operatorname{tr} f\left(a(e^{i\theta})\right) d\theta$$

for every function $f \in C_0(\mathbf{R})$.

This is Corollary 6.25 with $A = T(a)$.

Corollary 6.27. *Let* $A \in \mathbf{A}(L^\infty_{N\times N})$ *and let* $S_T(A) = T(a)$ $(a \in L^\infty_{N\times N})$. *Let further* $s_1^{(n)}, \ldots, s_{nN}^{(n)}$ *stand for the singular values of* P_nAP_n. *Then*

$$\frac{1}{nN}\sum_{j=1}^{nN} g((s_j^{(n)})^2) \to \frac{1}{N}\frac{1}{2\pi}\int_0^{2\pi} \operatorname{tr} g\big(a^*(e^{i\theta})a(e^{i\theta})\big)\, d\theta \tag{6.20}$$

for every $g \in C_0(\mathbf{R})$.

This follows from Theorem 6.24 with $A_n = (P_nA^*P_n)(P_nAP_n)$ and the fact that the symbol of $\{A_n\}$ is a^*a. The following result is immediate from Corollary 6.27.

Corollary 6.28 (the Avram-Parter theorem in the block case). *If* $a \in L^\infty_{N\times N}$ *and* $s_1^{(n)}, \ldots, s_{nN}^{(n)}$ *are the singular values of* $T_n(a)$, *then*

$$\frac{1}{nN}\sum_{j=1}^{nN} g((s_j^{(n)})^2) \to \frac{1}{N}\frac{1}{2\pi}\int_0^{2\pi} \operatorname{tr} g\big(a^*(e^{i\theta})a(e^{i\theta})\big)\, d\theta \tag{6.21}$$

for every $g \in C_0(\mathbf{R})$.

We remark that the previous corollaries also hold with (6.20) and (6.21) replaced by

$$\frac{1}{nN}\sum_{j=1}^{nN} g(s_j^{(n)}) \to \frac{1}{N}\frac{1}{2\pi}\int_0^{2\pi} \operatorname{tr} g\big(|a(e^{i\theta})|\big)\, d\theta,$$

where $|a| := (a^*a)^{1/2}$.

Note. The above results (and especially Corollary 6.26) are certainly well known to specialists, but we have not found them in the literature.

6.7 The Szegö-Widom Limit Theorem

In this section we establish the analogue of Theorem 5.2 for block Toeplitz matrices.

First of all we remark that if $a \in L^\infty_{N\times N}$, then $T(\tilde{a})$ is invertible if and only if $a \in GL^\infty_{N\times N}$ and $T(a^{-1})$ is invertible (Theorem 6.4 and (6.9)). Further, when proving Theorem 6.9, we already observed that the matrix analogue of Theorem 2.10 is true: if $a \in L^\infty_{N\times N}$ and $T(a)$ is invertible, then $\{P_nT^{-1}(a)P_n\}$ is stable. Hence, using Lemma 5.3 and proceeding as in Section 5.1, we arrive at the conclusion that if $a \in (W \cap B_2^{1/2})_{N\times N}$ and $T(\tilde{a})$ (sic!) is invertible, then

$$\lim_{n\to\infty} \frac{D_n(a)}{\det P_nT^{-1}(a^{-1})P_n} = \det T(a^{-1})T(a). \tag{6.22}$$

Here is the analogue of Proposition 5.4.

Proposition 6.29. *Let $a \in W_{N\times N}$ and suppose $T(\tilde{a})$ (and thus $T(a^{-1})$) is invertible. Then* $\det a$ *has a continuous logarithm* $\log\det a$ *and*

$$\det P_n T^{-1}(a^{-1}) P_n = G(a)^n \quad \textit{for all} \;\; n \geq 1,$$

where

$$G(a) := \exp(\log\det a)_0. \tag{6.23}$$

Proof. The existence of a continuous logarithm of $\det a$ is a consequence of Theorem 6.5. Let $a = a_+ a_-$ be a left Wiener-Hopf factorization of a (recall Section 6.1). We then have $T(a^{-1}) = T(a_-^{-1})T(a_+^{-1})$ and hence,

$$P_n T^{-1}(a^{-1}) P_n = P_n T(a_+) T(a_-) P_n = T_n(a_+) T_n(a_-),$$

which implies that

$$\det P_n T^{-1}(a^{-1}) P_n = \big(\det(a_+)_0 \det(a_-)_0\big)^n.$$

On the other hand, $(\log\det a)_0 = (\log\det a_+)_0 + (\log\det a_-)_0 + 2k\pi i$ with some $k \in \mathbf{Z}$. For $b = \sum_{n\geq 0} b_n \chi_n \in W^+_{N\times N}$, denote by $\hat{b}$ the harmonic extension of b into the unit disk $\mathbf{D}$:

$$\hat{b}(re^{i\theta}) := \sum_{n\geq 0} r^n b_n e^{in\theta}.$$

Clearly, $\hat{b}(0) = b_0$. Thus,

$$\begin{aligned} \exp(\log\det a_+)_0 &= \exp(\log\det a_+)\hat{}(0) \\ &= \exp\log(\det a_+)\hat{}(0) = (\det a_+)\hat{}(0) = \det\big(\hat{a}_+(0)\big) = \det(a_+)_0 \end{aligned}$$

and similarly, $\exp(\log\det a_-)_0 = \det(a_-)_0$. Consequently,

$$\exp(\log\det a)_0 = \det(a_+)_0 \det(a_-)_0. \blacksquare$$

The following theorem was established by Widom [185].

Theorem 6.30 (Szegö-Widom limit theorem). *Let a be a matrix function in $(W \cap B_2^{1/2})_{N\times N}$ and suppose $T(a)$ is Fredholm of index zero. Then* $\det a$ *has a continuous logarithm* $\log\det a$ *on* $\mathbf{T}$, *$T(a^{-1})T(a) - I$ is a trace class operator on l^2_N, and*

$$\lim_{n\to\infty} D_n(a)/G(a)^n = E(a),$$

where $G(a)$ is given by (6.23) and

$$E(a) = \det T(a^{-1})T(a). \tag{6.24}$$

Under the additional assumption that $T(\tilde{a})$ is invertible, this follows from (6.22) and Proposition 6.29. This assumption can be removed by a perturbation argument and a theorem by Widom which says that if $a \in L^\infty_{N\times N}$ and $T(a)$ is Fredholm of index zero, then there exists a trigonometric polynomial $p \in W_{N\times N}$ such that $T(a + \varepsilon p)$ is invertible for all $\varepsilon \in \mathbf{C}$ belonging to some sufficiently small punctured disk with center $\varepsilon = 0$. Details are in [185] (also see [39, Theorem 10.31]).

At the present moment, no expression like (5.4) is known for $E(a) = \det T(a^{-1})T(a)$ in case $N > 1$. Also notice that $E(a) \neq 0$ if and only if both $T(a)$ and $T(\tilde{a})$ are invertible.

Widom [185] extended Theorem 6.30 to symbols a in the matrix-valued Krein algebra $(L^\infty \cap B_2^{1/2})_{N\times N}$ (also see [39, Chapter 10]). A block case analogue of Theorem 5.6 can also be found in [185].

6.8 Rational Matrix Symbols

The purpose of this section is to acquaint the reader with the language developed by Bart, Gohberg, Kaashoek, and van Schagen to treat certain problems on Toeplitz matrices with rational matrix symbols. This theory is well presented in many works of these authors and in the book [81] (for example), by virtue of which we confine ourselves to citing a few sample results without proofs.

Realizations. If C, A, B are any matrices of the sizes $N\times m$, $m\times m$, $m\times N$, respectively, then $C(zI - A)^{-1}B$ is a rational matrix function of the size $N \times N$ which equals the zero matrix at infinity. It is well known from linear systems theory that the converse is also true: if b is a rational matrix function of the size $N \times N$ and $b(\infty)$ is the zero matrix, then there are $N \times m$, $m \times m$, $m \times N$ matrices C, A, B such that

$$b(z) = C(zI - A)^{-1}B.$$

Such a representation is called a *realization* of b. There are infinitely many different realizations of b, but there is an essentially unique realization in which m is minimal.

Suppose a is an $N \times N$ rational matrix function such that $a(\infty) = I$. Let

$$a(z) = I + C(zI - A)^{-1}B \tag{6.25}$$

be any realization, assume A is an $m \times m$ matrix, and suppose that m is minimal. One denotes by $A^\times$ the matrix $A - BC$; note that $A^\times$ depends not only on A but also on B and C. One can show that

$$a^{-1}(z) = I - C(zI - A^\times)^{-1}B. \tag{6.26}$$

Riesz projections. If a and a^{-1} are bounded on $\mathbf{T}$, then neither A nor $A^\times$ have eigenvalues on $\mathbf{T}$ and hence, the matrices (operators)

$$P := I - \frac{1}{2\pi i}\int_{\mathbf{T}} (zI - A)^{-1}\, dz, \qquad P^\times := I - \frac{1}{2\pi i}\int_{\mathbf{T}} (zI - A^\times)^{-1}\, dz$$

are well defined projections. They are referred to as the *Riesz projections* of A and $A^\times$ (associated with the unit disk $\mathbf{D}$).

Assumptions. Throughout the rest of this section we assume that a is a rational $N \times N$ matrix function, that a and a^{-1} are bounded on $\mathbf{T}$, and that we are given a realization (6.25) with an $m \times m$ matrix A in which m is minimal.

Theorem 6.31 (Gohberg-Kaashoek). *The operator $T(a)$ is invertible on l^2_N if and only if $\mathbf{C}^m$ decomposes into the direct sum*

$$\mathbf{C}^m = \operatorname{Ker} P \oplus \operatorname{Im} P^\times.$$

In that case a right Wiener-Hopf factorization $a = a_- a_+$ is given by

$$\begin{aligned} a_-(t) &= I + C(tI - A)^{-1}(I - S)B, \\ a_+(t) &= I + CS(tI - A)^{-1}B, \\ a_-^{-1}(t) &= I - C(I - S)(tI - A^\times)^{-1}B, \\ a_+^{-1}(t) &= I - C(tI - A^\times)^{-1}SB, \end{aligned}$$

where $t \in \mathbf{T}$ and S is the projection of $\mathbf{C}^m$ onto $\operatorname{Im} P^\times$ *parallel to* $\operatorname{Ker} P$.

For a proof, see, e.g., [81, Theorem XXIV.7.1].

This theorem along with the simultaneous invertibility of the two operators $T(\tilde{a})$ and $T(a^{-1})$ (recall (6.9)) implies that $T(a)$ and $T(\tilde{a})$ are both invertible on l^2_N if and only if $\mathbf{C}^m$ decomposes into the direct sums

$$\mathbf{C}^m = \operatorname{Ker} P \oplus \operatorname{Im} P^\times \quad \text{and} \quad \mathbf{C}^m = \operatorname{Ker} P^\times \oplus \operatorname{Im} P. \tag{6.27}$$

One can show that (6.27) is equivalent to the condition

$$\det\big((I - P)(I - P^\times) + PP^\times\big) \neq 0. \tag{6.28}$$

From Theorem 6.9 we therefore deduce that $\{T_n(a)\}$ is stable if and only if (6.28) holds.

Given an operator $T \in \mathcal{B}(\mathbf{C}^m)$, we denote by $PTP|\operatorname{Im} P$ the compression of T to $\operatorname{Im} P \subset \mathbf{C}^m$:

$$PTP|\operatorname{Im} P : \operatorname{Im} P \to \operatorname{Im} P, \; x \mapsto PTx.$$

Theorem 6.32 (Gohberg-Kaashoek-van Schagen). *For all $n \geq 1$, we have* $\det(PAP\,|\operatorname{Im} P) \neq 0$ *and*

$$D_n(a) = \frac{\det(P(A^\times)^n P|\operatorname{Im} P)}{(\det(PAP|\operatorname{Im} P))^n}.$$

A proof is in [83] and [81, Lemma XXIV.11.2].

Theorem 6.32 provides an "exact" formula for the Toeplitz determinants whose nature is entirely different from the formula given by Theorem 5.35. Note that Theorem 6.32 expresses the $nN \times nN$ determinant $D_n(a)$ in terms of the quotient of two determinants of order $\dim \operatorname{Im} P$ (which is independent of n).

Corollary 6.33 (Gohberg-Kaashoek-van Schagen). *In addition to the above assumptions, suppose that $T(a)$ is Fredholm of index zero. Then*

$$\lim_{n\to\infty} D_n(a)/G(a)^n = E(a),$$

where

$$G(a) := \frac{\det(P^\times A^\times P^\times|\operatorname{Im} P^\times)}{\det(PAP|\operatorname{Im} P)}, \tag{6.29}$$

$$E(a) := \det\big((I-P)(I-P^\times)+PP^\times\big) = \det(I-P-P^\times). \tag{6.30}$$

Proof. First of all we remark that, obviously,

$$\det(PTP|\operatorname{Im} P) = \det(I-P+PTP) = \det(I-P+PT)$$

for every $T \in \mathcal{B}(\mathbf{C}^m)$. Thus, from Theorem 6.32 we obtain

$$D_n(a)/G(a)^n = \det\big[\big(I-P+P(A^\times)^n\big)\big(I-P^\times+P^\times(A^\times)\big)^{-n}\big].$$

We have

$$\begin{aligned}
&\big[I-P+P(A^\times)^n\big]\big[I-P^\times+P^\times(A^\times)\big]^{-n} \\
&\quad = (I-P)\big[I-P^\times+P^\times(A^\times)\big]^{-n} + P\big[(I-P^\times)A^\times+P^\times\big]^n \\
&\quad = (I-P)\big[I-P^\times+(P^\times A^\times)^n\big]^{-1} + P\big[\big((I-P^\times)A^\times\big)^n+P^\times\big],
\end{aligned}$$

and this converges to $(I-P)(I-P^\times)+PP^\times$ because the eigenvalues of $P^\times A^\times$ and $(I-P^\times)A^\times$ are located in

$$\{\lambda \in \mathbf{C} : |\lambda| > 1\} \quad \text{and} \quad \{\lambda \in \mathbf{C} : |\lambda| < 1\},$$

respectively. ■

A direct proof of the fact that the right-hand sides of (6.23) and (6.29) coincide is in [83] and [81, Proposition XXIV.11.4]. Widom [187] showed in a direct way that the right-hand sides of (6.24) and (6.30) are equal to one another.

Exact formulas for $D_n(a)$ which look like the formulas of Sections 5.6 and 5.7 were established by Gorodetsky [90], [91] and Tismenetsky [168] (see also [22]).

6.9 Multilevel Toeplitz Matrices

Quarter-plane Toeplitz operators. Let $l^2(\mathbf{N}^2)$ denote the usual l^2 space of the discrete quarter-plane $\mathbf{N}^2 = \mathbf{N}\times\mathbf{N}$. A discrete Wiener-Hopf equation on the quarter-plane is an equation of the form

$$\sum_{k,l\geq 1} a_{i-k,j-l}\, x_{kl} = y_{ij} \qquad (i,j\geq 1); \tag{6.31}$$

here $\{a_{m,n}\}$ and $\{y_{ij}\}$ are given and $\{x_{kl}\}$ is sought. The operator generated by the left-hand side of (6.31) is bounded on $l^2(\mathbf{N}^2)$ if (and only if) there is a function a in L^∞ on the torus $\mathbf{T}^2 = \mathbf{T}\times\mathbf{T}$ such that

$$a_{m,n} = \frac{1}{(2\pi)^2}\int_0^{2\pi}\int_0^{2\pi} a(e^{i\theta}, e^{i\varphi})e^{-im\theta}e^{-in\varphi}\,d\theta\,d\varphi \quad \text{for all} \quad m,n\in\mathbf{Z}.$$

In that case the operator induced by the left-hand side of (6.31) is denoted by $T_{++}(a)$ and the function $a\in L^\infty(\mathbf{T}^2)$ is referred to as the *symbol* of $T_{++}(a)$. The operator $T_{++}(a)$ is called a *Toeplitz operator over the quarter-plane* and the "matrix"

$$T_{++}(a) = (a_{i-k,j-l})_{i,j,k,l=1}^\infty$$

is (nowadays and especially in numerical literature) called a *two-dimensional multilevel* or a *two-level Toeplitz matrix.*

Let $a\in L^\infty(\mathbf{T}^2)$. We then can write

$$a(\xi,\eta) = \sum_{m,n\in\mathbf{Z}} a_{m,n}\xi^m\eta^n = \sum_{m\in\mathbf{Z}}\xi^m b_m(\eta) = \sum_{n\in\mathbf{Z}}\eta^n c_n(\xi), \quad (\xi,\eta)\in\mathbf{T}^2,$$

where $\{b_m\}$ and $\{c_n\}$ are sequences of functions in $L^\infty := L^\infty(\mathbf{T})$. With respect to the decompositions

$$l^2(\mathbf{N}^2) = \bigoplus_{i=1}^\infty l^2(\mathbf{N}\times\{i\}), \qquad l^2(\mathbf{N}^2) = \bigoplus_{j=1}^\infty l^2(\{j\}\times\mathbf{N}),$$

the operator $T_{++}(a)$ is given by the operator matrices

$$T_{++}(a) = \big(T(c_{j-k})\big)_{j,k=1}^\infty, \qquad T_{++}(a) = \big(T(b_{j-k})\big)_{j,k=1}^\infty. \tag{6.32}$$

Thus, quarter-plane Toeplitz matrices may be interpreted as Toeplitz matrices whose entries are Toeplitz matrices.

The Douglas-Howe observation. Suppose $b\in L^\infty$ and define the function $a\in L^\infty(\mathbf{T}^2)$ by $a(\xi,\eta) = b(\xi\eta^{-1})$, $(\xi,\eta)\in\mathbf{T}^2$. The decomposition

$$l^2(\mathbf{N}^2) = \bigoplus_{n=1}^\infty l^2(H_n), \qquad H_n := \big\{(i,j)\in\mathbf{N}^2 : i+j = n+1\big\}$$

then induces the representation of $T_{++}(a)$ by the operator matrix

$$T_{++}(a) = \operatorname{diag}\big(T_1(b), T_2(b), T_3(b), \ldots\big), \tag{6.33}$$

where $T_n(b) = (b_{j-k})_{j,k=1}^n$. Equivalently, on denoting by $e_{ij} \in l^2(\mathbf{N}^2)$ the sequence whose ij entry is 1 and the other entries of which are zero, we see that (6.33) is the matrix representation of $T_{++}(a)$ with respect to the orthonormal basis $\{e_{ij}\}_{i,j=0}^\infty$ ordered as follows:

$$e_{11}, e_{12}, e_{21}, e_{13}, e_{22}, e_{31}, e_{14}, \ldots .$$

From (6.33) we infer that

$$T_{++}(a) \text{ is Fredholm} \iff \big\{T_n(b)\big\} \text{ is stable;}$$

$$T_{++}(a) \text{ is invertible} \iff \begin{cases} \big\{T_n(b)\big\} \text{ is stable, and} \\ \det T_n(b) \neq 0 \text{ for all } n \geq 1. \end{cases}$$

Consequently, at least for nice symbols b, we get an effectively verifiable Fredholm criterion for $T_{++}(a)$. However, the moral of this construction due to [61] is that invertibility of quarter-plane Toeplitz operators is a very subtle property.

Tensor products. In order to study quarter-plane Toeplitz operators we have to work with some (harmless) concrete tensor products.

The space $l^2(\mathbf{N}^2)$ is the Hilbert space tensor product of two copies of $l^2 = l^2(\mathbf{N})$,

$$l^2(\mathbf{N}^2) = l^2 \otimes l^2.$$

This means that finite sums $\sum_j x^{(j)} \otimes y^{(j)}$ are dense in $l^2(\mathbf{N}^2)$ where, for $x = \{x_k\} \in l^2$ and $y = \{y_k\} \in l^2$, the sequence $x \otimes y \in l^2(\mathbf{N}^2)$ is given by $x \otimes y := \{x_k y_l\}$.

Given $A, B \in \mathcal{B}(l^2)$, the operator $A \otimes B \in \mathcal{B}(l^2(\mathbf{N}^2))$ is defined as the linear and continuous extension to $l^2(\mathbf{N}^2)$ of the map given for $x, y \in l^2$ by

$$(A \otimes B)(x \otimes y) := Ax \otimes By.$$

One can show that $\|A \otimes B\| = \|A\|\,\|B\|$. Also notice that if $A \in \mathcal{B}(l^2)$ and $\gamma \in \mathbf{C}$, then

$$\gamma I \otimes A = I \otimes \gamma A, \qquad A \otimes \gamma I = \gamma A \otimes I.$$

In particular, if $\mathcal{A}$ is a subalgebra of $\mathcal{B}(l^2)$, then the sets $\{I \otimes A : A \in \mathcal{A}\}$ and $\{A \otimes I : A \in \mathcal{A}\}$ are subalgebras of $\mathcal{B}(l^2(\mathbf{N}^2))$.

For $b, c \in L^\infty$, we define $b \otimes c \in L^\infty(\mathbf{T}^2)$ by

$$(b \otimes c)(\xi, \eta) := b(\xi)c(\eta), \qquad (\xi, \eta) \in \mathbf{T}^2.$$

Given closed subalgebras $\mathcal{B}, \mathcal{C}$ of L^∞, we let $\mathcal{B} \otimes \mathcal{C}$ stand for the smallest closed subalgebra of $L^\infty(\mathbf{T}^2)$ which contains all finite sums

$$\sum_j b_j \otimes c_j, \qquad b_j \in \mathcal{B}, \qquad c_j \in \mathcal{C}.$$

We remark that $L^\infty \otimes L^\infty$ is a proper subalgebra of $L^\infty(\mathbf{T}^2)$ (see, e.g., [39, Section 8.7(d)]). It is easily seen that if $a \in L^\infty \otimes L^\infty$ is a finite sum $\sum_j b_j \otimes c_j$ with $b_j, c_j \in L^\infty$, then

$$T_{++}(a) = \sum_j T(b_j) \otimes T(c_j). \tag{6.34}$$

The algebra $C \otimes C$ coincides with $C(\mathbf{T}^2)$. Functions $a \in PC \otimes PC$ enjoy the property that at each point $(\xi, \eta) \in \mathbf{T}^2 = (e^{i\theta_0}, e^{i\varphi_0})$ the four one-sided limits

$$a(\xi \pm 0, \eta \pm 0) := \lim_{\substack{\theta \to \theta_0 \pm 0 \\ \varphi \to \varphi_0 \pm 0}} a(e^{i\theta}, e^{i\varphi})$$

exist (see, e.g., [39, Section 8.7(b)]).

Fredholm criteria. Given $a \in PC \otimes PC$ and $(\tau, \mu) \in \mathbf{T} \times [0,1]$, we define two functions $a^1_{\tau,\mu}$ and $a^2_{\tau,\mu}$ in PC by

$$a^1_{\tau,\mu}(t) = (1-\mu)a(\tau - 0, t) + \mu a(\tau + 0, t),$$
$$a^2_{\tau,\mu}(t) = (1-\mu)a(t, \tau - 0) + \mu a(t, \tau + 0), \qquad t \in \mathbf{T}.$$

Clearly, if $a \in C \otimes C = C(\mathbf{T}^2)$, then $a^1_{\tau,\mu}$ and $a^2_{\tau,\mu}$ are independent of μ and are simply given by

$$a^1_\tau(t) = a(\tau, t), \quad a^2_\tau(t) = a(t, \tau), \quad t \in \mathbf{T}.$$

Theorem 6.34. *Let $a \in PC \otimes PC$. The operator $T_{++}(a)$ is Fredholm on $l^2(\mathbf{N}^2)$ if and only if the operators $T(a^1_{\tau,\mu})$ and $T(a^2_{\tau,\mu})$ are invertible for all $(\tau, \mu) \in \mathbf{T} \times [0,1]$. In that case* $\operatorname{Ind} T_{++}(a) = 0$.

For $a \in C \otimes C$, this criterion was established by Simonenko [162] and by Douglas and Howe [61], for $a \in PC \otimes PC$, the theorem is Duduchava's [66]. Obviously, if $a \in C \otimes C$, the criterion can be rephrased as follows: $T_{++}(a)$ is Fredholm if and only if $a(\xi, \eta) \neq 0$ for all $(\xi, \eta) \in \mathbf{T}^2$ and if for each $t \in \mathbf{T}$ the functions

$$\eta \mapsto a(t, \eta) \quad \text{and} \quad \xi \mapsto a(\xi, t)$$

have zero winding number about the origin.

Intuitively, one usually thinks of the torus $\mathbf{T}^2$ as consisting of many copies of the circle $\mathbf{T}$. A continuous function $a : \mathbf{T}^2 \to \mathbf{C}$ deforms each of these circles into a continuous closed curve in the plane and the range of a, i.e., the set $a(\mathbf{T}^2)$, is the union of all these curves (see Figure 52). The situation is a little more complicated for symbols in $PC \otimes PC$. In Figure 53 we plotted (pieces of) the essential range of the function

$$a(\xi, \eta) = 3 - 3i + 4\psi_\delta(\xi) + \frac{1}{2}\psi_\beta(\xi)\psi_\gamma(\eta), \qquad (\xi, \eta) \in \mathbf{T}^2$$

for $\delta = 0.7$, $\beta = 0.5$, $\gamma = 1.7 + i$, and ψ_α as in Example 1.7. In Figure 54 we see (pieces of) the essential range of a function $a \in PC \otimes PC$ of the form

$$a(\xi, \eta) = \varphi(\xi) + c\psi_\beta(\xi)\psi_\gamma(\eta), \qquad (\xi, \eta) \in \mathbf{T}^2$$

where $\varphi \in C(\mathbf{T})$, $c \in \mathbf{C}$, $\beta = 5.5$, and $\gamma = 0.2$. In Figures 53 and 54 we can interpret the essential range of a as the set coated by a piece of a logarithmic spiral, degenerating to a circular arc in Figure 54, which is moving and rotating (by the angle $0.5 \times 2\pi$ in Figure 53 and by the angle $5.5 \times 2\pi$ in Figure 54) in the plane.

For our purposes, we need a generalization of Theorem 6.34. Given two closed subalgebras $\mathcal{C}, \mathcal{D}$ of $\mathcal{B}(l^2)$, we denote by $\mathcal{C} \otimes \mathcal{D}$ the smallest closed subalgebra of $\mathcal{B}(l^2(\mathbf{N}^2))$ which contains all operators $C \otimes D$ with $C \in \mathcal{C}$ and $D \in \mathcal{D}$. Let $\mathbf{A} := \mathbf{A}(PC)$ be as in Section 3.3. A dense subset of $\mathbf{A} \otimes \mathbf{A}$ is formed by the finite sums

$$A = \sum_j B_j \otimes C_j, \qquad B_j \in \mathbf{A}, \quad C_j \in \mathbf{A}. \tag{6.35}$$

Put $\mathbf{K} := \mathcal{K}(l^2)$. We know from Theorem 3.10 that the maximal ideal space of $\mathbf{A}^\pi := \mathbf{A}/\mathbf{K}$ can be identified with $\mathbf{T} \times [0, 1]$. Let $\Gamma : \mathbf{A}^\pi \to C(\mathbf{T} \times [0, 1])$ be the Gelfand map. For $(\tau, \mu) \in \mathbf{T} \times [0, 1]$ and a finite sum (6.35), we define $A^1_{\tau,\mu}$ and $A^2_{\tau,\mu}$ in $\mathbf{A}$ by

$$A^1_{\tau,\mu} := \sum_j (\Gamma B_j^\pi)(\tau, \mu) C_j, \qquad A^2_{\tau,\mu} := \sum_j (\Gamma C_j^\pi)(\tau, \mu) B_j.$$

One can show that the maps $A \mapsto A^i_{\tau,\mu}$ $(i = 1, 2)$ are well-defined and extend to C^*-algebra homomorphisms of $\mathbf{A} \otimes \mathbf{A}$ onto $\mathbf{A}$. Clearly, if A is the operator $T_{++}(a)$ then

$$A^1_{\tau,\mu} = T(a^1_{\tau,\mu}), \qquad A^2_{\tau,\mu} = T(a^2_{\tau,\mu}).$$

Theorem 6.35 (Duduchava). *An operator $A \in \mathbf{A} \otimes \mathbf{A}$ is Fredholm on $l^2(\mathbf{N}^2)$ if and only if the operators $A^1_{\tau,\mu}$ and $A^2_{\tau,\mu}$ are invertible on l^2 for all $(\tau, \mu) \in \mathbf{T} \times [0, 1]$.*

This theorem was established in [66]. A full proof is also in [39, Theorem 8.43]. We confine ourselves to outlining the idea behind the proof: bilocalization (which has its roots in the work of Douglas and Howe [61] and Pilidi [133]; also see Rodino's paper [146]).

Bilocalization. It is easily seen that $\mathbf{K} \otimes \mathbf{K} = \mathcal{K}(l^2(\mathbf{N}^2))$. Thus, an operator $A \in \mathbf{A} \otimes \mathbf{A}$ is Fredholm if and only if

$$A^\pi_{12} := A + \mathbf{K} \otimes \mathbf{K} \in G(\mathbf{A} \otimes \mathbf{A}/\mathbf{K} \otimes \mathbf{K}). \tag{6.36}$$

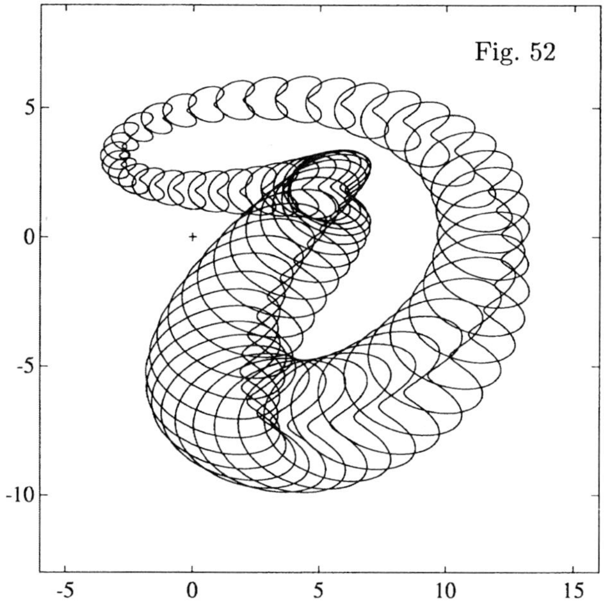

Figure 52 indicates the range of a continuous function $a : \mathbf{T}^2 \to \mathbf{C}$.

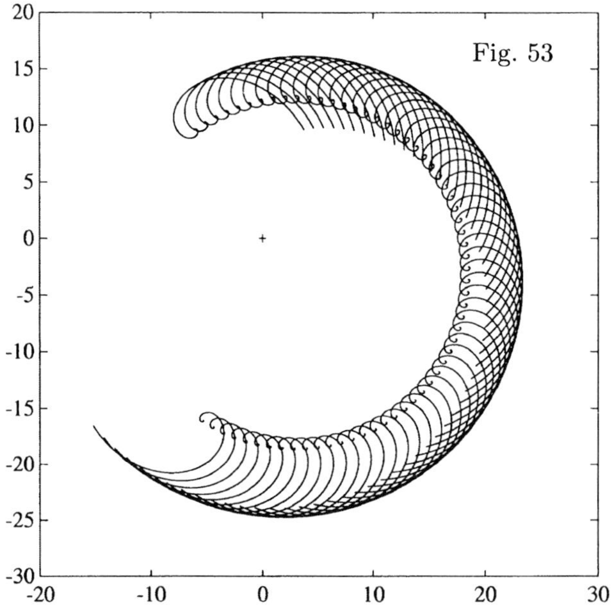

Figure 53 indicates the essential range of a discontinuous function belonging to the algebra $PC \otimes PC$.

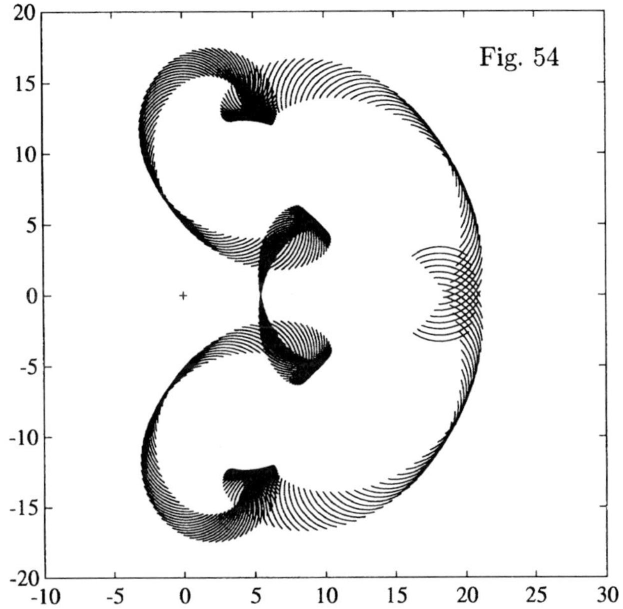

Figure 54 provides an idea of the essential range of another function in $PC \otimes PC$.

Condition (6.36) is in turn equivalent to the two conditions

$$A_1^\pi := A + \mathbf{K} \otimes \mathbf{A} \in G(\mathbf{A} \otimes \mathbf{A}/\mathbf{K} \otimes \mathbf{A}), \tag{6.37}$$

$$A_2^\pi := A + \mathbf{A} \otimes \mathbf{K} \in G(\mathbf{A} \otimes \mathbf{A}/\mathbf{A} \otimes \mathbf{K}); \tag{6.38}$$

note that if B_1^π and C_2^π are the inverses of A_1^π and A_2^π, then $(B+C-BAC)_{12}^\pi$ is the inverse of A_{12}^π.

The algebra $\mathbf{A} \otimes I/\mathbf{K} \otimes \mathbf{A}$ is isomorphic to $\mathbf{A}^\pi := \mathbf{A}/\mathbf{K}$ and is a central subalgebra of $\mathbf{A} \otimes \mathbf{A}/\mathbf{K} \otimes \mathbf{A}$. We can therefore localize over the maximal ideal space $\mathbf{T} \times [0,1]$ of $\mathbf{A}^\pi$. It results that if A is given by (6.35), then (6.37) is satisfied if and only if

$$\sum_j (\Gamma B_j^\pi)(\tau,\mu) \otimes C_j = \sum_j I \otimes (\Gamma B_j^\pi)(\tau,\mu) C_j = I \otimes A_{\tau,\mu}^1$$

is invertible for all $(\tau,\mu) \in \mathbf{T} \times [0,1]$ (see [39, pp. 363–369] for details). Analogously we obtain that (6.38) is valid if and only if $A_{\tau,\mu}^2 \otimes I$ is invertible for all $(\tau,\mu) \in \mathbf{T} \otimes [0,1]$. ■

Stability. We define the projections $P_n \otimes P_n$ on $l^2(\mathbf{N}^2)$ by

$$\big((P_n \otimes P_n)x\big)_{jk} = \begin{cases} x_{jk} & \text{if } 1 \le j \le n,\ 1 \le k \le n, \\ 0 & \text{otherwise,} \end{cases}$$

and we identify

$$T_{n,n}(a) := (P_n \otimes P_n)T_{++}(a)(P_n \otimes P_n)|\mathrm{Im}(P_n \otimes P_n)$$

with the operator on $\mathbf{C}^n \otimes \mathbf{C}^n$ given by the "matrix"

$$(a_{i-k,j-l})_{i,k,j,l=1}^n.$$

Note that if $T_{++}(a)$ is represented by (6.32), then $T_{n,n}(a)$ is given by

$$T_{n,n}(a) = \big(T_n(c_{j-k})\big)_{j,k=1}^n, \qquad T_{n,n}(a) = \big(T_n(b_{j-k})\big)_{j,k=1}^n.$$

A sequence $\{A_n\}_{n=1}^\infty$ of operators $A_n \in \mathcal{B}(\mathbf{C}^n \otimes \mathbf{C}^n)$ is said to be *stable* if

$$\limsup_{n\to\infty} \|A_n^{-1}\| < \infty.$$

Theorem 6.36. *Let $a \in PC \otimes PC$. The sequence $\{T_{n,n}(a)\}$ is stable if and only if the four operators*

$$T_{++}(a_{00}),\ T_{++}(a_{01}),\ T_{++}(a_{10}),\ T_{++}(a_{11}) \tag{6.39}$$

are invertible, where, for $(\xi,\eta) \in \mathbf{T}^2$,

$$\begin{aligned} a_{00}(\xi,\eta) &= a(\xi,\eta), & a_{01}(\xi,\eta) &= a(\xi,\eta^{-1}),\\ a_{10}(\xi,\eta) &= a(\xi^{-1},\eta), & a_{11}(\xi,\eta) &= a(\xi^{-1},\eta^{-1}). \end{aligned}$$

For $a \in C\otimes C$ this theorem is due to Kozak [108], [109], for $a \in PC\otimes PC$ it was first proved in our paper [36]. We remark that $T_{++}(a_{10})$ and $T_{++}(a_{11})$ are the transposed operators of $T_{++}(a_{01})$ and $T_{++}(a_{00})$, respectively, so that the theorem is actually true with the four operators (6.39) replaced by the two operators $T_{++}(a)$ and $T_{++}(a_{01})$. However, the theorem also holds in the block case, i.e., for $a \in (PC \otimes PC)_{N\times N}$, in which case none of the four operators (6.39) can be removed. Also see Figures 55 and 56.

Theorem 6.36 can again be generalized. Let $\mathbf{F}$ stand for the C^*-algebra of all sequences $\{A_n\}_{n=1}^\infty$ of operators

$$A_n \in \mathcal{B}(\mathbf{C}^n \otimes \mathbf{C}^n) = \mathcal{B}\big(\mathrm{Im}(P_n \otimes P_n)\big)$$

such that

$$\|\{A_n\}\| := \sup_{n\geq 1} \|A_n\| < \infty.$$

For closed subalgebras $\mathcal{C}$ and $\mathcal{D}$ of $\mathcal{F}$, define $\mathcal{C} \otimes \mathcal{D}$ as the smallest closed subalgebra of $\mathbf{F}$ containing all sequences $\{C_n \otimes D_n\}$ with $\{C_n\} \in \mathcal{C}$ and $\{D_n\} \in \mathcal{D}$. Let $\mathbf{S} := \mathbf{S}(PC)$ and $\mathbf{I} := \mathcal{J}$ be as in Section 3.3.

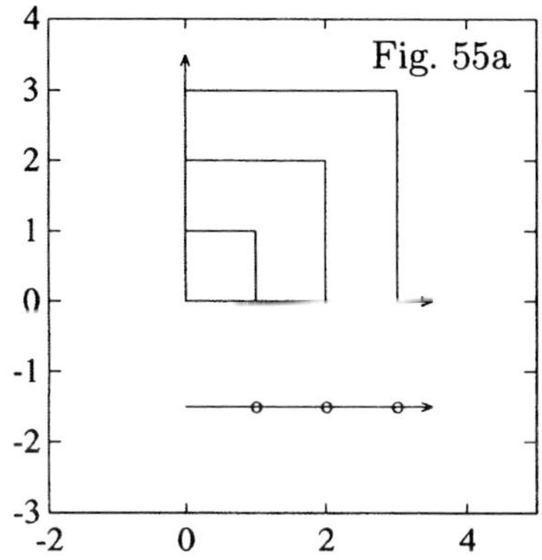

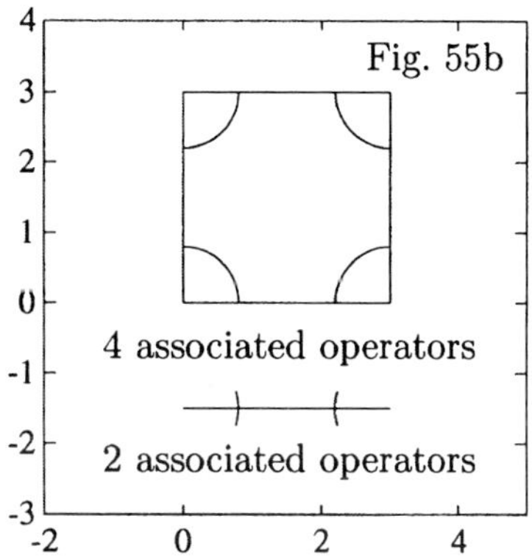

Figure 55 illustrates the background behind Theorems 6.9 and 6.36. Since a segment has two endpoints, the stability of $\{T_n(a)\}$ is determined by two convolution operators on the discrete half-line, i.e., by two Toeplitz operators (Theorem 6.9). The stability of $\{T_{n,n}(a)\}$ is equivalent to the invertibility of an infinite set of discrete convolution operators indexed by the points on the boundary of a square. The four edges yield the four quarter-plane operators (6.39), while the remaining boundary points yield half-space operators. The latter operators are automatically invertible if the former operators are invertible. Thus, in the end we are left with the four operators (6.39).

For $\{A_n\} \in \mathbf{S} \otimes \mathbf{S}$, the four strong limits

$$\begin{aligned}
A_{00} &:= \operatorname*{s\text{-lim}}_{n\to\infty} (P_n \otimes P_n)A_n(P_n \otimes P_n),\\
A_{01} &:= \operatorname*{s\text{-lim}}_{n\to\infty} (P_n \otimes W_n)A_n(P_n \otimes W_n),\\
A_{10} &:= \operatorname*{s\text{-lim}}_{n\to\infty} (W_n \otimes P_n)A_n(W_n \otimes P_n),\\
A_{11} &:= \operatorname*{s\text{-lim}}_{n\to\infty} (W_n \otimes W_n)A_n(W_n \otimes W_n)
\end{aligned}$$

exist and belong to $\mathbf{A} \otimes \mathbf{A}$. From (6.34) we infer that if $a \in PC \otimes PC$, then

$$\{T_{n,n}(a)\} \in \mathbf{S} \otimes \mathbf{S}$$

and the above four strong limits are just the four operators (6.39).

Theorem 6.37. *If $\{A_n\} \in \mathbf{S} \otimes \mathbf{S}$, then $\{A_n\}$ is stable if and only if the four operators $A_{00}, A_{01}, A_{10}, A_{00}$ are invertible.*

This theorem is (implicitly) in [39]. The proofs of Theorems 6.36 and 6.37 are based on the following two-dimensional analogue of Theorem 2.23:

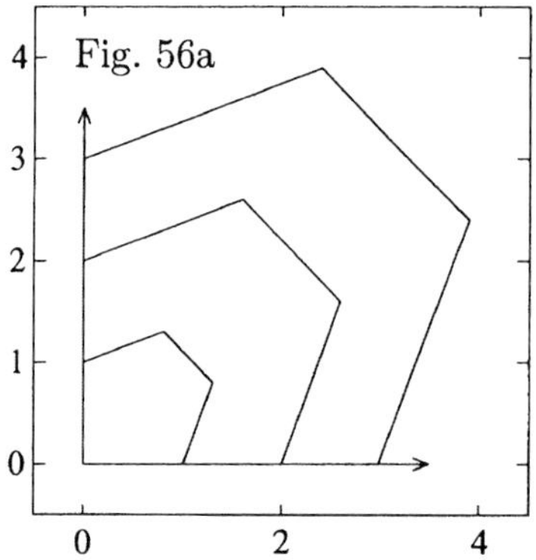

Fig. 56a

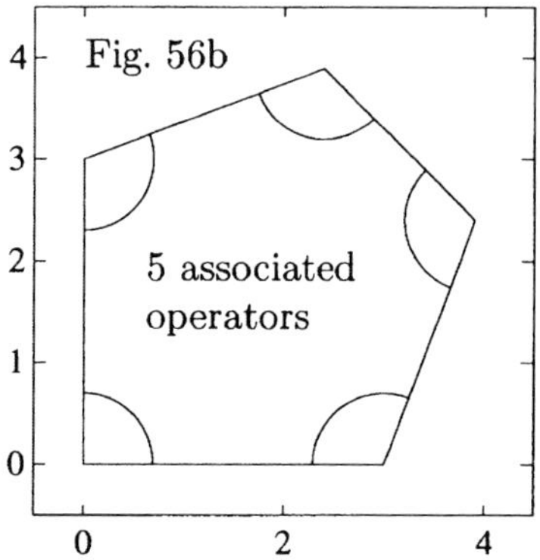

Fig. 56b

One can also consider the compressions of $T_{++}(a)$ to other increasing finite sets in the discrete quarter-plane. Suppose the symbol a is in $C(\mathbf{T}^2)$. Then in the case of homothetic pentagons, the corresponding sequence of truncated operators is stable if and only if five associated convolution operators over certain angles are invertible [108].

if $\{A_n\} \in \mathbf{S} \otimes \mathbf{S}$, then $\{A_n\}$ is stable if and only if $A_{00}, A_{01}, A_{10}, A_{11}$ are invertible and

$$\{A_n\}_{12}^{\pi} := \{A_n\} + \mathbf{I} \otimes \mathbf{I} \in G(\mathbf{S} \otimes \mathbf{S}/\mathbf{I} \otimes \mathbf{I}). \tag{6.40}$$

Using bilocalization and Theorems 3.10 and 6.35 one can show that (6.40) is automatically satisfied if the above four operators are Fredholm. ■

Norms of inverses and pseudospectra. Theorem 6.37 tells us that the map

$$\mathbf{S} \otimes \mathbf{S}/\mathcal{N} \otimes \mathcal{N} \to \bigoplus_{j=1}^{4} \mathcal{B}\big(l^2(\mathbf{N}^2)\big),$$
$$\{A_n\} + \mathcal{N} \otimes \mathcal{N} \mapsto (A_{00}, A_{01}, A_{10}, A_{11})$$

is a C^*-algebra homomorphism which preserves spectra. Proceeding as in Chapter 3, we therefore arrive at the following conclusions (see [41]).

Theorem 6.38. *If $\{A_n\} \in \mathbf{S} \otimes \mathbf{S}$ then*

$$\lim_{n\to\infty} \|A_n\| = \max\big\{\|A_{00}\|, \|A_{01}\|, \|A_{10}\|, \|A_{11}\|\big\}$$

and for each $\varepsilon > 0$,

$$\begin{aligned} u\text{-}\lim_{n\to\infty} \operatorname{sp}_\varepsilon A_n &= p\text{-}\lim_{n\to\infty} \operatorname{sp}_\varepsilon A_n \\ &= \operatorname{sp}_\varepsilon A_{00} \cup \operatorname{sp}_\varepsilon A_{01} \cup \operatorname{sp}_\varepsilon A_{10} \cup \operatorname{sp}_\varepsilon A_{11}. \end{aligned}$$ ■

Corollary 6.39. *Let $a \in PC \otimes PC$, $K \in \mathcal{K}(l^2(\mathbf{N}^2))$, and put*

$$A_n := T_{n,n}(a) + (P_n \otimes P_n)K(P_n \otimes P_n).$$

Then

$$\lim_{n\to\infty} \|A_n^{-1}\|$$
$$= \max\left\{\left\|\left(T_{++}(a)+K\right)^{-1}\right\|, \left\|T_{++}^{-1}(a_{01})\right\|, \left\|T_{++}^{-1}(a_{10})\right\|, \left\|T_{++}^{-1}(a_{11})\right\|\right\}$$

and for each $\varepsilon > 0$,

$$u\text{-}\lim_{n\to\infty} \operatorname{sp}_\varepsilon A_n = p\text{-}\lim_{n\to\infty} \operatorname{sp}_\varepsilon A_n$$
$$= \operatorname{sp}_\varepsilon \left(T_{++}(a)+K\right) \cup \operatorname{sp}_\varepsilon T_{++}(a_{01}) \cup \operatorname{sp}_\varepsilon T_{++}(a_{10}) \cup \operatorname{sp}_\varepsilon T_{++}(a_{11}). \blacksquare$$

Block case. The Fredholm criteria of Theorems 6.34 and 6.35 remain true under the assumption that $a \in (PC \otimes PC)_{N\times N}$ and $A \in (\mathbf{A} \otimes \mathbf{A})_{N\times N}$. Note, however, that in the case $N > 1$ the index of $T_{++}(a)$ need not be zero (see [66]). Theorems 6.36 to 6.38 as well as Corollary 6.39 hold literally in the block case.

Higher dimensions. All the above results extend to higher dimensions. For example, if $a \in C \otimes C \otimes C$ then $T_{+++}(a)$ is Fredholm on $l^2(\mathbf{N}^3)$ if and only if the three operators

$$T_{++}\left(a(\cdot,\cdot,\tau)\right), \quad T_{++}\left(a(\cdot,\tau,\cdot)\right), \quad T_{++}\left(a(\tau,\cdot,\cdot)\right)$$

are invertible for all $\tau \in \mathbf{T}$. In Theorems 6.36 to 6.38 and Corollary 6.39 it is 2^d operators which play the same role in d dimensions as four operators in two dimensions (see [41]).

Selfadjoint operators. One can easily show that if a is any function in $L^\infty(\mathbf{T}^2)$, then $\|T_{++}(a)\| = \|a\|_\infty$ and thus

$$\operatorname{sp} T_{++}(a) \subset \operatorname{conv} \mathcal{R}(a), \qquad \operatorname{sp} T_{n,n}(a) \subset \operatorname{conv} \mathcal{R}(a)$$

(see the proofs of Theorem 1.18 and Proposition 2.17). Theorem 1.25 also extends to higher dimensions:

$$\mathcal{R}(a) \subset \operatorname{sp}_{\text{ess}} T_{++}(a).$$

The operator $T_{++}(a)$ is selfadjoint if and only if a is real-valued. We do not know whether for real-valued symbols Theorem 1.27 holds in the multidimensional case, i.e., whether

$$\operatorname{sp}_{\text{ess}} T_{++}(a) = \operatorname{sp} T_{++}(a) = \operatorname{conv} \mathcal{R}(a) \tag{6.41}$$

for every real-valued $a \in L^\infty$. Theorem 6.34 implies that (6.41) is valid for real-valued symbols $a \in PC \otimes PC$. For such symbols, we also have the following result.

Theorem 6.40. *If $a \in PC \otimes PC$ is real-valued, then*

$$u\text{-}\lim_{n\to\infty} \operatorname{sp} T_{n,n}(a) = p\text{-}\lim_{n\to\infty} \operatorname{sp} T_{n,n}(a) = \operatorname{conv} \mathcal{R}(a).$$

Proof. This follows without difficulty from Corollary 6.39 and (6.41) along with the fact that

$$\operatorname{sp}_\varepsilon A = \{\lambda \in \mathbf{C} : \operatorname{dist}(\lambda, \operatorname{sp} A) \le \varepsilon\}$$

whenever A is selfadjoint. ■

The best higher-dimensional analogues of the Avram-Parter theorem and of Szegö's first limit theorem we are aware of were recently established by Tyrtyshnikov.

Let $\{s_j^{(n,n)}\}_{j=1}^{n^2}$ and $\{\lambda_j^{(n,n)}\}_{j=1}^{n^2}$ denote the singular values and the eigenvalues of $T_{n,n}(a)$, respectively.

Theorem 6.41 (Tyrtyshnikov). *If $a \in L^2(\mathbf{T}^2)$, then*

$$\lim_{n\to\infty} \frac{1}{n^2} \sum_{j=1}^{n^2} f\big(s_j^{(n,n)}\big) = \frac{1}{(2\pi)^2} \int_0^{2\pi} \int_0^{2\pi} f\big(|a(e^{i\theta}, e^{i\varphi})|\big)\, d\theta\, d\varphi$$

for every $f \in C_0(\mathbf{R})$. If $a \in L^2(\mathbf{T}^2)$ is real-valued, then

$$\lim_{n\to\infty} \frac{1}{n^2} \sum_{j=1}^{n^2} f\big(\lambda_j^{(n)}\big) = \frac{1}{(2\pi)^2} \int_0^{2\pi} \int_0^{2\pi} f\big(a(e^{i\theta}, e^{i\varphi})\big)\, d\theta\, d\varphi$$

for every $f \in C_0(\mathbf{R})$.

The d-dimensional version of this theorem was first published in [176]; a full proof is in [177]. ■

Extensions of Szegö's strong limit theorem. A higher-dimensional version of Szegö's strong limit theorem was probably first established by Linnik [119], who studied the determinants of $T_{n\Omega}(a)$ as $n \to \infty$ where $\Omega = G \cap \mathbf{Z}^d$, $G \subset \mathbf{R}^d$ is a domain with a smooth boundary, and $n\Omega := nG \cap \mathbf{Z}^d$. Continuous analogues of this result (i.e., determinants of truncated convolution integral operators) were considered by Widom [186]. Doktorski [57] was able to establish a second-order result for the determinants of $T_{n\Omega}(a)$ in the case where Ω is a d-dimensional polyhedron. The first d terms of the asymptotics of $\det T_{n\Omega}(a)$ were recently determined by Thorsen [167] provided Ω is the cube $[0,1]^d$.

We confine ourselves to citing Doktorski's result for $\Omega = [0,1]^2$. We call a function $a \in C(\mathbf{T}^2)$ *weakly sectorial* if there exists a path in the complex plane which joins the origin to infinity and, for all sufficiently large

n, lies outside some ε-neighborhood of the union of $\mathcal{R}(a)$ and $\operatorname{sp} T_{n,n}(a)$. For example, if $0 \notin \operatorname{conv} \mathcal{R}(a)$, then a is weakly sectorial. Without loss of generality let us also assume that this path does not pass through the point 1. Let $\log z$ be the branch of the logarithm which is analytic in the plane cut along this path and takes the value 0 at 1.

Theorem 6.42 (Doktorski). *Let $a \in C(\mathbf{T}^2)$ be weakly sectorial and suppose*

$$\sum_{j,k\in\mathbf{Z}} |a_{jk}| + \sum_{j,k\in\mathbf{Z}} \big(|j| + |k|\big)|a_{jk}|^2 < \infty.$$

Then as $n \to \infty$,

$$\log \det T_{n,n}(a) = n^2(\log a)_{00} + \frac{n}{2} \sum_{j,k\in\mathbf{Z}} \big(|j| + |k|\big)(\log a)_{jk}(\log a)_{-j,-k} + o(n). \quad \blacksquare$$

7
Banach Space Phenomena

7.1 Boundedness

For $1 \leq p < \infty$, $\mu \in \mathbf{R}$, and $J \in \{\mathbf{N}, \mathbf{Z}\}$, we denote by $l^p_\mu(J)$ the Banach space of all sequences $x = \{x_j\}_{j \in J}$ such that

$$\|x\|_{p,\mu} := \left(\sum_{j \in J} (|j| + 1)^{p\mu} |x_j|^p \right)^{1/p} < \infty.$$

Put $l^p(J) := l^p_0(J)$ and $\|\cdot\|_p := \|\cdot\|_{p,0}$.

Multiplier algebras. One can show that if the Laurent matrix $L(a) = (a_{j-k})_{j,k \in \mathbf{Z}}$ induces a bounded operator on $l^p_\mu(\mathbf{Z})$, then there is a function $a \in L^\infty$ whose Fourier coefficients form the sequence $\{a_n\}_{n \in \mathbf{Z}}$. Let M^p_μ denote the collection of all $a \in L^\infty$ for which $L(a)$ induces a bounded operator on $l^p_\mu(\mathbf{Z})$. The set M^p_μ is a Banach algebra with pointwise algebraic operations and the norm

$$\|a\|_{M^p_\mu} := \|L(a)\|_{\mathcal{B}(l^p_\mu(\mathbf{Z}))}$$

and is referred to as the *multiplier algebra* of $l^p_\mu(\mathbf{Z})$.

In what follows we abbreviate $l^p_\mu(\mathbf{N})$ to l^p_μ. Clearly, if $a \in M^p_\mu$, then the Toeplitz matrix $T(a) = (a_{j-k})_{j,k=1}^\infty$ generates a bounded operator on l^p_μ. Note that the converse is only true for $\mu = 0$: if $T(a)$ is bounded on l^p, then $L(a)$ is necessarily bounded on $l^p(\mathbf{Z})$, but if $\mu \neq 0$, then there exist $a \in W$ such that $T(a)$ is bounded on l^p_μ while $L(a)$ is unbounded on $l^p_\mu(\mathbf{Z})$ (see, e.g., [39, Remark 6.3]).

Properties of multipliers. Several properties of the algebras M_μ^p are recorded in [39, Section 2.5 and Chapter 6]. We here confine ourselves to citing a few facts without proofs.

For $\nu \geq 0$, denote by W^ν the weighted Wiener algebra of all functions $a \in W$ such that

$$\|a\|_{W^\nu} := \sum_{n \in \mathbf{Z}} \left(|n|+1\right)^\nu |a_n| < \infty;$$

notice that we encountered W^ν already in Theorem 2.15.

Proposition 7.1. (a) $M_\mu^1 = W^{|\mu|}$ *for all* $\mu \in \mathbf{R}$.

(b) *Let* $1 < p < \infty$ *and* $1/p + 1/q = 1$. *If* $\mu > 1/q$, *then*

$$a \in M_\mu^p \iff \sum_{n \in \mathbf{Z}} (|n|+1)^{\mu p} |a_n|^p < \infty,$$

and if $\mu < -1/p$, *then*

$$a \in M_\mu^p \iff \sum_{n \in \mathbf{Z}} (|n|+1)^{|\mu| q} |a_n|^q < \infty.$$

(c) *If* $1 < p < \infty$, $1/p + 1/q = 1$, *and* $\mu \in \{-1/p, 1/q\}$, *then functions in* M_μ^p *cannot have jumps.*

(d) *If* $1 < p < \infty$, $1/p + 1/q = 1$, *and* $\mu \in (-1/p, 1/q)$, *then every function* $a \in L^\infty$ *with finite total variation* $V_1(a)$ *belongs to* M_μ^p; *there exists a constant* $c_{p,\mu}$ *such that*

$$\|a\|_{M_\mu^p} \leq c_{p,\mu}\big(\|a\|_\infty + V_1(a)\big) \quad (\textit{Stechkin's inequality}).$$

Given an integer $k \geq 1$, we define the set PC^k as in Section 5.3. In this chapter we need the following results.

Corollary 7.2. (a) *If* $a \in W^{|\mu|}$, *then* $T(a)$ *is bounded on* l_μ^p *for all* $p \in [1, \infty)$.

(b) *If* $a \in PC^1$, *then* $T(a)$ *is bounded on* l_μ^p *if* $p \in (1, \infty)$ *and* $\mu \in (-1/p, 1/q)$ *where* $1/p + 1/q = 1$.

Part (a) is simply the inequality $\|x * y\|_{p,\mu} \leq \|x\|_{1,\mu} \|y\|_{p,\mu}$, and part (b) follows from Proposition 7.1(d) along with the fact that functions in PC^1 have finite total variation. ∎

7.2 Fredholmness and Invertibility

An operator $A \in \mathcal{B}(l_\mu^p)$ is said to be *Fredholm* if it is normally solvable (i.e., if $\operatorname{Im} A$ is a closed subspace of l_μ^p) and the kernel $\operatorname{Ker} A := \{x \in l_\mu^p : Ax = 0\}$

as well as the cokernel Coker $A := l^p_\mu/\text{Im}\, A$ have finite dimensions. Equivalently, $A \in \mathcal{B}(l^p_\mu)$ is Fredholm if and only the coset $A + \mathcal{K}(l^p_\mu)$ is invertible in the Calkin algebra $\mathcal{B}(l^p_\mu)/\mathcal{K}(l^p_\mu)$. The index of a Fredholm operator is defined by

$$\text{Ind}\, A = \dim \text{Ker}\, A - \dim \text{Coker}\, A.$$

A self-contained presentation of the Fredholm theory of operators on Banach spaces is in [87], for example.

Two general results. The Hartman-Wintner and Coburn theorems extend to Toeplitz operators on l^p_μ.

Theorem 7.3 (Duduchava). *Let $1 \le p < \infty$ and $\mu \in \mathbf{R}$.*

(a) *If $a \in M^p_\mu$ and $T(a)$ is Fredholm on l^p_μ, then a is invertible in L^∞.*

(b) *If $a \in M^p_\mu$, then $T(a)$ is invertible on l^p_μ if and only if $T(a)$ is Fredholm of index zero on l^p_μ.*

A full proof is in [39, Theorems 6.5 and 6.6]. Also see Duduchava's book [67] and his articles [64], [65]. ■

Continuous symbols. The identity

$$T(ab) = T(a)T(b) + H(a)H(\tilde{b})$$

(Proposition 1.12) holds for $a, b \in M^p_\mu$ on l^p_μ. The Hankel operators $H(a)$ and $H(\tilde{a})$ are easily seen to be compact on l^p_μ if $a \in W^{|\mu|}$.

Theorem 7.4 (Gohberg-Duduchava). *Let $1 \le p < \infty$, $\mu \in \mathbf{R}$, and $a \in W^{|\mu|}$. The operator $T(a)$ is Fredholm on l^p_μ if and only if $a(t) \neq 0$ for all $t \in \mathbf{T}$. In that case* $\text{Ind}\, T(a) = -\text{wind}(a, 0)$.

In the case $\mu = 0$ this theorem is Gohberg's, for general μ it was first explicitly stated by Duduchava (see [64] and [65]). The proof is the same as the one of Theorem 1.17. ■

Piecewise continuous symbols. Considering Toeplitz operators on l^p_μ is heavily motivated by phenomena caused by piecewise continuous symbols.

Throughout the rest of this section, let

$$1 < p < \infty, \qquad 1/p + 1/q = 1, \qquad \mu \in (-1/p, 1/q).$$

Given two points $z, w \in \mathbf{C}$ and a number $r \in (1, \infty)$, we denote by $\mathcal{A}_r(z, w)$ the *circular arc* at the points of which the line segment $[z, w]$ is seen at the angle

$$\frac{2\pi}{\max\{r, s\}} \qquad (1/r + 1/s = 1)$$

and which lies on the left (resp., right) of the straight line passing first z and then w if $1 < r < 2$ (resp., $2 < r < \infty$). For $r = 2$, $\mathcal{A}_r(z, w)$ is nothing but the line segment $[z, w]$ itself. See Figure 57.

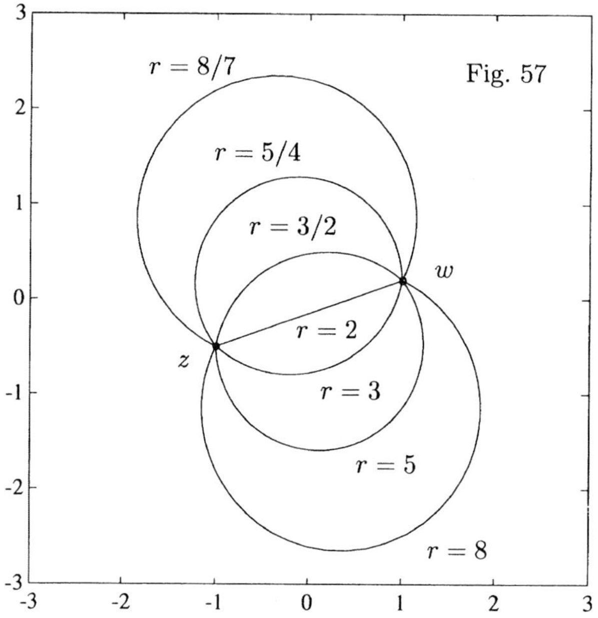

Figure 57 shows the circular arcs $\mathcal{A}_r(z, w)$ between the two points z and w for several values of the parameter r. Half-circles arise in the cases $r = 4/3$ or $r = 4$.

For $a \in PC$, we let $a_r^\#$ stand for the closed continuous and naturally oriented curve which results from the essential range of a by filling in the arc $\mathcal{A}_r(a(t-0), a(t+0))$ at each point $t \in \mathbf{T}$ where a has a jump. Figure 58 shows an example.

Here is one of the most beautiful results of Toeplitz theory.

Theorem 7.5 (Duduchava). *Let $1 < p < \infty$, $1/p + 1/q = 1$, and $\mu \in (-1/p, 1/q)$. Put $r := (1/q - \mu)^{-1}$. If $a \in PC^2$, then $T(a)$ is Fredholm on l^p_μ if and only if $0 \notin a_r^\#$. In that case* $\operatorname{Ind} T(a) = -\operatorname{wind}(a_r^\#, 0)$.

This theorem was established by Duduchava [65] and is illustrated by Figure 58. The appearance of circular arcs in the spectra of Toeplitz operators on H^p was previously discovered by Widom [182] and Gohberg and Krupnik [86] (also see [31]). A full proof of Theorem 7.5 is in [39, Proposition 6.32]. Here is an outline of the *basic steps of the proof.*

Fix $\tau \in \mathbf{T}$ and $\beta \in \mathbf{C} \setminus \mathbf{Z}$. Define $\varphi_\beta := \varphi_{\beta,\tau}$ by (5.65). Also define functions $\xi_\beta := \xi_{\beta,\tau}$ and $\eta_\beta := \eta_{\beta,\tau}$ by

$$\xi_\beta(t) := (1 - \tau/t)^\beta := \exp\big(\beta \log|1 - \tau/t| + i\beta \arg(1 - \tau/t)\big),$$
$$\eta_\beta(t) := (1 - t/\tau)^\beta := \exp\big(\beta \log|1 - t/\tau| + i\beta \arg(1 - t/\tau)\big),$$

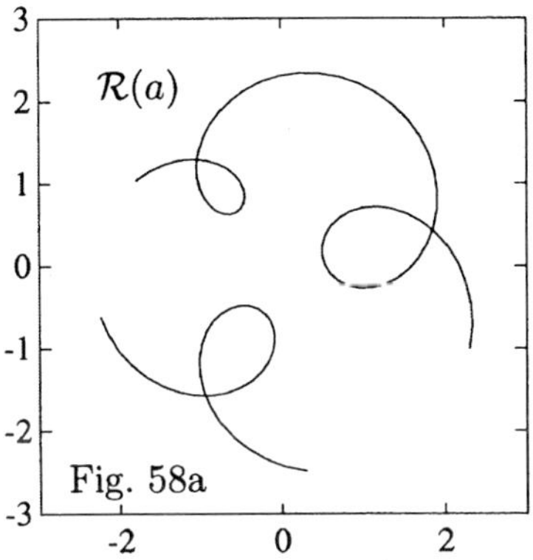

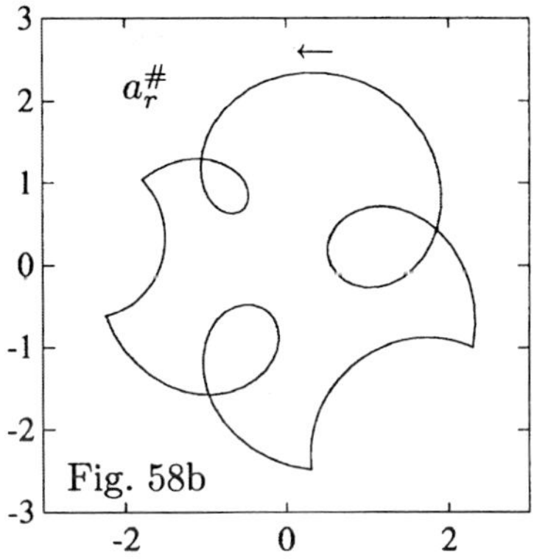

The essential range $\mathcal{R}(a)$ of a piecewise continuous function is seen in Figure 58a. The curve $a_r^{\#}$ corresponding to $p = 3$ and thus to $r = 3/2$ is shown in Figure 58b.

where $\arg z \in (-\pi, \pi]$. It can be verified without difficulty that

$$\varphi_\beta = \xi_{-\beta}\, \eta_\beta.$$

We know from Section 5.11 that φ_β has the Fourier coefficients

$$(\varphi_\beta)_n = (-1/\tau)^n \frac{\sin \pi\beta}{\pi} \frac{1}{\beta - n}.$$

Put

$$(\xi_\beta)_{-n} \quad := \quad \begin{cases} (-\tau)^n \binom{\beta}{n} & \text{for} \quad n \geq 0, \\ 0 & \text{for} \quad n < 0, \end{cases}$$

$$(\eta_\beta)_n \quad := \quad \begin{cases} (-1/\tau)^n \binom{\beta}{n} & \text{for} \quad n \geq 0, \\ 0 & \text{for} \quad n < 0, \end{cases}$$

and denote by $T(\xi_\beta)$ and $T(\eta_\beta)$ the Toeplitz matrices

$$T(\xi_\beta) := \big((\xi_\beta)_{j-k}\big)_{j,k=1}^{\infty}, \qquad T(\eta_\beta) := \big((\eta_\beta)_{j-k}\big)_{j,k=1}^{\infty}.$$

The functions ξ_β and η_β belong to L^1 if and only if $\operatorname{Re}\beta > -1$. In that case $(\xi_\beta)_n$ and $(\eta_\beta)_n$ are just the Fourier coefficients of ξ_β and η_β. Finally, let

$$\mu_n^{(\beta)} := (-1)^n \binom{-1-\beta}{n-1} \qquad (n \geq 1)$$

and

$$M_\beta := \operatorname{diag}\big(\mu_1^{(\beta)}, \mu_2^{(\beta)}, \mu_3^{(\beta)}, \ldots\big).$$

It is easily seen that

$$c_\beta^{-1} n^{\operatorname{Re}\beta} \leq |\mu_n^{(\beta)}| \leq c_\beta n^{\operatorname{Re}\beta} \tag{7.1}$$

with some constant $c_\beta \in (0,\infty)$ independent of n (see [39, Lemma 6.21]). Consequently, $M_\beta : l^p_\mu \to l^p_{\mu-\mathrm{Re}\,\beta}$ is a bounded and invertible operator.

The following "factorization" result is the key to the entire theory of Toeplitz operators with piecewise continuous symbols on l^p_μ.

Lemma 7.6 (Duduchava). *If $\beta \in \mathbf{C}\setminus\mathbf{Z}$, then*

$$T(\eta_\beta)T(\xi_{-\beta}) = \Gamma_\beta M_{-\beta} T(\varphi_\beta) M_\beta,$$

where $\Gamma_\beta := \Gamma(1+\beta)\Gamma(1-\beta) = \pi\beta/\sin\pi\beta$.

This is an identity for binomial coefficients which can be checked without difficulty *once* it has been guessed. Two different proofs are in [65] and [39, Theorem 6.20]. ∎

Lemma 7.7. *Let $\beta \in \mathbf{C}\setminus\mathbf{Z}$, $1 < p < \infty$, $1/p + 1/q = 1$, $\mu \in (-1/p, 1/q)$. Then the following are equivalent:*

(i) *$T(\varphi_\beta)$ is Fredholm of index $-\varkappa$ on l^p_μ;*

(ii) $\varkappa - 1/p < \mathrm{Re}\,\beta + \mu < \varkappa + 1/q$;

(iii) $0 \notin (\varphi_\beta)^\#_r$ *and* $\mathrm{wind}((\varphi_\beta)^\#_r, 0) = \varkappa$ *where* $r := (1/q - \mu)^{-1}$.

Proof. The equivalence (ii) $\Leftrightarrow$ (iii) is elementary plane geometry (though it takes some trouble to convince oneself of its validity). To prove the implication (ii) $\Leftrightarrow$ (i), put $\alpha := \beta - \varkappa$. Then $|\mathrm{Re}\,\alpha| < 1$. Consider the operator

$$A_\alpha = T(\eta_{-\alpha})T(\xi_\alpha).$$

If $\mathrm{Re}\,\alpha = 0$, then $\eta_{\pm\alpha}$ and $\xi_{\pm\alpha}$ are bounded, and since $\varphi_\alpha = \xi_{-\alpha}\eta_\alpha$, it is easily seen that

$$A_\alpha T(\varphi_\alpha) = T(\varphi_\alpha)A_\alpha = I.$$

By an analyticity argument (see [39, p. 259]) we can show that this identity also holds for $|\mathrm{Re}\,\alpha| < 1$. From Lemma 7.6 we infer that A_α can also be written in the form

$$A_\alpha = \Gamma_{-\alpha} M_\alpha T(\varphi_{-\alpha}) M_{-\alpha}.$$

The operators

$$M_{-\alpha} : l^p_\mu \to l^p_{\mu+\mathrm{Re}\,\alpha}, \qquad M_\alpha : l^p_{\mu+\mathrm{Re}\,\alpha} \to l^p_\mu$$

are bounded, and since $-1/p < \mu + \mathrm{Re}\,\alpha < 1/q$ by assumption, Corollary 7.2(b) implies that $T(\varphi_{-\alpha})$ is bounded on $l^p_{\mu+\mathrm{Re}\,\alpha}$. This yields the boundedness of A_α on l^p_μ and therefore proves that $T(\varphi_\alpha)$ is invertible. As

$$T(\varphi_\beta) = \begin{cases} T(\varphi_\alpha)T(\chi_\varkappa) & \text{if} \quad \varkappa > 0, \\ T(\chi_\varkappa)T(\varphi_\beta) & \text{if} \quad \varkappa < 0, \end{cases}$$

we finally arrive at (i). The reverse implication (i) $\Rightarrow$ (iii) can be shown by an index perturbation argument. ■

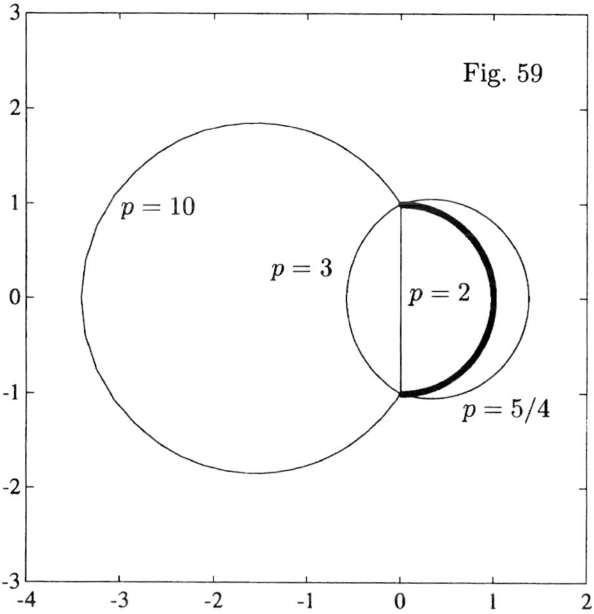

The essential spectrum of $T(\psi_{1/2})$ on l^p consists of the bold circular arc in Figure 59 and of another circular arc, which depends on p. The other arc is shown in Figure 59 for several p. In the case $p = 4/3$, both arcs coincide. Thus, $T(\psi_{1/2})$ is invertible on l^p for $1 < p < 2$ and is Fredholm of index -1 on l^p for $2 < p < \infty$. The operator $T(\psi_{1/2})$ is not normally solvable on l^2.

Corollary 7.8. *Let $\beta \in \mathbf{C}\setminus\mathbf{Z}$, $1 < p < \infty$, $1/p + 1/q = 1$, $\mu \in (-1/p, 1/q)$. Suppose $b \in C \cap PC^2$ has no zeros on $\mathbf{T}$ and* wind$(b, 0) = 0$. *Then $T(\varphi_\beta b)$ is invertible on l^p_μ if and only if*

$$-1/p < \operatorname{Re}\beta + \mu < 1/q.$$

Proof. We have $T(\varphi_\beta b) = T(\varphi_\beta)T(b) + H(\varphi_\beta)H(\tilde{b})$ and since $b_n = O(1/n^2)$ by (5.26), it follows that $b \in W^{|\mu|}$. Therefore $H(\tilde{b})$ is compact, and the assertion results from Theorem 7.4 and Lemma 7.7. ■

If a is as in Theorem 7.5 and $0 \notin a_r^\#$, we can write

$$a = \varphi_{\beta_1,\tau_1}\varphi_{\beta_2,\tau_2}\cdots\varphi_{\beta_m,\tau_m} b$$

with $-1/p < \operatorname{Re}\beta_j < 1/q$ for all j and a function $b \in C \cap PC^2$ which has no zeros on $\mathbf{T}$ and has winding number wind$(a_r^\#, 0)$. One can show that $T(\varphi\psi) - T(\varphi)T(\psi)$ is compact on l^2_μ whenever $\varphi, \psi \in PC^2$ have no common discontinuities. This observation in conjunction with Theorem 7.4 and Lemma 7.7 gives Theorem 7.5.

We remark that the smoothness condition in Theorem 7.5 can be relaxed. For example, the theorem is true if only $a \in PC^1$ is required. Note, however, that Theorem 7.4 is no longer valid under the sole assumption that $a \in C \cap M^p_\mu$ and that, in the same vein, Theorem 7.5 cannot be applied to general symbols $a \in PC \cap M^p_\mu$ (see [39, p. 255]).

7.3 Continuous Symbols

The purpose of this section is to illustrate that some of the results we derived with C^*-algebra techniques can also be established without such devices.

To avoid unnecessary complications, let us consider operators on l^p, i.e., let $\mu = 0$. Put $M^p := M^p_0$. We define P_n and W_n on l^p in the usual way,

$$P_n : \{x_1, x_2, \ldots\} \mapsto \{x_1, \ldots, x_n, 0, 0, \ldots\},$$
$$W_n : \{x_1, x_2, \ldots\} \mapsto \{x_n, \ldots, x_1, 0, 0, \ldots\}.$$

Notice that the operators W_n are uniformly bounded on l^p_μ if and only if $\mu = 0$. The norm of an operator A on l^p or on $\mathbf{C}^n_p := \operatorname{Im} P_n \subset l^p$ (with the l^p norm) will be denoted by $\|A\|_p$. Throughout this section we suppose that $1 \le p < \infty$.

The Toeplitz algebra. Let C_p denote the closure of the Wiener algebra W in M^p and let $\mathbf{A}(C_p)$ stand for the smallest closed subalgebra of $\mathcal{B}(l^p)$ containing the set $\{T(c) : c \in W\}$.

Theorem 7.9. *We have*

$$\mathbf{A}(C_p) = \{T(c) + K : c \in C_p,\ K \in \mathcal{K}(l^p)\}. \tag{7.2}$$

Since $\|T(b)\| = \|b\|_{M^p}$ for all $b \in W$, it results that $T(c) \in \mathbf{A}(C_p)$ whenever $c \in C_p$. One can show that $\mathcal{K}(l^p) \subset \mathbf{A}(C_p)$ (see, e.g., [39, Proposition 4.51]). Thus, the equality (7.2) will follows as soon as we have shown that the set on the right is closed. But this is a consequence of the fact that

$$\|a\|_{M^p} = \|T(a)\|_{\text{ess}} =: \operatorname{dist}(T(a), \mathcal{K}(l^p)) \tag{7.3}$$

for every $a \in M^p$ (see, e.g., [39, Proposition 4.4(d)]). ■

We remark that Theorem 7.4 remains valid for Toeplitz operators on l^p with symbols in C_p.

Stability. A sequence $\{A_n\} = \{A_n\}_{n=1}^{\infty}$ of operators $A_n \in \mathcal{B}(\mathbf{C}_p^n)$ is said to be *stable* on l^p if $\limsup_{n\to\infty} \|A_n^{-1}\|_p < \infty$.

Theorem 7.10. *Let $a \in C_p$. The sequence $\{T_n(a)\}$ is stable on l^p if and only if $T(a)$ is invertible on l^p.*

The "if portion" can be proved in the same way as the sufficiency part of Theorem 2.11 (note that Lemma 2.8 remains literally true for Banach space operators). Conversely, if $\{T_n(a)\}$ is stable and $\|T_n^{-1}(a)\|_p \le M$ for all sufficiently large n, then

$$\|x\|_p = \lim_{n\to\infty} \|P_n x\|_p \le M \lim_{n\to\infty} \|T_n(a)x\|_p = M\|T(a)x\|_p$$

for all $x \in l^p$, which shows that $T(a)$ is normally solvable and injective. We have $l^p = (l^q)^*$ $(1/p + 1/q = 1)$ for $p \in (1,\infty)$ and $l^1 = c_0^*$, where c_0 is the Banach space of all zero sequences. If $\{T_n(a)\}$ is stable on l^p, then $\{T_n(\overline{a})\}$ is stable on l^q and c_0, respectively. Since $P_n \to I$ strongly on l^q and c_0, this implies as above that $T(\overline{a})$ is injective on l^q and c_0, respectively. Thus, $T(a) = T^*(\overline{a})$ is invertible. ■

Note that if $T(a)$ is invertible and $\{T_n(a)\}$ is stable on l^p, then $T_n^{-1}(a) \to T^{-1}(a)$ strongly on l^p (see the proof of Propositions 2.2 and 2.4).

The finite section algebras. Let $\mathcal{F}_p$ be the Banach algebra of all sequences $\{A_n\} = \{A_n\}_{n=1}^{\infty}$ of the operators $A_n \in \mathcal{B}(\mathbf{C}_p^n)$ such that

$$\|\{A_n\}\|_p := \sup_{n\ge 1} \|A_n\|_p < \infty,$$

and let $\mathcal{N}_p$ denote the sequences $\{C_n\} \in \mathcal{F}_p$ for which $\|C_n\|_p \to 0$ as $n \to \infty$. We define $\mathbf{S}(C_p)$ as the smallest closed subalgebra of $\mathcal{F}_p$ which contains the set $\{\{T_n(c)\} : c \in W\}$.

Theorem 7.11. *The algebra $\mathbf{S}(C_p)$ equals*

$$\big\{\{T_n(c) + P_nKP_n + W_nLW_n + C_n\} : c \in C_p;\ K, L \in \mathcal{K}(l^p), \{C_n\} \in \mathcal{N}_p\big\}.$$

Moreover, $\mathbf{S}(C_p)$ is inverse closed in $\mathcal{F}_p$ and $\mathbf{S}(C_p)/\mathcal{N}_p$ is inverse closed in $\mathcal{F}_p/\mathcal{N}_p$.

Denote the above set by X. In [39, Proposition 7.27(b)] it is proved that X is a subset of $\mathbf{S}(C_p)$. Since, by the Banach-Steinhaus theorem,

$$\begin{aligned}
\|T(c) + K\| &\le \liminf_{n\to\infty} \|T_n(c) + P_nKP_n + W_nLW_n + C_n\|_p, \\
\|T(\tilde{c}) + L\| &\le \liminf_{n\to\infty} \|T_n(\tilde{c}) + W_nKW_n + P_nLP_n + W_nC_nW_n\|_p \\
&= \liminf_{n\to\infty} \|T_n(c) + P_nKP_n + W_nLW_n + C_n\|_p,
\end{aligned}$$

we obtain from (7.3) and Theorem 7.9 that the set X is closed and thus coincides with $\mathbf{S}(C_p)$. The inverse closedness of the algebras can be shown by computations as in Section 2.3. For example, if $\{T_n(a)\}$ is stable, then

$$\begin{aligned} T_n^{-1}(a) &= T_n(a^{-1}) + T_n^{-1}(a)P_nMP_n + T_n^{-1}(a)W_nNW_n \\ &= T_n(a^{-1}) + P_nT^{-1}(a)MP_n + W_nT^{-1}(\tilde{a})NW_n + C_n \end{aligned}$$

with $M, N \in \mathcal{K}(l^p)$ and $\{C_n\} \in \mathcal{N}_p$ (again recall that Lemma 2.8 also holds in the Banach space case). ■

Norms of inverses. We equip the direct sum $\mathbf{A}(C_p) \oplus \mathbf{A}(C_p)$ with the norm

$$\|(A,B)\|_p := \max\{\|A\|_p, \|B\|_p\}.$$

Here is the Banach space analogue of Theorem 3.6.

Theorem 7.12. *The map* $\operatorname{Sym}_p : \mathbf{S}(C_p)/\mathcal{N}_p \to \mathbf{A}(C_p) \oplus \mathbf{A}(C_p)$ *given by*

$$\{A_n\} + \mathcal{N}_p \mapsto (A, \tilde{A}); \qquad A := s\text{-}\lim_{n\to\infty} A_n, \qquad \tilde{A} := s\text{-}\lim_{n\to\infty} W_nA_nW_n,$$

is an isometric (and thus injective) Banach algebra homomorphism. Furthermore, if $\{A_n\} \in \mathbf{S}(C_p)$, *then*

$$\limsup_{n\to\infty} \|A_n\|_p = \lim_{n\to\infty} \|A_n\|_p. \tag{7.4}$$

Proof. It is clear that Sym_p is a Banach algebra homomorphism and that

$$\|(A, \tilde{A})\|_p \le \liminf_{n\to\infty} \|A_n\|_p \tag{7.5}$$

(note that $\|W_n\|_p = 1$). Since the only compact Toeplitz operator on l^p is the zero operator, we obtain from Theorem 7.11 that Sym_p is injective. In the case $p = 2$ this is all we need to conclude that Sym_p is isometric (see Section 3.2). For $p \ne 2$, this conclusion requires hand-work.

A simple approximation argument and Theorem 7.11 reveal that it suffices to show that

$$\lim_{n\to\infty} \|A_n\|_p = \max\{\|T(a) + K\|_p, \|T(\tilde{a}) + L\|_p\}, \tag{7.6}$$

where $A_n = T_n(a) + P_nKP_n + W_nLW_n$, $a \in W$ has a finitely supported sequence of Fourier coefficients, $a_n = 0$ for $|n| > s \ge 1$, and the compact operators K, L are subject to the condition

$$P_{n_0}KP_{n_0} = K, \qquad P_{n_0}LP_{n_0} = L \tag{7.7}$$

for some $n_0 \ge 1$.

Pick $n > 4\max\{n_0, s\}$ and put $l := n/4$. Let $x \in \operatorname{Im} P_n$ be any element such that $\|x\|_p = 1$. We claim that there exists an integer $N \in (l+s, 3l-s]$ such that

$$\sum_{k=N-s+1}^{N+s} |x_k|^p < [l/s]^{-1}, \tag{7.8}$$

where $[l/s]$ stands for the integral part of l/s. Indeed, since

$$1 = \sum_{k=1}^{n} |x_k|^p \geq \sum_{k=l+1}^{3l} |x_k|^p \geq \sum_{d=0}^{d_0-1} \sum_{k=l+2ds+1}^{l+2(d+1)s} |x_k|^p$$

with $d_0 = [l/s]$, there exists a d_1 such that $0 \leq d_1 \leq d_0 - 1$ and

$$\sum_{k=l+2d_1s+1}^{l+2(d_1+1)s} |x_k|^p \leq [l/s]^{-1},$$

which proves our claim with $N := l + (2d_1 + 1)s$.

Given $1 \leq n_1 < n_2$, we put $P_{(n_1,n_2]} = P_{n_2} - P_{n_1}$. With N as above, we have

$$A_n = P_N A_n P_N + P_{(N,n]} A_n P_{(N,n]} + P_{(N,n]} A_n P_N + P_N A_n P_{(N,n]}.$$

Since $N + n_0 \leq 3l - s + l < 4l = n$ and $P_{n_0} L P_{n_0} = L$ (recall (7.7)), the equality $P_N W_n L W_n P_N = 0$ holds. Hence $P_N A_n P_N = P_N(T(a) + K)P_N$, which implies that

$$\|P_N A_n P_N\|_p \leq \|T(a) + K\|_p =: M_1. \tag{7.9}$$

As $N > n_0$ and $P_{n_0} K P_{n_0} = K$ (by (7.7)), we see that $P_{(N,n]} P_n K P_n P_{(N,n]} = 0$. Therefore,

$$\begin{aligned} P_{(N,n]} A_n P_{(N,n]} &= P_{(N,n]}\big(T_n(a) + W_n L W_n\big)P_{(N,n]} \\ &= P_{(N,n]} W_n \big(T(\tilde{a}) + L\big) W_n P_{(N,n]} \\ &= W_n P_{n-N}\big(T(\tilde{a}) + L\big)P_{n-N} W_n, \end{aligned}$$

whence

$$\|P_{(N,n]} A_n P_{(N,n]}\|_p \leq \|T(\tilde{a}) + L\|_p =: M_2. \tag{7.10}$$

Again taking into account (7.7) we get

$$P_{(N,n]} A_n P_N + P_N A_n P_{(N,n]} = P_{(N,n]} T_n(a) P_N + P_N T_n(a) P_{(N,n]}.$$

Now recall that $a_n = 0$ for $|n| > s$. Thus, if $N < k \leq n$, then

$$\big(T_n(a) P_N x\big)_k = \sum_{j=1}^{N} a_{k-j} x_j = \sum_{j=N-s+1}^{N} a_{k-j} x_j$$

and if $1 < k \le N$, then

$$\left(T_n(a)P_{(N,n]}x\right)_k = \sum_{j=N+1}^{n} a_{k-j}\,x_j = \sum_{j=N+1}^{N+s} a_{k-j}\,x_j.$$

Consequently,

$$\begin{aligned}
&\left\|P_{(N,n]}A_nP_Nx + P_NA_nP_{(N,n]}x\right\|_p^p \\
&= \sum_{k=N+1}^{n}\left|\sum_{j=N-s+1}^{N} a_{k-j}\,x_j\right|^p + \sum_{k=1}^{N}\left|\sum_{j=N+1}^{N+s} a_{k-j}\,x_j\right|^p \\
&\le \|a\|_1^p\left(\sum_{j=N-s+1}^{N}|x_j|^p + \sum_{j=N+1}^{N+s}|x_j|^p\right) \\
&= \|a\|_1^p \sum_{j=N-s+1}^{N+s}|x_j|^p \le \|a\|_1^p\,[l/s]^{-1},
\end{aligned} \tag{7.11}$$

the estimates resulting from the inequality $\|a * x\|_p \le \|a\|_1\,\|x\|_p$ and from (7.8), respectively. Put

$$u := P_NA_nP_Nx, \quad v := P_{(N,n]}A_nP_{(N,n]},$$
$$w := (P_{(N,n]}A_nP_N + P_NA_nP_{(N,n]})x.$$

Then $A_nx = u + v + w$ and thus,

$$\|A_nx\|_p \le \|u+v\|_p + \|w\|_p = \left(\sum_{j=1}^{n}|u_j+v_j|^p\right)^{1/p} + \|w\|_p$$
$$= \left(\sum_{j=1}^{N}|u_j|^p + \sum_{j=N+1}^{n}|v_j|^p\right)^{1/p} + \|w\|_p = \left(\|u\|_p^p + \|v\|_p^p\right)^{1/p} + \|w\|_p.$$

From (7.9), (7.10), (7.11) we therefore obtain

$$\begin{aligned}
\|A_nx\|_p &\le \left(M_1^p\|P_Nx\|_p^p + M_2^p\|P_{(N,n]}x\|_p^p\right)^{1/p} + \|w\|_p \\
&\le \max\{M_1,M_2\}\left(\|P_Nx\|_p^p + \|P_{(N,n]}x\|_p^p\right)^{1/p} + \|w\|_p \\
&= \max\{M_1,M_2\}\|x\|_p + \|w\|_p \\
&\le \max\{M_1,M_2\} + \|a\|_1\,[l/s]^{-1}.
\end{aligned}$$

As $n \to \infty$, we have $[l/s]^{-1} = [n/(4s)]^{-1} \to 0$, which gives the estimate

$$\limsup_{n\to\infty}\|A_n\|_p \le \max\{M_1,M_2\} = \|(A,\tilde{A})\|_p. \tag{7.12}$$

Combining (7.12) and (7.5) we arrive at (7.6). ■

Corollary 7.13. *If $a \in C_p$ and $T(a)$ is invertible on l^p, then*

$$\lim_{n\to\infty} \|T_n^{-1}(a)\|_p = \max\{\|T^{-1}(a)\|_p, \|T^{-1}(\tilde{a})\|_p\}. \qquad (7.13)$$

This is immediate from Theorems 7.10, 7.11, 7.12. ∎

We remark that $T(\tilde{a})$ is the transposed operator of $T(a)$. Hence,

$$\|T^{-1}(\tilde{a})\|_p = \|T^{-1}(a)\|_q \qquad (1/p + 1/q = 1),$$

where in the case $p = 1$ we have to interpret $\|T^{-1}(a)\|_q$ as the norm of $T^{-1}(a)$ on l^∞ or c_0. Thus, (7.13) may also be written in the form

$$\lim_{n\to\infty} \|T_n^{-1}(a)\|_p = \max\{\|T^{-1}(a)\|_p, \|T^{-1}(a)\|_q\}. \qquad (7.14)$$

Example 7.14 (Grudsky-Kozak). Let

$$a_+(t) = 2 - t, \qquad a_-(t) = 4 + 4t^{-1} + 3t^{-2} \qquad (t \in \mathbf{T}).$$

Since $a_+(z)$ has no zeros in $\{|z| \le 1\}$ and $a_-(z)$ has no zeros in $|z| \ge 1$, we see that $a_+ \in GW_+$ and $a_- \in GW_-$. For $a = a_-^{-1}a_+^{-1}$ we therefore have

$$\begin{aligned} T^{-1}(a) &= T(a_+)T(a_-) \\ &= \begin{pmatrix} 2 & 0 & 0 & \dots \\ -1 & 2 & 0 & \dots \\ 0 & -1 & 2 & \dots \\ \dots & \dots & \dots & \dots \end{pmatrix} \begin{pmatrix} 4 & 4 & 3 & 0 & \dots \\ 0 & 4 & 4 & 3 & \dots \\ 0 & 0 & 4 & 4 & \dots \\ \dots & \dots & \dots & \dots & \dots \end{pmatrix} \\ &= \begin{pmatrix} 8 & 8 & 6 & 0 & 0 & \dots \\ -4 & 4 & 5 & 6 & 0 & \dots \\ 0 & -4 & 4 & 5 & 6 & \dots \\ \dots & \dots & \dots & \dots & \dots & \dots \end{pmatrix}. \end{aligned}$$

The norm of a matrix on l^1 is the maximum of the l^1 norms of the columns. Consequently,

$$\|T^{-1}(a)\|_1 = \max\{12, 16, 19\} = 19, \qquad \|T^{-1}(\tilde{a})\|_1 = \max\{22, 19\} = 22,$$

which shows that $\|T^{-1}(a)\|_1 < \|T^{-1}(\tilde{a})\|_1$. It follows that if $p \in (1, \infty)$ is sufficiently close to 1, then $\|T^{-1}(a)\|_p < \|T^{-1}(\tilde{a})\|_p$. Moral: the maximum in (7.13) and (7.14) cannot be removed in general. ∎

Spaces with weight. One can show that if $a \in W^{|\mu|}$ then $\{T_n(a)\}$ is stable on l^p_μ $(1 \le p < \infty, \mu \in \mathbf{R})$ if and only if $T(a)$ is invertible on l^p_μ (see, e.g., [136], [137]). However, in that case it turns out that

$$\lim_{n\to\infty} \|T_n^{-1}(a)\|_{p,\mu} = \max\{\|T^{-1}(a)\|_{p,\mu}, \|T^{-1}(\tilde{a})\|_p\}, \qquad (7.15)$$

i.e., when taking the norm of $T^{-1}(\tilde{a})$ we have to ignore the weight.

Pseudospectra. For $\varepsilon > 0$, the *ε-pseudospectrum* of an operator A on $\mathbf{C}_p^n$ or l^p is defined by

$$\mathrm{sp}_\varepsilon^{(p)} A := \{\lambda \in \mathbf{C} : \|(A - \lambda I)^{-1}\|_p \geq 1/\varepsilon\}.$$

In order to employ Corollary 7.13 to determine the limiting set of the pseudospectra $\mathrm{sp}_\varepsilon^{(p)} T_n(a)$, we need a Banach space analogue of Theorem 3.14. Here it is.

Theorem 7.15. *Let (X, dm) be a space with a measure and let $1 < p < \infty$. Suppose $A \in \mathcal{B}(L^p(X, dm))$ and $A - \lambda I$ is invertible for all λ in some open subset U of $\mathbf{C}$. If $\|(A-\lambda I)^{-1}\|_p \leq M$ for all $\lambda \in U$, then $\|(A-\lambda I)^{-1}\|_p < M$ for all $\lambda \in U$.*

The only proof of this theorem we know is much more complicated than the proof of Theorem 3.14. This proof can be found in our paper [30] with Grudsky. ■

Theorem 7.16. *Let $1 < p < \infty$ and $\varepsilon > 0$. If $\{A_n\} \in \mathbf{S}(C_p)$, then*

$$u\text{-}\lim_{n\to\infty} \mathrm{sp}_\varepsilon^{(p)} A_n = p\text{-}\lim_{n\to\infty} \mathrm{sp}_\varepsilon^{(p)} A_n = \mathrm{sp}_\varepsilon^{(p)} A \cup \mathrm{sp}_\varepsilon^{(p)} \tilde{A}.$$

In particular, if $a \in C_p$, then

$$u\text{-}\lim_{n\to\infty} \mathrm{sp}_\varepsilon^{(p)} T_n(a) = p\text{-}\lim_{n\to\infty} \mathrm{sp}_\varepsilon^{(p)} T_n(a) = \mathrm{sp}_\varepsilon^{(p)} T(a) \cup \mathrm{sp}_\varepsilon^{(p)} T(\tilde{a}).$$

Once Theorems 7.12 and 7.15 are available, the proof is analogous to the proof of Theorem 3.17 (also see [30]). ■

Approximation numbers. Given an operator $A_n \in \mathcal{B}(\mathbf{C}_p^n)$, the kth *approximation number* $s_k^{(p)}(A_n)$ $(k = 0, 1, \ldots, n)$ is defined by

$$s_k^{(p)}(A_n) = \inf\{\|A_n - F_{n-k}\|_p : F_{n-k} \in \mathcal{F}_{n-k}^{(n)}\},$$

where $\mathcal{F}_{n-k}^{(n)}$ stands for the $n \times n$ matrices of rank at most $n - k$. Clearly,

$$s_0^{(p)}(A_n) = 0, \qquad s_n^{(p)}(A_n) = \|A_n\|_p.$$

One can show that

$$s_1^{(p)}(A_n) = \begin{cases} 1/\|A_n^{-1}\|_p & \text{if } A_n \text{ is invertible,} \\ 0 & \text{if } A_n \text{ is not invertible.} \end{cases}$$

(see, e.g., [27]).

Several results of Chapter 4 can be extended to Toeplitz operators on l^p.

Theorem 7.17. *Let $1 < p < \infty$ and $a \in C_p$.*

(a) *For each k, $s_{n-k}^{(p)}(T_n(a)) \to \|T(a)\|_p$ as $n \to \infty$.*

(b) *If $T(a)$ is not Fredholm, i.e., if a has a zero on the unit circle $\mathbf{T}$, then $s_k^{(p)}(T_n(a)) \to 0$ as $n \to \infty$ for each k.*

(c) *If $T(a)$ is Fredholm of index $k \in \mathbf{Z}$, then the approximation numbers of $\{T_n(a)\}$ on l^p have the $|k|$-splitting property:*

$$\lim_{n\to\infty} s_{|k|}^{(p)}\big(T_n(a)\big) = 0, \qquad \liminf_{n\to\infty} s_{|k|+1}^{(p)}\big(T_n(a)\big) > 0.$$

The proof is similar to the proofs of Theorems 4.5 and 4.13 (see [27]). ■

Speed of approximation. Let $a \in W$ and suppose $T(a)$ is invertible on the space l^1. By Corollary 7.13,

$$\lim_{n\to\infty} \|T_n^{-1}(a)\|_1 = \max\big\{\|T^{-1}(a)\|_1, \|T^{-1}(\tilde{a})\|_1\big\} =: N_1(a) < \infty. \tag{7.16}$$

Equivalently,

$$\lim_{n\to\infty} s_1^{(1)}\big(T_n(a)\big) = 1/N_1(a) > 0. \tag{7.17}$$

Here are two sample results about the speed of convergence in (7.16) and (7.17). Recall the definition of C^α given in Section 5.3.

Theorem 7.18 (Grudsky-Kozak). *If $a \in C^\alpha$ with $\alpha \in (1/2, 1]$ and $T(a)$ is invertible on l^1 then*

$$\begin{aligned} \|T_n^{-1}(a)\|_1 &= N_1(a) + O(1/n^{\alpha-1/2}), \\ s_1^{(1)}\big(T_n(a)\big) &= 1/N_1(a) + O(1/n^{\alpha-1/2}). \end{aligned}$$

If $a \in C^{k+\beta}$ with $k \in \mathbf{N}$ and $\beta \in (0,1)$ and if $T(a)$ is invertible on l^1, then

$$\begin{aligned} \|T_n^{-1}(a)\|_1 &= N_1(a) + O(1/n^{k-1+\beta}), \\ s_1^{(1)}\big(T_n(a)\big) &= 1/N_1(a) + O(1/n^{k-1+\beta}). \end{aligned}$$

Theorem 7.19 (Grudsky-Kozak). *Suppose $a \in W$ can be analytically extended onto some annulus $\{z \in \mathbf{C} : r_1 \le |z| \le r_2\}$ where $r_1 \in (1, \infty)$ and $r_2 \in (1,\infty)$. Put $q := \max\{r_1, 1/r_2\}$. If $T(a)$ is invertible on l^1 then*

$$\begin{aligned} &\|T_n^{-1}(a)\|_1 = N_1(a) + O(q^n), \\ &s_1^{(1)}\big(T_n(a)\big) = 1/N_1(a) + O(q^n). \end{aligned}$$

Proofs are in [93]. ■

Eigenvalues. Unlike norms, pseudospectra, or approximation numbers, the eigenvalues of finite matrices do not depend on the space we let the matrix act on.

Notes. Baxter [16] and Reich [138] showed that $\{T_n(a)\}$ is stable on l^1 if $a \in W$ and $T(a)$ is invertible on l^1. Theorem 7.10 as it is stated and Theorem 7.9 are due to Gohberg and Feldman [80]. Theorem 7.11 was established in [39, Proposition 7.27].

Due to the success of C^*-algebra techniques (see Chapter 3 and also Arveson's works [3], [4]), the results of this section needed some time to come to light. Theorem 7.12 was de facto already predicted in [142, Remark 3 on p. 303] and this theorem is, in disguised form, also contained in the results of [94, pp. 186–205]. Grudsky and Kozak were the first to explicitly state and prove Corollary 7.13 for $p = 1$; they used techniques different from ours. Theorem 7.12 and the proof given in the text appeared explicitly in [30] for the first time. Formula (7.15) is also from [30]; its proof makes use of [137, Remark 4.51, 2^0]. Theorems 7.15 and 7.16 were established in [30], too. Theorem 7.17 was obtained in [27]. Example 7.14 and Theorems 7.18 and 7.19 are Grudsky and Kozak's [93].

7.4 Piecewise Continuous Symbols

Theorem 7.5 tells us that the spectrum of a Toeplitz operator with a piecewise continuous symbol depends heavily on the underlying space. The same is true for stability.

Theorem 7.20 (Verbitsky-Krupnik). *Let* $\beta \in \mathbf{C} \setminus \mathbf{Z}$, $1 < p < \infty$, $1/p + 1/q = 1$, $\mu \in (-1/p, 1/q)$. *Assume* $b \in C \cap PC^2$ *has no zeros on* $\mathbf{T}$ *and* $\operatorname{wind}(b, 0) = 0$. *Let* $\varphi_\beta \in PC^2$ *be given by* (5.64). *Then the following are equivalent:*

(i) $\{T_n(\varphi_\beta b)\}$ *is stable on* l^p_μ;

(ii) $-1/p < \operatorname{Re}\beta + \mu < 1/q$, $-1/q < \operatorname{Re}\beta < 1/p$.

Proofs are in Verbitsky and Krupnik's original paper [179] and also in the books [35, Proposition 3.11] ($\mu = 0$) and [39, Theorem 7.37]. Figure 60 illustrates the theorem. In what follows we will make use of the implication (ii) $\Rightarrow$ (i) of this theorem. Here is a proof of this implication.

Suppose first that $b = 1$. If $-1/p < \operatorname{Re}\beta + \mu < 1/q$, then $T(\varphi_\beta)$ is invertible on l^p_μ by virtue of Lemma 7.7. From Lemma 7.6 we infer that

$$\Gamma_\beta M_{-\beta} T(\varphi_\beta) M_\beta = T(\eta_\beta) T(\xi_{-\beta}) \tag{7.18}$$

whence

$$\Gamma_\beta P_n M_{-\beta} P_n T_n(\varphi_\beta) P_n M_\beta P_n = T_n(\eta_\beta) T_n(\xi_{-\beta}).$$

It follows that $\det T_n(\varphi_\beta) \neq 0$ for all $n \geq 1$. Thus, by Lemma 2.9, the operators $Q_n T^{-1}(\varphi_\beta) Q_n | \operatorname{Im} Q_n$ are invertible for all $n \geq 1$. From (7.18) we

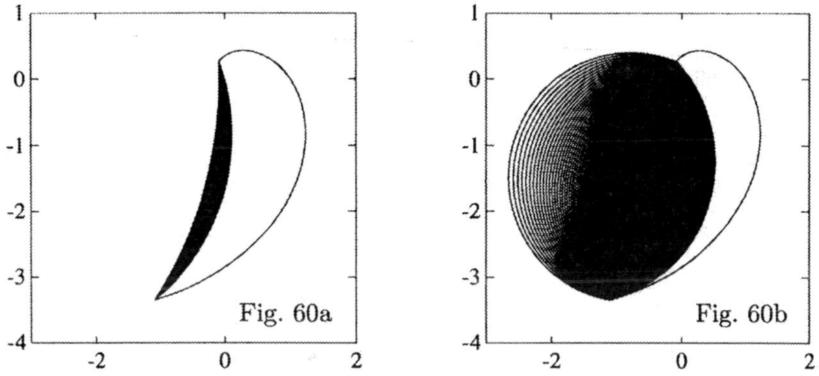

In Figures 60a and 60b we see a set X which is the union of the essential range of ψ_β for $\beta = 0.6 + 0.4i$ and a heavy set between the endpoints of the jump of ψ_β. In geometric language, Theorem 7.20 says that $\{T_n(\psi_\beta) - \lambda P_n\}$ is stable on l^p_μ if and only if λ is located in the unbounded component of the set $\mathbf{C} \setminus X$. Figure 60a corresponds to the case $p = 2.5$, $\mu = 0.15$, while in Figure 60b we have $p = 3$ and $\mu = -0.1$.

get

$$T^{-1}(\varphi_\beta) = \Gamma_\beta M_\beta T(\xi_\beta) T(\eta_{-\beta}) M_{-\beta} = \Gamma_\beta M_\beta T(\varphi_{-\beta}) M_{-\beta} \tag{7.19}$$

and, consequently,

$$Q_n T^{-1}(\varphi_\beta) Q_n = \Gamma_\beta Q_n M_{\beta,n} T(\varphi_{-\beta}) M_{-\beta,n} Q_n,$$

where $M_{\alpha,n} := \operatorname{diag}(\mu_n^{(\alpha)}, \mu_{n+1}^{(\alpha)}, \ldots)$. We therefore see that the operator $M_{\beta,n} T(\varphi_{-\beta}) M_{-\beta,n}$ is invertible on the weighted l^p space

$$l^p_{\varrho_n} := \left\{ x = \{x_k\}_{k=1}^\infty : \|x\|^p_{p,\varrho_n} := \sum_{k=1}^\infty (n+k)^{p\mu} |x_k|^p < \infty \right\}$$

for each $n \ge 1$. Furthermore,

$$\begin{aligned} \left\| \left(Q_n T^{-1}(\varphi_\beta) Q_n\right)^{-1} Q_n \right\|_{p,\mu} &= \left\| \Gamma_\beta^{-1} M_{-\beta,n}^{-1} T^{-1}(\varphi_{-\beta}) M_{\beta,n}^{-1} \right\|_{p,\varrho_n} \\ &= \left\| \Gamma_\beta^{-1} \Gamma_{-\beta} M_{-\beta,n}^{-1} M_{-\beta} T(\varphi_\beta) M_\beta M_{\beta,n}^{-1} \right\|_{p,\varrho_n} \quad \text{(by (7.19))}, \end{aligned}$$

and taking into account (7.1), this is easily seen to be at most $c\|T(\varphi_\beta)\|_{p,r_n}$ where $c < \infty$ is some constant independent of n and $l^p_{r_n}$ is the weighted l^p space

$$l^p_{r_n} := \left\{ x = \{x_k\}_{k=1}^\infty : \|x\|^p_{p,r_n} := \sum_{k=1}^\infty \frac{(n+k)^{(\mu+\alpha)p}}{k^{\alpha p}} |x_k|^p < \infty \right\}$$

with $\alpha := \operatorname{Re}\beta$. Without loss of generality assume $\mu + \alpha \geq 0$; otherwise consider adjoints. Then $(n+k)^{(\mu+\alpha)p} \leq d\big(n^{(\mu+\alpha)p} + k^{(\mu+\alpha)p}\big)$ with some $d < \infty$ independent of n and k, which implies that

$$\|T(\varphi_\beta)\|_{p,r_n}^p \leq d\big(\|T(\varphi_\beta)\|_{p,-\alpha}^p + \|T(\varphi_\beta)\|_{p,\mu}^p\big).$$

Since $-1/q < \alpha < 1/p$ and $-1/p < \mu < 1/q$, we deduce from Corollary 7.2(b) that

$$\|T(\varphi_\beta)\|_{p,-\alpha} < \infty, \qquad \|T(\varphi_\beta)\|_{p,\mu} < \infty.$$

In summary, $\|(Q_n T^{-1}(\varphi_\beta)Q_n)^{-1}Q_n\|_{p,\mu}$ is uniformly bounded. Lemma 2.8 therefore proves that $\{T_n(\varphi_\beta)\}$ is stable.

From (5.26) we get $b_n = O(1/n^2)$. Hence $b \in W^{|\mu|}$, and we can show that b admits a Wiener-Hopf factorization $b = b_- b_+$ with $b_-^{\pm 1} \in W_-^{|\mu|}$ and $b_+^{\pm 1} \in W_+^{|\mu|}$. Consequently, $T(\varphi_\beta b) = T(b_+)T(\varphi_\beta)T(b_-) + K$ with $K \in \mathcal{K}(l_\mu^p)$. The reasoning of Section 2.4 now gives the stability of $T(\varphi_\beta b)$ (note that Theorem 2.16 remains true with l^2 replaced by l_μ^p). ■

Theorem 5.49 implies that if $a \in PC^2\backslash C$ has exactly one jump and (5.82) holds, then

$$p\text{-}\lim_{n\to\infty} \operatorname{sp} T_n(a) \subset \mathcal{R}(a). \tag{7.20}$$

The proof of this inclusion given in Chapter 5 has recourse to several deep results, in particular to Refinement 5.46. Employing the implication (ii) ⇒ (i) of Theorem 7.20, we will give a very simple proof of (7.20), which demonstrates the *usefulness of Banach space methods for getting Hilbert space results.*

Corollary 7.21. *Let $a \in PC^2\backslash C$ have exactly one jump, for $\lambda \in \mathbf{C}\backslash\mathcal{R}(a)$ define $\{\arg(a-\lambda)\}$ as in Section* 5.12, *and suppose*

$$\frac{1}{2\pi}|\{\arg(a-\lambda)\}| < 1 \quad \textit{for all} \quad \lambda \in \mathbf{C}\backslash\mathcal{R}(a).$$

Then p-$\lim_{n\to\infty} \operatorname{sp} T_n(a) \subset \mathcal{R}(a)$.

Proof. Fix $\lambda \in \mathbf{C}\backslash\mathcal{R}(a)$. As in the proof of Theorem 5.49, we can write $a - \lambda = \varphi_{\beta(\lambda)} b_\lambda$ where $b_\lambda \in C \cap PC^2$, b_λ has no zeros on $\mathbf{T}$, $\operatorname{wind}(b_\lambda, 0) = 0$, and $|\operatorname{Re}\beta(\lambda)| < 1$. We may assume that $0 \leq \operatorname{Re}\beta(\lambda) < 1$, since otherwise we can pass to adjoints.

If $0 \leq \operatorname{Re}\beta(\lambda) < 1/2$, then $\{T_n(\varphi_{\beta(\lambda)} b_\lambda)\}$ is stable on l^2 (Corollary 2.19 or the implication (ii) ⇒ (i) of Theorem 7.20 with $p = q = 2$ and $\mu = 0$). In case $1/2 < \operatorname{Re}\beta(\lambda) < 1$, put

$$1/p := \operatorname{Re}\beta(\lambda) + \varepsilon, \qquad \mu := -\operatorname{Re}\beta(\lambda),$$

where $0 < \varepsilon < 1 - \operatorname{Re}\beta(\lambda)$. Then $p \in (1,\infty)$, $\mu \in (-1/p, 1/q)$, and

$$-1/p < \operatorname{Re}\beta(\lambda) + \mu < 1/q, \qquad -1/q < \operatorname{Re}\beta(\lambda) < 1/p.$$

Consequently, $\{T_n(\varphi_{\beta(\lambda)}b_\lambda)\}$ is stable on l^p_μ by virtue of the implication (ii) $\Rightarrow$ (i) of Theorem 7.20.

The rest of the proof is the proof of Proposition 5.26. Namely, in either case there exists a constant $M < \infty$ such that

$$\|T_n^{-1}(a-\lambda)\| = \|T_n^{-1}(\varphi_{\beta(\lambda)}b_\lambda)\| \le M \quad \text{for all} \quad n \ge n_0,$$

where $\|\cdot\|$ is $\|\cdot\|_{2,0}$ for $0 \le \operatorname{Re}\beta(\lambda) < 1/2$ and $\|\cdot\|_{p,\mu}$ for $1/2 < \operatorname{Re}\beta(\lambda) < 1$. Hence, if $|\mu - \lambda| < 1/(2M)$, then $\|T_n^{-1}(a-\mu)\| \le 2M$ for all $n \ge n_0$, which shows that λ has a neighborhood $U(\lambda)$ such that $U(\lambda) \cap \operatorname{sp} T_n(a) = \emptyset$ for all $n \ge n_0$. Thus, $\lambda \notin p\text{-}\lim T_n(a)$. ■

We finally extend Theorem 7.20 to the case of an arbitrary number of jumps. Recall the definition of the circular arc $\mathcal{A}_r(z,w)$ (Section 7.2). Given two numbers $r_1, r_2 \in (1,\infty)$, we put $[r_1, r_2] := [\min\{r_1, r_2\}, \max\{r_1, r_2\}]$ and

$$\mathcal{O}_{r_1,r_2}(z,w) = \bigcup_{r\in[r_1,r_2]} \mathcal{A}_r(z,w).$$

Thus, if $r_1 \ne r_2$, then $\mathcal{O}_{r_1,r_2}(z,w)$ is the closed domain bounded by the two circular arcs $\mathcal{A}_{r_1}(z,w)$ and $\mathcal{A}_{r_2}(z,w)$. The set $\mathcal{O}_{r_1,r_2}(z,w)$ may look like a crescent (Figure 60a), like a football (Figure 60b), or even like an Egyptian scarab (Figure 61a). For $a \in PC$, let $a^\#_{r_1,r_2}$ be the set resulting from the essential range of a by filling in the set $\mathcal{O}_{r_1,r_2}(a(t-0), a(t+0))$ at each jump. The sets plotted in Figure 60 arise in this way from a function with a single jump. In Figure 61 we consider the case of two jumps. If ϱ is any number between r_1 and r_2, then the curve $a^\#_\varrho$ (defined in Section 7.2) lies entirely in $a^\#_{r_1,r_2}$. Clearly, if $0 \notin a^\#_{r_1,r_2}$, then the winding number of $a^\#_\varrho$ about the origin is independent of the choice of ϱ between r_1 and r_2 (canonical choices are $\varrho = r_1$, $\varrho = r_2$, or $\varrho = (r_1+r_2)/2$); we denote this winding number by $\operatorname{wind}(a^\#_{r_1,r_2}, 0)$.

Theorem 7.22. *Let $1 < p < \infty$, $1/p + 1/q = 1$, and $\mu \in (-1/p, 1/q)$. Put*

$$r_1 := (1/q - \mu)^{-1} \quad \textit{and} \quad r_2 := (1/p)^{-1} = p.$$

If $a \in PC^2$, then the following are equivalent:

(i) *$\{T_n(a)\}$ is stable on l^p_μ;*

(ii) *$T(a)$ is invertible on l^p_μ and $T(\tilde{a})$ is invertible on l^p;*

(iii) *$T(a)$ is invertible on both l^p_μ and l^q;*

(iv) *$0 \notin a^\#_{r_1,r_2}$ and $\operatorname{wind}(a^\#_{r_1,r_2}, 0) = 0$.*

The proof is based on the Banach space version of the approach of Sections 2.5 and 2.7. Full proofs are in [35, Theorem 3.19] ($\mu = 0$), [39, Theorem 7.42] ($\mu = 0$), and [141] (general μ). ■

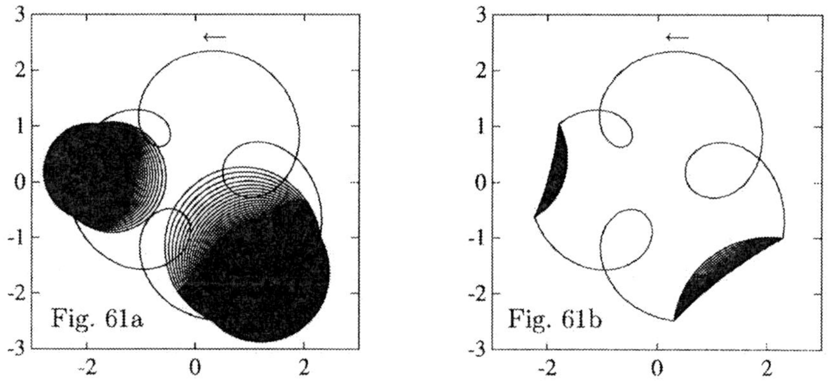

In Figures 61a and 61b we plotted the sets $a^{\#}_{r_1,r_2}$ for a symbol with two jumps for two different choices of p and μ. We clearly see the Egyptian scarabs and the crescents.

We remark that this theorem holds under weaker hypotheses concerning the symbol a; for instance, it is sufficient to require that $a \in PC^1$. The theorem is also true for symbols with countably many jumps, and the equivalences (i) ⇔ (ii) ⇔ (iii) are valid for block Toeplitz operators (see the references cited above).

Concluding note. The problem of studying the stability of $\{T_n(a)\}$ on l^p for piecewise continuous symbols a had been intensively studied since the early 1970s. The 1977 Verbitsky-Krupnik result (Theorem 7.20) solved the problem in the case of a single jump. Only at the turn of the 1980s, were we able to dispose of the case of a finite number of jumps [34], and a few months later, to dispose of the case of countably many jumps [154]. In a sense, it was the search for a proof of Theorem 7.22 for countably many discontinuities which resulted in the paper [154] and thus in the approach of Sections 2.5 and 2.7, which in one way or another prevails everywhere in this book.

7.5 Loss of Symmetry

In Chapter 4, we observed that some central results on the asymptotic distribution of the singular values (= approximation numbers) of Toeplitz matrices on l^2 can be derived with the help of Theorem 4.1, the singular value decomposition, which reduces problems for $\{A_n\}$ to questions about the infinite diagonal operator

$$\operatorname{diag}\big(s_1^{(2)}(A_1), s_1^{(2)}(A_2), s_2^{(2)}(A_2), s_1^{(2)}(A_3), s_2^{(2)}(A_3), s_3^{(2)}(A_3), \dots\big).$$

It would therefore be very nice to have an analogous result for l^p.

For example, we could ask the following: given $A_n \in \mathcal{B}(\mathbf{C}_p^n)$, are there invertible isometries $U_n, V_n \in \mathcal{B}(\mathbf{C}_p^n)$ and a diagonal matrix $S_n \in \mathcal{B}(\mathbf{C}_p^n)$ such that $A_n = U_n S_n V_n$? If the answer were "yes", the approximation numbers of A_n would coincide with those of S_n, and a theorem by Pietsch [131, Theorem 11.11.3] would tell us that the latter numbers are the absolute values of the diagonal elements of S_n.

However, the answer to the above question is "no". Looking at the (real) unit spheres

$$\partial B_1^{(p)} := \big\{(x,y) \in \mathbf{R}^2 : |x|^p + |y|^p = 1\big\}$$

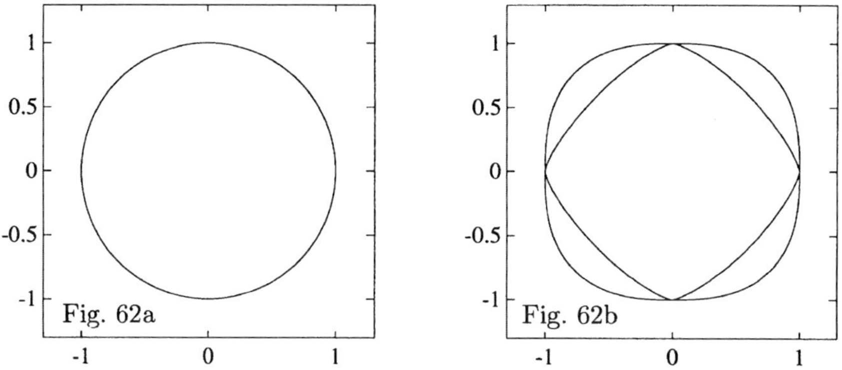

Fig. 62a Fig. 62b

(Figure 62) we see that $\partial B_1^{(2)}$ has the symmetry group $O(2)$ while the symmetry group of $\partial B_1^{(p)}$ $(p \neq 2)$ is the dieder group D_4, which contains only eight elements. Equivalently, the invertible isometries of $\mathcal{B}(\mathbf{C}_2^2)$ are the 2×2 unitary matrices, whereas a matrix $U_2 \in \mathcal{B}(\mathbf{C}_2^2)$ $(p \neq 2)$ is an invertible isometry if and only if

$$U_2 = \begin{pmatrix} \lambda & 0 \\ 0 & \mu \end{pmatrix} \quad \text{or} \quad U_2 = \begin{pmatrix} 0 & \lambda \\ \mu & 0 \end{pmatrix} \quad \text{with } (\mu, \lambda) \in \mathbf{T}^2.$$

Hence, a matrix $A_2 \in \mathcal{B}(\mathbf{C}_p^2)$ $(p \neq 2)$ is of the form $A_2 = U_2 S_2 V_2$ with invertible isometries U_2, V_2 and a diagonal matrix S_2 if and only if

$$A_2 = \begin{pmatrix} a & 0 \\ 0 & b \end{pmatrix} \quad \text{or} \quad A_2 = \begin{pmatrix} 0 & a \\ b & 0 \end{pmatrix} \quad \text{with } (a, b) \in \mathbf{C}^2.$$

In the end, it is this dramatic loss of symmetry which causes the precipice between the Hilbert and Banach space cases. In view of this fact, the circumstance that so many results about Toeplitz operators on l^2 can be carried over to Toeplitz operators on l^p is something like a miracle.

References

[1] G.R. Allan: Ideals of vector-valued functions. *Proc. London Math. Soc., 3rd ser.* **18** (1968), 193–216.

[2] P.M. Anselone and I.H. Sloan: Spectral approximations for Wiener-Hopf operators, II. *J. Integral Equations Appl.* **4** (1992), 465–489.

[3] W. Arveson: C^*-algebras in numerical linear algebra. *J. Funct. Anal.* **122** (1994), 333–360.

[4] W. Arveson: The role of C^*-algebras in infinite dimensional numerical linear algebra. *Contemp. Math.* **167** (1994), 115–129.

[5] F. Avram: On bilinear forms in Gaussian random variables and Toeplitz matrices. *Probab. Theory Related Fields* **79** (1988), 37–45.

[6] S. Axler: Paul Halmos and Toeplitz operators. In: *Paul Halmos Celebrating 50 Years of Mathematics* (J.H. Ewing and F.W. Gehring, eds.), pp. 257–263, Springer-Verlag, New York, 1991.

[7] E. Azoff and K.F. Clancey: Toeplitz operators with sectorial matrix-valued symbols. *Indiana Univ. Math. J.* **26** (1977), 933–938.

[8] J. Barria and P. Halmos: Asymptotic Toeplitz operators. *Trans. Amer. Math. Soc.* **273** (1982), 621–630.

[9] H. Bart, I. Gohberg, and M.A. Kaashoek: The state space method in problems of analysis. In: *Proceedings of the ICIAM* 87 (A.H.P. van der Burgh and R.M.M Mattheij, eds.), pp. 1–15, La Villette, Paris, 1987.

[10] E. Basor: Asymptotic formulas for Toeplitz determinants. *Trans. Amer. Math. Soc.* **239** (1978), 33–65.

[11] E. Basor: A localization theorem for Toeplitz determinants. *Indiana Univ. Math. J.* **28** (1979), 975–983.

[12] E. Basor: Review of "Invertibility and Asymptotics of Toeplitz Matrices". *Linear Algebra Appl.* **68** (1985), 275–278.

[13] E. Basor and K.E. Morrison: The Fisher-Hartwig conjecture and Toeplitz eigenvalues. *Linear Algebra Appl.* **202** (1994), 129–142.

[14] E. Basor and K.E. Morrison: The extended Fisher-Hartwig conjecture for symbols with multiple jump discontinuities. *Operator Theory: Adv. and Appl.* **71** (1994), 16–28.

[15] E. Basor and C.A. Tracy: The Fisher-Hartwig conjecture and generalizations. *Phys. A* **177** (1991), 167–173.

[16] G. Baxter: A norm inequality for a finite-section Wiener-Hopf equation. *Illinois J. Math.* **7** (1963), 97–103.

[17] R.M. Beam and R.F. Warming: The asymptotic spectra of banded Toeplitz and quasi-Toeplitz matrices. *SIAM J. Sci. Comput.* **14** (1993), 971–1006.

[18] L. Berg: Über eine Identität von W. Trench zwischen der Toeplitzschen und einer verallgemeinerten Vandermondeschen Determinante. *Z. Angew. Math. Mech.* **66** (1986), 314–315.

[19] L. Berg: *Lineare Gleichungssysteme mit Bandstruktur.* Deutscher Verlag d. Wiss., Berlin, 1986.

[20] P. M. Blekher: On the conjecture of Fisher and Hartwig in the theory of Toeplitz determinants. *Funct. Anal. Appl.* **16** (1982), 59–83.

[21] A. Böttcher: Toeplitz determinants with piecewise continuous generating function. *Z. Anal. Anwendungen* **1** (1982), 23–39.

[22] A. Böttcher: Status report on rationally generated block Toeplitz and Wiener-Hopf determinants. Unpublished manuscript, 34 pages, 1989 (available from the author on request).

[23] A. Böttcher: Pseudospectra and singular values of large convolution operators. *J. Integral Equations Appl.* **6** (1994), 267–301.

[24] A. Böttcher: Magnete, Determinanten und Fourier-Summen. *Spektrum der Wissenschaft*, März 1994, 25–27.

[25] A. Böttcher: The Onsager formula, the Fisher-Hartwig conjecture, and their influence on research into Toeplitz operators. *J. Statist. Phys.* (Lars Onsager Festschrift) **78** (1995), 575–584.

[26] A. Böttcher: Infinite matrices and projection methods. In: *Lectures on Operator Theory and Its Applications* (P. Lancaster, ed.), pp. 1–72, Fields Institute Monographs, Vol. 3, American Mathematical Society, Providence, RI, 1996.

[27] A. Böttcher: On the approximation numbers of large Toeplitz matrices. *Documenta Mathematica* **2** (1997) 1–29.

[28] A. Böttcher and S.M. Grudsky: Toeplitz operators with discontinuous symbols – phenomena beyond piecewise continuity. *Operator Theory: Adv. and Appl.* **90** (1996), 55–118.

[29] A. Böttcher and S.M. Grudsky: On the condition numbers of large semidefinite Toeplitz matrices. *Linear Algebra Appl.* (to appear).

[30] A. Böttcher, S.M. Grudsky, and B. Silbermann: Norms of inverses, spectra, and pseudospectra of large truncated Wiener-Hopf operators and Toeplitz matrices. *New York J. Math.* **3** (1997), 1–31.

[31] A. Böttcher and Yu.I. Karlovich: *Carleson Curves, Muckenhoupt Weights, and Toeplitz Operators.* Birkhäuser Verlag, Basel, 1997.

[32] A. Böttcher, Yu.I. Karlovich, and V.S. Rabinovich: Emergence, persistence, and disappearance of logarithmic spirals in the spectra of singular integral operators. *Integral Equations and Operator Theory* **25** (1996), 406–444.

[33] A. Böttcher and B. Silbermann: The asymptotic behavior of Toeplitz determinants for generating functions with zeros of integral orders. *Math. Nachr.* **102** (1981), 79–105.

[34] A. Böttcher and B. Silbermann: Über das Reduktionsverfahren für diskrete Wiener-Hopf-Gleichungen mit unstetigem Symbol. *Z. Anal. Anwendungen* **1** (1982), 1–5.

[35] A. Böttcher and B. Silbermann: *Invertibility and Asymptotics of Toeplitz Matrices.* Akademie-Verlag, Berlin, 1983.

[36] A. Böttcher and B. Silbermann: The finite section method for Toeplitz operators on the quarter-plane with piecewise continuous symbols. *Math. Nachr.* **110** (1983), 279–291.

[37] A. Böttcher and B. Silbermann: Toeplitz matrices and determinants with Fisher-Hartwig symbols. *J. Funct. Anal.* **62** (1985), 178–214.

[38] A. Böttcher and B. Silbermann: Toeplitz operators and determinants generated by symbols with one Fisher-Hartwig singularity. *Math. Nachr.* **127** (1986), 95–124.

[39] A. Böttcher and B. Silbermann: *Analysis of Toeplitz Operators.* Akademie-Verlag, Berlin, 1989 and Springer-Verlag, Berlin, 1990.

[40] A. Böttcher and H. Widom: Two remarks on spectral approximations for Wiener-Hopf operators. *J. Integral Equations Appl.* **6** (1994), 31–36.

[41] A. Böttcher and H. Wolf: Spectral approximation for Segal-Bargmann space Toeplitz operators. In: *Linear Operators, Banach Center Publ.*, Vol. 38, pp. 25–48, PAN, Warsaw, 1997.

[42] N. Bourbaki: The most mysterious shape of all. *Quantum*, March/April 1994, 32–35.

[43] A. Brown and P. Halmos: Algebraic properties of Toeplitz operators. *J. Reine Angew. Math.* **231** (1963), 89–102.

[44] R.H. Chan and G. Strang: Toeplitz equations by conjugate gradients with circulant preconditioner. *SIAM J. Sci. Statist. Comput.* **10** (1989), 104–119.

[45] R.H. Chan and Xiao-Qing Jin: A family of block preconditioners for block systems. *SIAM J. Sci. Statist. Comput.* **13** (1992), 1218–1235.

[46] K.F. Clancey: A local result for systems of Riemann-Hilbert barrier problems. *Trans. Amer. Math. Soc.* **200** (1974), 315–325.

[47] K.F. Clancey: Exact sequences of algebras generated by singular integral operators. *Integral Equations and Operator Theory* **4** (1981), 185–205.

[48] K. Clancey and I. Gohberg: *Factorization of Matrix Functions and Singular Integral Operators.* Birkhäuser Verlag, Basel, 1981.

[49] L.A. Coburn: Weyl's theorem for non-normal operators. *Michigan Math. J.* **13** (1966), 285–286.

[50] L.A. Coburn: The C^*-algebra generated by an isometry. Part I: *Bull. Amer. Math. Soc.* **73** (1967), 722–726; Part II: *Trans. Amer. Math. Soc.* **137** (1969), 211–217.

[51] K.M. Day: Toeplitz matrices generated by the Laurent series expansion of an arbitrary rational function. *Trans. Amer. Math. Soc.* **206** (1975), 224–245.

[52] K.M. Day: Measures associated with Toeplitz matrices generated by the Laurent expansion of rational functions. *Trans. Amer. Math. Soc.* **209** (1975), 175–183.

[53] A. Devinatz: Toeplitz operators on H^2 spaces. *Trans. Amer. Math. Soc.* **112** (1964), 304–317.

[54] A. Devinatz: An extension of a limit theorem of G. Szegö. *J. Math. Anal. Appl.* **14** (1966), 499–510.

[55] A. Devinatz: The strong Szegö limit theorem. *Illinois J. Math.* **11** (1967), 160–175.

[56] J. Dixmier: *Les C^*-algèbres et leurs représentations.* Gauthier-Villars, Paris, 1969 [Russian transl.: Nauka, Moscow, 1974].

[57] R.Ya. Doktorski: A generalization of the limit theorem of G. Szegö to the multidimensional case. *Sibirsk. Mat. Zh.* **25** (1984), 20–29 [Russian].

[58] R.G. Douglas: Toeplitz and Wiener-Hopf operators in $H^\infty + C$. *Bull. Amer. Math. Soc.* **74** (1968), 895–899.

[59] R.G. Douglas: *Banach Algebra Techniques in Operator Theory.* Academic Press, New York 1972.

[60] R.G. Douglas: Local Toeplitz operators. *Proc. London Math. Soc., 3rd ser.,* **36** (1978), 243–272.

[61] R.G. Douglas and R. Howe: On the C^*-algebra of Toeplitz operators on the quarter-plane. *Trans. Amer. Math. Soc.* **158** (1971), 203–217.

[62] R.G. Douglas and D. Sarason: Fredholm Toeplitz operators. *Proc. Amer. Math. Soc.* **26** (1970), 117–120.

[63] R.G. Douglas and H. Widom: Toeplitz operators with locally sectorial symbols. *Indiana Univ. Math. J.* **20** (1970), 385–388.

[64] R. Duduchava: Discrete Wiener-Hopf equations on l^p spaces with weight. Soobshzh. *Akad. Nauk Gruz. SSR* **67** (1972), 17–20 [Russian].

[65] R. Duduchava: On discrete Wiener-Hopf equations. *Trudy Tbiliss. Mat. Inst.* **50** (1975), 42–59. [Russian].

[66] R. Duduchava: Discrete convolution operators on the quarter-plane and their indices. *Math. USSR Izv.* **11** (1977), 1072–1084.

[67] R. Duduchava: *Integral Equations with Fixed Singularities.* Teubner-Verlag, Leipzig, 1979.

[68] R.E. Edwards: *Fourier Series. A Modern Introduction*, Vol. 1. Springer-Verlag, New York, 1979 [Russian transl.: Mir, Moscow, 1985].

[69] T. Ehrhardt: *Toeplitz determinants with several Fisher-Hartwig singularities.* Dissertationsschrift, TU Chemnitz, 1997.

[70] T. Ehrhardt and B. Silbermann: Toeplitz determinants with one Fisher-Hartwig singularity. *J. Funct. Anal.* **148** (1997), 229–256.

[71] P.A. Fillmore: *A User's Guide to Operator Algebras.* Wiley, New York 1996.

[72] M.E. Fisher and R.E. Hartwig: Toeplitz determinants – some applications, theorems, and conjectures. *Adv. Chem. Phys.* **15** (1968), 333–353.

[73] M.E. Fisher and R.E. Hartwig: Asymptotic behavior of Toeplitz matrices and determinants. *Arch. Rational Mech. Anal.* **32** (1969), 190–225.

[74] D. Gaier: *Lectures on Complex Approximation.* Birkhäuser Verlag, Basel, 1987.

[75] F.D. Gakhov: On Riemann's boundary value problem. *Mat. Sb.* **2 (44)** (1937), 673–683 [Russian].

[76] F.D. Gakhov: *Boundary Values Problems.* Pergamon Press, London, 1966 [extended Russian edition: Nauka, Moscow, 1977].

[77] I. Gohberg: On an application of the theory of normed rings to singular integral equations. *Uspekhi Matem. Nauk* **7** (1952), 149–156 [Russian].

[78] I. Gohberg: On the number of solutions of homogeneous singular equations with continuous coefficients. *Dokl. Akad. Nauk SSSR* **112** (1958), 327–330 [Russian].

[79] I. Gohberg: The factorization problem in normed rings, functions of isometric and symmetric operators, and singular integral equations. *Uspekhi Mat. Nauk* **19** (1964), 71–124 [Russian].

[80] I. Gohberg and I.A. Feldman: *Convolution Equations and Projection Methods for Their Solution.* American Mathematical Society, Providence, RI, 1974 [Russian original: Nauka, Moscow, 1971; German transl.: Akademie-Verlag, Berlin, 1974].

[81] I. Gohberg, S. Goldberg, and M.A. Kaashoek: *Classes of Linear Operators,* Vol. I: Birkhäuser Verlag, Basel, 1990; Vol. II: Birkhäuser Verlag, Basel, 1993.

[82] I. Gohberg, S. Goldberg, and N. Krupnik: Traces and determinants of linear operators. *Integral Equations and Operator Theory* **26** (1996), 136–187.

[83] I. Gohberg, M.A. Kaashoek, and F. van Schagen: Szegö-Kac-Achiezer formulas in terms of realization of the symbol. *J. Funct. Anal.* **74** (1987), 24–51.

[84] I. Gohberg and M.G. Krein: *Introduction to the Theory of Linear Nonselfadjoint Operators in Hilbert Space.* American Mathematical Society, Providence, RI, 1969 [Russian original: Nauka, Moscow, 1965].

[85] I. Gohberg and M.G. Krein: Systems of integral equations on a half-line with kernel depending upon the difference of the arguments. *Amer. Math. Soc. Transl.* (2) **14** (1960), 217–287 [Russian original: *Uspekhi Mat. Nauk* **13** (1958), 3–72].

[86] I. Gohberg and N. Krupnik: Singular integral operators with piecewise continuous coefficients. *Izv. Akad. Nauk SSSR* **35** (1971), 940–964 [Russian].

[87] I. Gohberg and N. Krupnik: *One-Dimensional Linear Singular Integral Equations,* Vols. I and II. Birkhäuser Verlag, Basel, 1992 [Russian original: Shtiintsa, Kishinev, 1973; German transl.: Birkhäuser Verlag, Basel, 1979].

[88] I. Gohberg and N. Krupnik: On the algebra generated by Toeplitz matrices. *Funktsional. Anal. Prilozhen.* **3** (1969), 46–52.

[89] B.L. Golinski and I.A. Ibragimov: On the limit theorem of G. Szegö. *Math. USSR Izv.* **5** (1971), 421–446 [Russian original: *Izv. Akad Nauk SSSR, Ser. Mat.* **35** (1971), 408–427].

[90] M. Gorodetsky: Toeplitz determinants generated by rational functions. In: *Integral Differential Equations and Approximate Solutions* (V. Dybin, ed.), pp. 49–54, Rostov-on-Don University Press, Elista, 1985 [Russian].

[91] M. Gorodetsky: On block Toeplitz matrices with analytic symbols. *Linear Algebra Appl.* **116** (1989), 41–51.

[92] U. Grenander and G. Szegö: *Toeplitz Forms and Their Applications.* University of California Press, Berkeley, 1958 [Russian transl.: Izd. Inostr. Lit., Moscow, 1961].

[93] S.M. Grudsky and A.V. Kozak: On the convergence speed of the norms of the inverses of truncated Toeplitz operators. In: *Integro-Differential Equations Applications*, pp. 45–55, Rostov-on-Don Univeersity Press, Rostov-on-Don, 1995 [Russian].

[94] R. Hagen, S. Roch, and B. Silbermann: *Spectral Theory of Approximation Methods for Convolution Equations.* Birkhäuser Verlag, Basel, 1995.

[95] P. Halmos: *A Hilbert Space Problem Book.* D. van Nostrand, Princeton 1967.

[96] R. Harte and M. Mbekhta: On generalized inverses in C^*-algebras. Part I: *Studia Math.* **103** (1992), 71–77; Part II: *ibid.* **106** (1993), 129–138.

[97] P. Hartman: On completely continuous Hankel operators. *Proc. Amer. Math. Soc.* **9** (1958), 862–866.

[98] P. Hartman and A. Wintner: The spectra of Toeplitz's matrices. *Amer. J. Math.* **76** (1954), 867–882.

[99] G. Heinig: Endliche Toeplitzmatrizen und zweidimensionale diskrete Wiener-Hopf-Operatoren mit homogenem Symbol. *Math. Nachr.* **82** (1978), 29–68.

[100] G. Heinig and F. Hellinger: The finite section method for Moore-Penrose inversion of Toeplitz operators. *Integral Equations and Operator Theory* **19** (1994), 419–446.

[101] G. Heinig and K. Rost: *Algebraic Methods for Toeplitz-Like Matrices and Operators.* Akademie-Verlag, Berlin, 1984 and Birkhäuser Verlag, Basel, 1984.

[102] I.I. Hirschman, Jr.: On a formula of Kac and Achiezer. *J. Math. Mech.* **16** (1966), 167–196.

[103] I.I. Hirschman, Jr.: The spectra of certain Toeplitz matrices. *Illinois J. Math.* **11** (1967), 145–159.

[104] T. Høholdt and J. Justesen: Determinants of a class of Toeplitz matrices. *Math. Scand.* **43** (1978), 250–258.

[105] Xiao-Qing Jin: Fast iterative solvers for symmetric Toeplitz systems - a survey and an extension. *J. Comput. Appl. Math.* **66** (1996), 315–321.

[106] K. Johansson: On Szegö's asymptotic formula for Toeplitz determinants and generalizations. *Bull. Sci. Math.* **112** (1988), 257–304.

[107] K. Johansson: On random matrices from the compact classical groups. *Ann. of Math.* **145** (1997), 519–545.

[108] A.V. Kozak: On the reduction method for multidimensional discrete convolutions. *Mat. Issled.* **8** (29) (1973), 157–160 [Russian].

[109] A.V. Kozak: A local principle in the theory of projection methods. *Soviet Math. Dokl.* **14** (1974), 1580–1583.

[110] M.G. Krein: Integral equations on a half-line with kernel depending upon the difference of the arguments. *Amer. Math. Soc. Transl.* (2) **22** (1962), 163–288.

[111] M.G. Krein: On some new Banach algebras and Wiener-Lévy type theorems for Fourier series and integrals. *Amer. Math. Soc. Transl.* **93** (1970), 177–199 [Russian original: *Mat. Issled.* **1** (1966), 82–109].

[112] M.G. Krein and I. Spitkovsky: On some generalizations of Szegö's first limit theorem. *Anal. Math.* **9** (1983), 23–41 [Russian].

[113] Ta-Kang Ku and C.-C. Jay: Design and analysis of Toeplitz preconditioners. *IEEE Trans. Signal Process.* **40** (1992), 129–141.

[114] H. Landau: On Szegö's eigenvalue distribution theorem and non-Hermitian kernels. *J. Analyse Math.* **28** (1975), 335–357.

[115] H. Landau: Loss in unstable resonators. *J. Opt. Soc. Amer.* **66** (1976), 525–529.

[116] H. Landau: The notion of approximate eigenvalues applied to an integral equation of laser theory. *Quart. Appl. Math.*, April 1977, 165–171.

[117] L.M. Libkind: Asymptotics of the eigenvalues of Toeplitz forms. *Math. Notes* **11** (1972), 97–101.

[118] R. Libby: *Asymptotics of determinants and eigenvalue distribution for Toeplitz matrices associated with certain discontinuous symbols.* Ph.D. Thesis, University of California, Santa Cruz, 1990.

[119] I.Yu. Linnik: The multidimensional analogue of a limit theorem of G. Szegö. *Math. USSR Izv.* **9** (1975), 1323–1332.

[120] G.S. Litvinchuk and I. Spitkovsky: *Factorization of Measurable Matrix Functions.* Akademie-Verlag, Berlin, 1987 and Birkhäuser Verlag, Basel, 1987.

[121] B.M. McCoy and T.T. Wu: *The Two-Dimensional Ising Model.* Harvard University Press, Cambridge, MA, 1973.

[122] E.W. Montroll, R.B. Potts, and J.C. Ward: Correlations and spontaneous magnetization of the two-dimensional Ising model. *J. Math. Phys.* **4** (1963), 308–326.

[123] R.H. Moore and M.Z. Nashed: Approximation of generalized inverses of linear operators. *SIAM J. Appl. Math.* **27** (1974), 1–16.

[124] K.E. Morrison: Spectral approximation of multiplication operators. *New York J. Math.* **1** (1995), 75–96.

[125] M.Z. Nashed: Perturbations and approximations for generalized inverses and linear operator equations. In: *Generalized Inverses and Applications* (M.Z. Nashed, ed.), pp. 325–396, Academic Press, New York, 1976.

[126] Z. Nehari: On bounded bilinear forms. *Ann. of Math.* **65** (1957), 153–162.

[127] N.K. Nikolski: *Treatise on the Shift Operator.* Springer-Verlag, Berlin, 1986 [Russian original: Nauka, Moscow, 1980].

[128] S.V. Parter: On the distribution of the singular values of Toeplitz matrices. *Linear Algebra Appl.* **80** (1986), 115–130.

[129] J.R. Partington: *An Introduction to Hankel Operators.* London Mathematical Society Student Texts, Vol. 13, Cambridge University Press, Cambridge, UK, 1988.

[130] V.V. Peller: Hankel operators and their applications (the ideals $\mathcal{K}_p$, Besov classes, random processes). *Soviet Math. Dokl.* **21** (1980), 683–688.

[131] A. Pietsch: *Operator Ideals.* Deutscher Verlag d. Wiss., Berlin, 1978 [Russian transl.: Mir, Moscow 1982].

[132] A. Pietsch: *Eigenvalues and s-Numbers.* Geest & Portig, Leipzig, 1987.

[133] V.S. Pilidi: On multidimensional bisingular operators. *Soviet Math. Dokl.* **12** (1971), 1723–1726.

[134] J. Plemelj: Ein Ergänzungssatz zur Cauchy'schen Integraldarstellung analytischer Funktionen, Randwerte betreffend. *Monatsh. Math. Phys.* **19** (1908), 105–210.

[135] S.C. Power: *Hankel Operators on Hilbert Space.* Pitman Research Notes, No. 64, Pitman, Boston, 1982.

[136] S. Prössdorf and B. Silbermann: *Projektionsverfahren und die näherungsweise Lösung singulärer Gleichungen.* Teubner-Verlag, Leipzig, 1977.

[137] S. Prössdorf and B. Silbermann: *Numerical Analysis for Integral and Related Operator Equations.* Akademie-Verlag, Berlin, 1991 and Birkhäuser Verlag, Basel, 1991.

[138] E. Reich: On non-Hermitian Toeplitz matrices. *Math. Scand.* **10** (1962), 145–152.

[139] M. Reed and B. Simon: *Methods of Modern Mathematical Physics,* Vol. 1: Functional Analysis. Academic Press, New York, 1971 [Russian transl.: Mir, Moscow, 1977].

[140] L. Reichel and L.N. Trefethen: Eigenvalues and pseudo-eigenvalues of Toeplitz matrices. *Linear Algebra Appl.* **162** (1992), 153–185.

[141] S. Roch and B. Silbermann: Toeplitz-like operators, quasicommutator ideals, numerical analysis. Part I: *Math. Nachr.* **120** (1985), 141–173; Part II: *Math. Nachr.* **134** (1987), 381-391.

[142] S. Roch and B. Silbermann: Limiting sets of eigenvalues and singular values of Toeplitz matrices. *Asymptotic Anal.* **8** (1994), 293–309.

[143] S. Roch and B. Silbermann: Index calculus for approximation methods and singular value decomposition. Preprint TU Chemnitz-Zwickau, 1996.

[144] S. Roch and B. Silbermann: A note on the singular values of Cauchy-Toeplitz matrices. Preprint TU Chemnitz-Zwickau, 1996.

[145] S. Roch and B. Silbermann: C^*-algebra techniques in numerical analysis. *J. Operator Theory* **35** (1996), 241–280.

[146] L. Rodino: Polysingular integral operators. *Ann. Mat. Pura Appl.* (IV) **124** (1980), 59–106.

[147] D. Sarason: *Function Theory on the Unit Circle.* Virginia Polytechnic Institute and State University, Blacksburg, 1978.

[148] P. Schmidt and F. Spitzer: The Toeplitz matrices of an arbitrary Laurent polynomial. *Math. Scand.* **8** (1960), 15–38.

[149] D. SeLegue: A C^*-algebraic extension of the Szegö trace formula. Talk given at the GPOTS, Arizona State University, Tempe, May 22, 1996.

[150] V.N. Semenyuta and A.V. Khevelev: A local principle for special classes of Banach algebras. *Izv. Severo-Kavkaz. Nauchn. Z. Vyssh. Shkoly, Ser. Estestv. Nauk,* **1/1977** (1977), 15–17 [Russian].

[151] S. Serra: On the extreme spectral properties of Toeplitz matrices generated by L^1 functions with several minima (maxima). *BIT* **34** (1996), 135–142.

[152] S. Serra: On the extreme eigenvalues of Hermitian (block) Toeplitz matrices. *Linear Algebra Appl.* **270** (1997), 109–129.

[153] B. Silbermann: The strong Szegö limit theorem for a class of singular generating functions, I. *Demonstratio Math.* **14** (1981), 647–667.

[154] B. Silbermann: Lokale Theorie des Reduktionsverfahrens für Toeplitzoperatoren. *Math. Nachr.* **104** (1981), 137–146.

[155] B. Silbermann: On the limiting set of singular values of Toeplitz matrices. *Linear Algebra Appl.* **182** (1993), 35–43.

[156] B. Silbermann: Asymptotic Moore-Penrose inversion of Toeplitz operators. *Linear Algebra Appl.* **256** (1996), 219–234.

[157] B. Simon: Notes on infinite determinants of Hilbert space operators. *Adv. in Math.* **24** (1977), 244-273.

[158] I.B. Simonenko: The Riemann boundary value problem with measurable coefficients. *Dokl. Akad. Nauk SSSR* **135** (1960), 538–541 [Russian].

[159] I.B. Simonenko: The Riemann boundary value problem for n pairs of functions with measurable coefficients and its application to the investigation of singular integrals in the spaces L^p with weight. *Izv. Akad. Nauk SSSR, Ser. Mat.,* **28** (1964), 277–306.

[160] I.B. Simonenko: A new general method of studying linear operator equations of the type of singular integral equations. Part I: *Izv. Akad. Nauk SSSR, Ser. Mat.*, **29** (1965), 567–586; Part II: *ibid.* **29** (1965), 757–782 [Russian].

[161] I.B. Simonenko: Some general questions of the theory of the Riemann boundary value problem. *Math. USSR Izv.* **2** (1968), 1091–1099.

[162] I.B. Simonenko: On multidimensional discrete convolutions. *Mat. Issled.* **3** (1968), 108–122.

[163] I. Spitkovsky: On the asymptotic behavior of determinants of block Toeplitz matrices in the locally sectorial case. *Zap. Nauchn. Sem. LOMI* **149** (1986), 76–92 [Russian].

[164] V.V. Strela and E.E. Tyrtyshnikov: Which circulant preconditioner is better? *Math. Comput.* **65** (1996), 137–150.

[165] G. Szegö: Ein Grenzwertsatz über die Toeplitzschen Determinanten einer reellen positiven Funktion. *Math. Ann.* **76** (1915), 490–503.

[166] G. Szegö: On certain Hermitian forms associated with the Fourier series of a positive function. *Festschrift Marcel Riesz*, pp. 222–238, Lund, 1952.

[167] B. Thorsen: An N-dimensional analogue of Szegö's limit theorem, *J. Math. Anal. Appl.* **198** (1996), 137–165.

[168] M. Tismenetsky: Determinant of block Toeplitz band matrices. *Linear Algebra Appl.* **85** (1987), 165–184.

[169] O. Toeplitz: Zur Theorie der quadratischen und bilinearen Formen von unendlichvielen Veränderlichen. *Math. Ann.* **70** (1911), 351–376.

[170] L.N. Trefethen: *Non-Normal Matrices and Pseudospectra.* To appear.

[171] L.N. Trefethen: Pseudospectra of matrices. In: *Numerical Analysis 1991* (D.F. Griffiths and G.A. Watson, eds.), pp. 234–266, Longman, London, 1992.

[172] S.R. Treil: Invertibility of Toeplitz operators does not imply applicability of the finite section method. *Dokl. Akad. Nauk SSSR* **292** (1987), 563–567 [Russian].

[173] W.F. Trench: Solution of systems with Toeplitz matrices generated by rational functions. *Linear Algebra Appl.* **74** (1986), 191–211.

[174] E.E. Tyrtyshnikov: Cauchy-Toeplitz matrices and some applications. *Linear Algebra Appl.* **149** (1991), 1–18.

[175] E.E. Tyrtyshnikov: Singular values of Cauchy-Toeplitz matrices. *Linear Algebra Appl.* **161** (1992), 99–116.

[176] E.E. Tyrtyshnikov: New theorems on the distribution of eigenvalues and singular values of multilevel Toeplitz matrices. *Dokl. Akad. Nauk* **333** (1993), 300–303 [Russian].

[177] E.E. Tyrtyshnikov: A unifying approach to some old and new theorems on distribution and clustering. *Linear Algebra Appl.* **232** (1996), 1–43.

[178] J.L. Ullman: A problem of Schmidt and Spitzer. *Bull. Amer. Math. Soc.* **73** (1967), 883-885.

[179] I.E. Verbitsky and N. Krupnik: On the applicability of the reduction method to discrete Wiener-Hopf equations with piecewise continuous symbol. *Mat. Issled.* **45** (1977), 17–28 [Russian].

[180] H. Widom: On the eigenvalues of certain Hermitean operators. *Trans. Amer. Math. Soc.* **88** (1958), 491–522.

[181] H. Widom: Inversion of Toeplitz matrices. III. *Notices Amer. Math. Soc.* **7** (1960), p. 63.

[182] H. Widom: Singular integral equations on L^p. *Trans. Amer. Math. Soc.* **97** (1960), 131–160.

[183] H. Widom: Toeplitz operators on H^p. *Pacific J. Math.* **19** (1966), 573–582.

[184] H. Widom: Toeplitz determinants with singular generating functions. *Amer. J. Math.* **95** (1973), 333–383.

[185] H. Widom: Asymptotic behavior of block Toeplitz matrices and determinants. II. *Adv. in Math.* **21** (1976), 1–29.

[186] H. Widom: Szegö's limit theorem – the higher-dimensional matrix case. *J. Funct. Anal.* **39** (1980), 182–198.

[187] H. Widom: On Wiener-Hopf determinants. *Operator Theory: Adv. and Appl.* **41** (1989), 519–543.

[188] H. Widom: On the singular values of Toeplitz matrices. *Z. Anal. Anwendungen* **8** (1989), 221–229.

[189] H. Widom: Eigenvalue distribution of nonselfadjoint Toeplitz matrices and the asymptotics of Toeplitz determinants in the case of nonvanishing index. *Operator Theory: Adv. and Appl.* **48** (1990), 387–421.

[190] H. Widom: Eigenvalue distribution for nonselfadjoint Toeplitz matrices. *Operator Theory*: Advances and Applications **71** (1994), 1–8.

[191] K.G. Wilson: Die Renormierungsgruppe. In: *Teilchen, Felder und Symmetrien*, pp. 50–67, Spcktrum der Wissenschaft Verlagsgesellschaft mbH, Heidelberg, 1988.

[192] A. Wintner: Zur Theorie der beschränkten Bilinearformen. *Math. Z.* **30** (1929), 228–282.

[193] N.L. Zamarashkin and E.E. Tyrtyshnikov: Distribution of the eigenvalues and singular numbers of Toeplitz matrices under weakened requirements on the generating function. *Mat. Sb.* **188** (1997), 83–92 [Russian].

Index

Symbol Index

Universitext *(continued)*

Rotman: Galois Theory
Rubel/Colliander: Entire and Meromorphic Functions
Sagan: Space-Filling Curves
Samelson: Notes on Lie Algebras
Schiff: Normal Families
Shapiro: Composition Operators and Classical Function Theory
Simonnet: Measures and Probability
Smith: Power Series From a Computational Point of View
Smoryński: Self-Reference and Modal Logic
Stillwell: Geometry of Surfaces
Stroock: An Introduction to the Theory of Large Deviations
Sunder: An Invitation to von Neumann Algebras
Tondeur: Foliations on Riemannian Manifolds
Wong: Weyl Transforms
Zong: Strange Phenomena in Convex and Discrete Geometry